INTRA-OPERATIVE MONITORING

A Comprehensive Approach

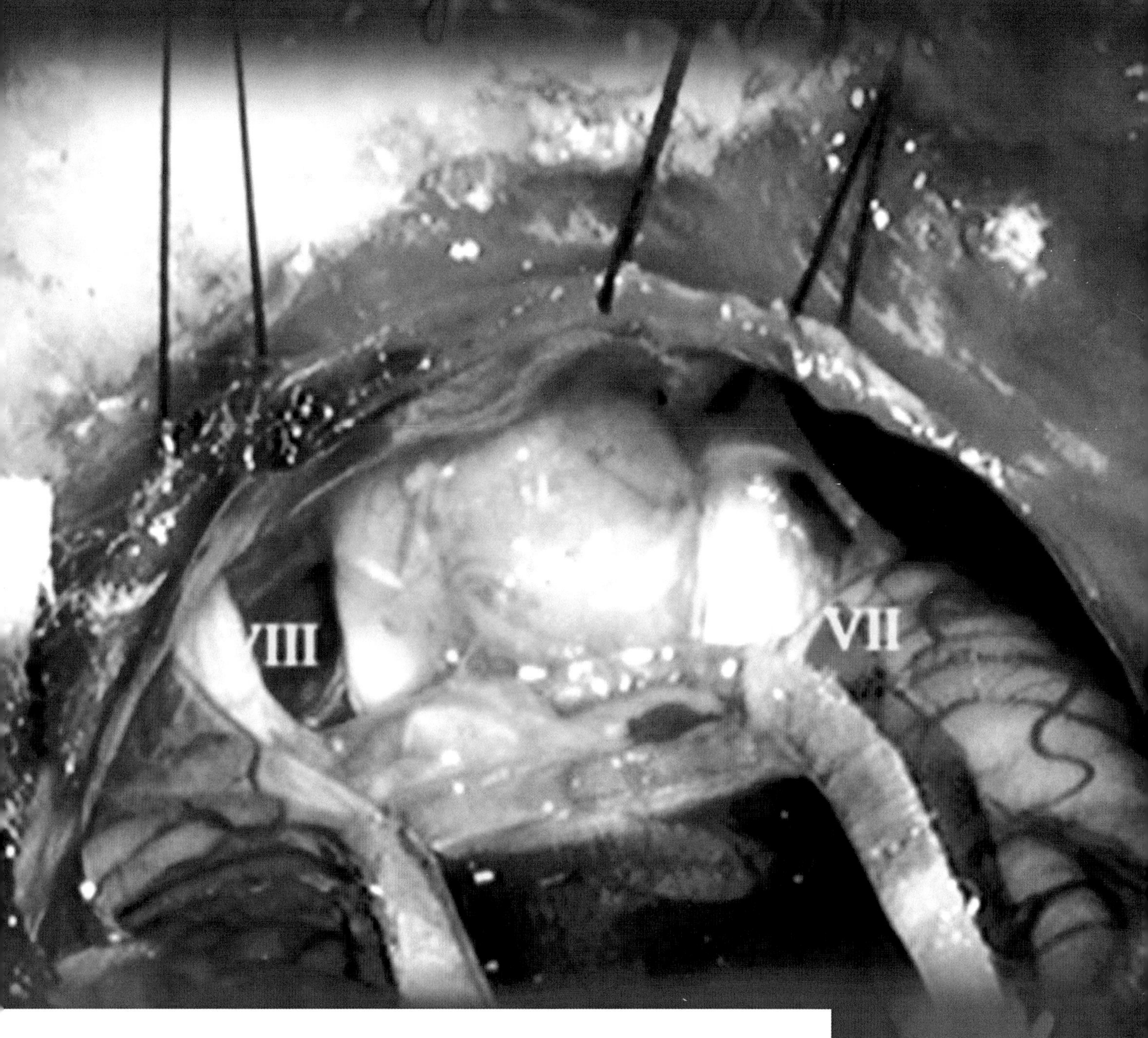

Donald Farrell M.D.

ISBN: Softcover 978-1-7960-3098-3
EBook 978-1-7960-3097-6

Print information available on the last page

Rev. date: 05/15/2019

To order additional copies of this book, contact:
Xlibris
1-888-795-4274
www.Xlibris.com
Orders@Xlibris.com

Clinical Evoked Potentials and Intra-operative Monitoring: A Comprehensive Approach

Donald F. Farrell, M.D.
Professor Emeritus, Neurology
Former Director, EEG and Clinical Neurophysiology
University of Washington Medical Center
University of Washington School of Medicine

Stephanie Ferri, R. EEG/EPT., CNIM, CLTM
Senior END Technologist
University of Washington Medical Center
Seattle, Washington 98195

The purpose of this book is to introduce the basics of evoked potentials and their use as a diagnostic tool and as an intraoperative tool to reduce injury during surgery on the central and peripheral nervous systems. There has been a reduction in the use of evoked potentials since the advent of magnetic resonance imaging, but in certain clinical situations evoked potentials remain more sensitive than the MRI scan. In many situations the physiologic information gathered from evoked potential studies complement the anatomic based MRI scan. In recent years there has been a marked proliferation in "intra-operative monitoring". Electro-physiologic intra-operative monitoring utilizes electro-diagnostic tools that continually monitor the sensory and motor systems. Sensory systems typically monitored with brainstem auditory evoked potentials, somatosensory evoked potentials, and occasionally visual evoked potentials, although use of the latter has been controversial. Motor systems are monitored by motor evoked potentials for assessment of the corticospinal or corticobulbar pathways and by electromyographic studies of the muscles for assessment of the various cranial, spinal, or peripheral nerves.

Evoked potential studies have become much easier to accomplish with more modern equipment. No longer must the monitoring personnel program the equipment for each test but is able to utilize standard program packages that come with the equipment, only needing to modify these programs for local use and then electronically select appropriate tests. Because of this equipment "improvement" monitoring personnel, while still having to understand the basics required to program an evoked potential study from scratch, no longer need to perform this function on a daily basis and can focus instead on the data collection necessary for either diagnosis or intra-operative monitoring.

Since the operating room is an "electrically unfriendly environment" where artifacts abound and the monitoring personnel must be able to troubleshoot and correct the artifact. It is imperative that monitoring personnel must excel at troubleshooting and eliminating artifacts that can interfere with reliable data collection as most issues in neuro-monitoring are of a of a technical nature. Troubleshooting must be accomplished in a rapid and efficient manner so as not to interfere with the surgery.

Service models for neuro-monitoring personnel range from technologist only with no professional supervision to technologist with professional supervision. The more ideal situation is a team of professional neurophysiologist and certified technologist. The professional neurophysiologist may be either a physician or non physician. A physician professional must be a licensed MD or DO and preferably certified by the American Board of Neurophysiological Monitoring, the American Board of Clinical Neurophysiology, American Board of Electroneurodiagnositc Medicine, or American Board of Psychiatry and Neurology, added qualification exam in clinical neurophysiology. A non-physician professional should be certified by the American Board of Neurophysiological Monitoring. Technologists should be registered by American Board of Registered Electroneurodiagnostic Technologists in intraoperative monitoring. Registry in electroencephalography or evoked potentials, while not necessary, is clearly preferable.

Professional supervision of technologists should be a nearly continuous, real time function throughout a surgical case. Ideally, the professional neurophysiologist should be in the operating room during critical periods of a surgery that requires rapid decision making and communication with the surgeons. During non-critical times or during cases that are not as demanding, professional supervision should preferable done with real time, with on line communication with the neuromonitoring equipment and with the technologists in the operating room.

This book will be divided into two sections. The first section will cover the different types of short latency evoked potentials, including chapters on pattern reversal evoked potentials, brainstem auditory evoked potentials, somatosensory evoked potentials and finally dermatomal sensory evoked potentials.

The second section will cover intra-operative monitoring including, skull-base surgery, spinal surgery, peripheral nerve surgery, pediatric surgery, electrocorticography for epilepsy surgery and brain tumor surgery, and finally monitoring during carotid artery surgery.

The senior author wishes to thank the outstanding END technologists that he has had the privilege of working with over the years. All of these outstanding technologists are triple registered, some quadriple registered in electroencephalography, evoked potentials, intra-operative monitoring, and long term epilepsy monitoring: Adele Wirch, Stephanie Ferri, Cindy Francke, Lara Henderson, Sara Wright and Debra Rollevson.

The senior author would also like to thank my post-doctoral fellows for their constant devotion to becoming outstanding neurologists and clinical neurophysiologists: Drs. Roman Kutsy, Morris Chang, Nathaniel Watson, Michael Doherty, Andrea Cheng-Hakimian, Ednea Simon, Shehim Hakimian, and John Oakley. I would like to thank the neurosurgical and neuro-otological surgeons that I have worked with over the years. As a neurologist, I have been impressed not only with their technical skills, but with the compassion that they direct towards their patients. Thanks to: Drs. George Ojemann, Basil Harris, Mitch Berger, Mark Mayberg, Dan Silbergeld, Bob Rostomily, Rich Ellenbogen, Larry Duckertt and George Gates. Finally, I would like to thank Watson B. Smith, Clinical Ultrasonographer, for providing figures and reviewing the chapter on "Carotid Endarterectomy".

Dedication (Dr. Farrell)

The senior author has a series of individuals to whom he would like to dedicate this work. First, the late Dr. Richmond S. Paine and late Dr. Harold Stevens. These two fine neurologists inspired a young medical student to pursue Neurology as a career.

Secondly, the late Dr. Frank Morrell and Dr. Guy M. McKhann, mentors during my residency and fellowship training who instilled in me the scientific approach to disorders of the nervous system and how we all have an obligation to advance our field of knowledge.

Finally, to my wife Eleanor who for over 50 years has provided encouragement and support. It would not have possible without you.

TABLE OF CONTENTS

Section 1: Short Latency Evoked Potentials

Introduction – General Principles

Modern computer based evoked potential equipment makes it relatively easy to carry out these types of studies. Processors are now very fast and powerful enough that one can anticipate accurate waveforms without the fear of aliasing when the waveforms are reconstructed. All commercial processors far exceed the standard of 2.5 times the fastest frequency being analyzed (Nyquist frequency). The processors are also powerful enough that the high amplitude fast peaks will not be truncated. A-D converters have improved in speed and capacity so that they are not a limiting factor in data analysis. However, care must still be observed so as not to allow artifacts to interfere with data collection. The desired waveforms are generated in a time-locked fashion to a stimulus. The number of trials necessary to identify a given waveform depends upon the size of the response in relationship to the background activity (noise). For example, the flash response recorded over the occipital cortex is large and can be seen in some individuals with a single flash. With amplitude of approximately 10 µV, only a small number of trials are necessary to be able to identify the response. PREPs require about 100 stimuli to clearly identity the various waveforms.

The somatosensory evoked responses are much smaller than the visual responses so the number of trials needs to be increased. The desired response being measured is closer in amplitude to the noise, and between 500 and 1000 trials are necessary to be able to identify the waveforms of interest.

Finally, BAEPs (brainstem auditory evoked potentials) are even smaller in amplitude compared to the noise, so that 2000 to 4000 trials are frequently necessary to be able to identify waves I through V.

To carry out these studies with a minimum amount of risk from a hostile electrical environment it is imperative that each placed electrode have a similar low impedance level. If the electrodes have dissimilar impedance values, it is very likely that a 60 Hz artifact will be introduced and this may interfere greatly with the collection of reliable data.

Stimulation rates are chosen to maximize the evoked response and reduce the chance of 60 Hz or harmonics or sub-harmonics of 60Hz to be introduced. For example, in stimulating the visual system one would not pick a stimulation rate that is a sub-harmonic of 60 Hz such as 1Hz, but would select a stimulation rate such as 1.1 Hz, not 1.0 Hz. In stimulating the somatosensory system one would not select 5.0 Hz, but would utilize one such as 5.4 Hz. Finally, in stimulating the auditory system, one would not use 10.0 Hz, but might more appropriately pick 10.1 Hz as the rate.

It is important to realize that evoked potentials are not real time events, but are processed data. The amplifiers enlarge the waveform, and then points of data are held in a bin of the A-D converter, and then digitized. The response is then overlaid. The time-locked response increasing in amplitude while the background (noise) being random cancels itself. Large single electrical artifacts do not greatly modify the evoked potential as they are averaged out of the response fairly rapidly. Smaller, recurrent artifacts such as a sub-harmonic of 60 Hz are more difficult to average out. The technologist running a given test should be watching the raw data as it is generated and as the evoked potential trails are accumulating. A judgment can be made to increase the number of trials to eliminate the artifact, or to abort the study, remove the cause of the artifact if possible and restart the study in a better, less hostile environment. In the operating room, there are a myriad of artifact generators. Some of the equipment in the operating room which is necessary for the surgery generates artifacts, others do not. To reduce artifacts, all cables necessary to accomplish the evoked potential studies in the operating room should be bundled. This will reduce some large artifacts arriving simultaneously at both Input 1 and Input 2 of the differential amplifiers and will cancel. This is one of the important properties of a differential amplifier; simultaneous electrical events arriving at the 2 inputs of the amplifier will not influence the output of the amplifier.

General Bibliography

Evoked Potentials:

Chiappa, K.H. and Yiannikas, Con. Evoked Potentials in Clinical Medicine. Raven Press, 1983.

Cracco, R.Q. and Bodus-Wollner, J. Evoked Potentials. Liss, 1986.

Halliday, A.M. (Editor). Evoked Potentials in Clinical Testing (Clinical Testing in Neurology and Neurosurgery Monographs). Churchill Livingstone, 1993.

Mauguiere, F. and Garcia-Iarrea, L. (Editors) Electroencephalography and Clinical Neurophysiology (Evoked Potentials). Elsevier, 1993.

ACNS Guideline 9A: Guidelines on Evoked Potentials. *J. Clin Neurophysiol.* 2006: 23: 125-137.

Intraoperative Monitoring:

Zouridakis, G. and Papanicolaou, A.C. A concise guide to Intraoperative Monitoring. CRC, 2000.

Moller, A.R. Intraoperative Neurophysiological Monitoring. Humana Press, 2005.

CHAPTER 1

Pattern Reversal Evoked Potentials (PREPs)

THE EARLIEST STUDIES using simple flash evoked potentials of the visual system were of little to no value clinically. It wasn't until the advent of pattern reversal evoked potentials in the early 1970s that this type of study became clinically useful. Its principle use has been to identify demyelinative lesions affecting the optic nerve and chiasm. It was recognized early that PREPs had little to no value in evaluating lesions in the optic tracts. It was also recognized early that large tumors and other mass lesions involving the posterior parietal lobes and posterior temporal lobes would not routinely change the evoked potential responses. Routine bedside visual field examination or formal visual field evaluations were found to be more reliable than evoked potential studies, and are the recommended method of evaluation for suspected lesions in these cortical locations.

On the other hand, PREPs are extremely valuable in the diagnosis of optic neuritis and/or optic nerve demyelination in patients with multiple sclerosis. In fact, it is superior to magnetic resonance imaging in the study of the optic nerve.

Guidelines for Pattern Reversal Evoked Potentials.

The electrode placement utilized for obtaining PREPs is different from the standard 10-20 electrode placement used in electroencephalography and other evoked potential studies. The most frequently used electrode of electrode application is the Queen Square System. This System is quite good for the majority of cases being studied, but may have to be modified slightly for rare individuals. Electrode placement includes: MO (mid-occipital) which is placed 5 centimeters above the inion in the midline. Lateral electrodes include LO (left occipital) and RO (right occipital) each placed 5 centimeters lateral to MO. The reference electrode is MF (mid-frontal) which is placed 12 centimeters superior to the nasion in the midline. A three channel recording can be done with these electrode combinations, but this author prefers a 5 channel recording which includes extra derivatives using linked ear electrodes (A1, A2). The five channels would include MF-A1, A2; MO-A1, A2, MO-MF, RO-MF and LO-MF. A 7 channel system which includes the above electrodes plus an extra electrode 5 centimeters above and below MO reduces the possibility of additional studies if the p100 is not well developed at MO. **(Figure 1).**

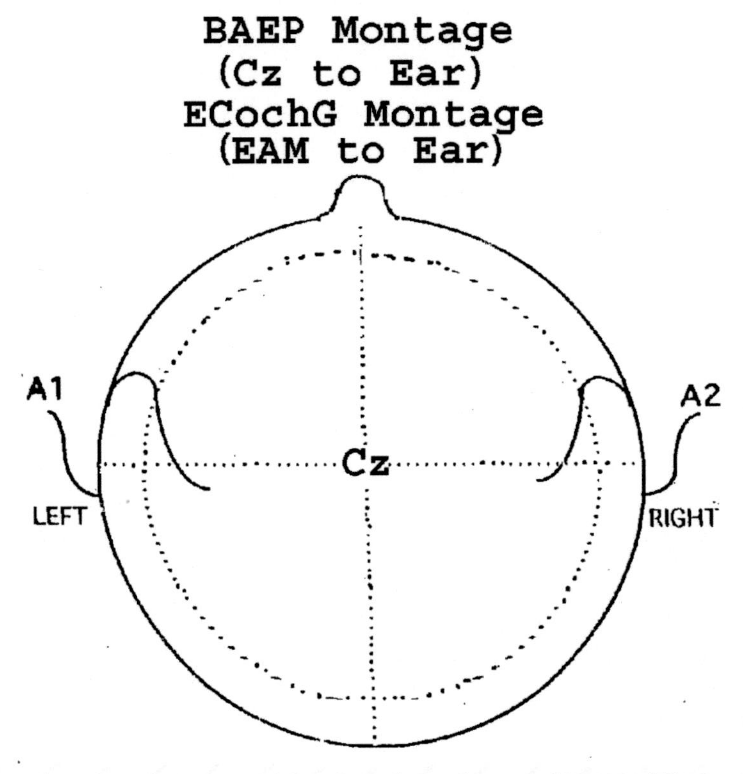

Figure 1. *This schematic shows the location of the Electrodes used to record PREPs. The Queen Square Montage.*

The 5 channel PREP allows one to observe the contribution of the frontal components to the overall waveforms, especially the p100 at MO. **(Fig. 2 and Fig. 3**

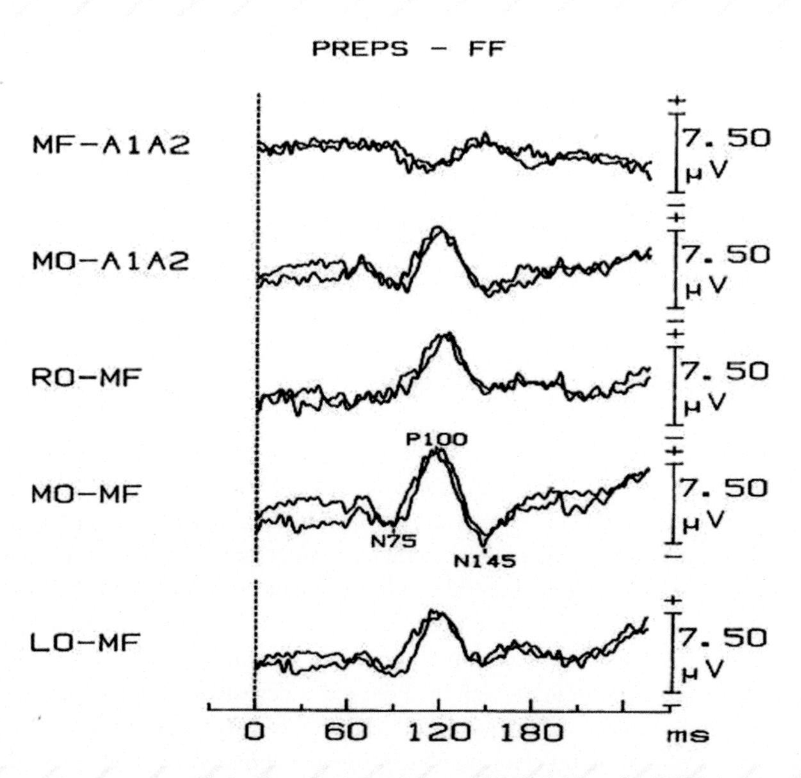

Figure 2. *This sample shows a well developed N75, P100, and N145.Note the P100 is an upward wave in our laboratory.*

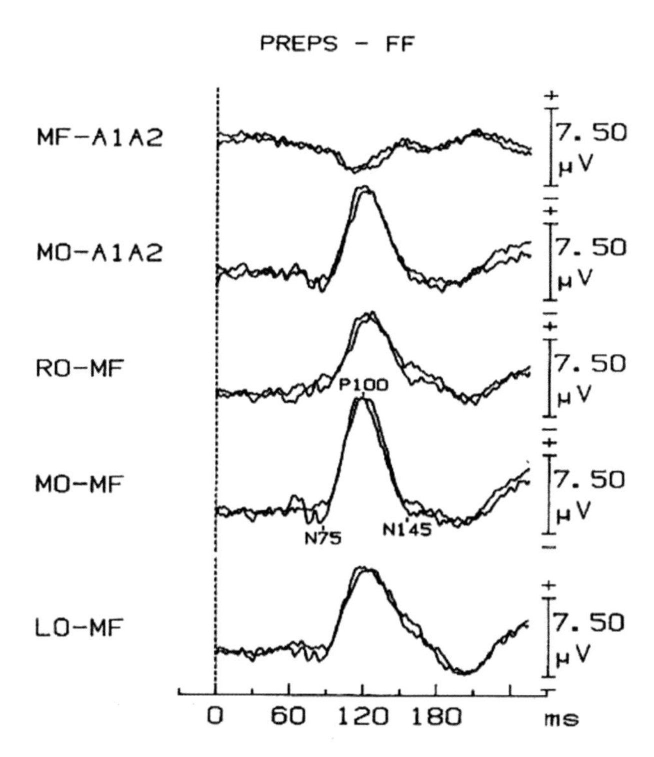

Figure 3. *This figure shows a different normal response. The N75 and N145 are clearly not present.*

Note that the amplitude of the p100 at MO - MF in Figures 2 and 3 is the same as MO-A1, A2 + MF-A1, A2. This particular combination of electrodes allows one to see one of the causes of a BIFID p100 when the N100 is shifted to the right **(Fig. 4).**

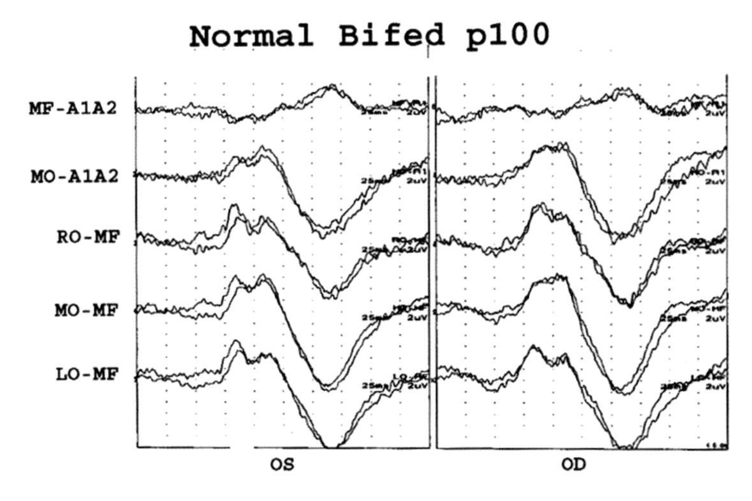

Figure 4. *This example demonstrates a bifid or double peaked p100. The Bifid n100 are seen in the fromtal derivatives (MF – A1, A2). as well as, all of the posterior derivatives.*

The latest standards for Pattern Reversal Evoked Potentials include:

1. Stimulation. A high-contrast black and white checkerboard pattern is delivered by a pattern generator using a television screen. The contrast must be greater than 50 percent. The checks should reverse at a rate of 4/ second or less. Typically, we use a turn around time of 1 second. At least 2 check sizes should be used for each patient. Small checks of 12 to 16 minutes with a field of 2-4 degrees and large checks of 40 to 50 minutes with

a field of 16-32 degrees. Some laboratories utilize one check size for their studies and these are medium checks of 24 to 32 minutes with a field of 6 to 12 degrees. Whichever check size and field size is used; no variation from these sizes should be carried out, as the visual cortex handles these different stimuli differently. The small check size is appropriate for examination of the most central macular fibers, and may identify a much localized central scotomata. The direct cortical response to the small check size stimulates the smallest area of retina (macula) and activates the smallest cortical area. The large check size is appropriate for examining the peri-macular regions of the retina and is best for picking up the typical centro-cecal scotomata found in optic neuritis and multiple sclerosis.

Checkerboard Pattern Generator
Pattern Reversal Evoked Potential

2. Recording: Band pass filters should be set at 1 to 100 Hz (-3dB). The low frequency filter should have a rollover of < 12 dB/octave. The high frequency filter should have a rollover of < 24 dB/octave.
3. Analysis time: 250 milliseconds are standard, but longer times may be necessary to measure markedly delayed wave forms.
4. Replication: Each study must be done a minimum of 2 times, with a replication of the results showing the peak of the p100 to be within 2.5 milliseconds of each other. The peak to peak amplitude of the p100 in the replicated trials should be less than 15 percent. A cooperative patient is necessary to carry out this study. The patient also must have a visual acuity better the 20/200 in order to see and fixate on the target. Uncooperative individuals who cannot or will not fixate on a central target will show variation from run to run.
5. Electrode placement: As mentioned earlier the Queen Square System is utilized for PREPs. The reference electrode MF is located in the midline 12 centimeters superior to the nasion. MO is placed in the midline 5 centimeters superior to the inion. LO and RO are placed 5 centimeters lateral to MO. If hemi-fields are to be done, additional electrodes are placed 5 centimeters lateral to LO and RO. These are named LT and RT. MO is the appropriate location for the majority of patients, but in that small minority the electrode may have to be moved up towards MP (mid-parietal) or down to the inion, each location being 5 centimeters superior and inferior to MO.
6. Recording montages: The guidelines suggest that 4 channels be recorded, but as mentioned earlier we always utilize between 5 and 7 channels.

Waveform Generators

The PREP as seen on the scalp over the occipital cortex is very variable amongst individuals. In some individuals there are a series of positive and negative waves characterized by their approximate latency. The series starts with a p55 followed by an N75 then the p100 and finally an N145. Unfortunately, the earliest cortical response, the p55 is not a consistent finding, nor is the N75 and N145. The only absolute waveform that is universally present is the p100. There is no single area of visual cortex that accounts for these different waveforms. They are instead generated in different areas of the occipital cortex. Direct cortical recordings show that the p55 is generated very close to the calcarine fissure and in the majority of cases is buried deep in between the hemispheres and is oriented to the midline, therefore not seen with scalp electrodes. Anatomically, there is a great deal of variation in the calcarine cortex, not only from person to person, but from side to side as well. When the calcarine cortex is exposed posteriorly over the occipital tip, one is likely to see this initial cortical response. The N75 is generated over a wider area of occipital cortex and is seen in the majority of individuals, but not all. The p100 is generated over much of the occipital cortex in both primary visual cortex (V1), and over the visual association cortex (V2 and V3). The p100, because of its universal presence, has been selected for use in analyzing the conduction latency for the visual system to determine if a delay is present. The N145 is generated from much of the same cortex that generates the p100.

When one records directly from the visual cortex there is no single area of wave generation. Adjacent electrodes 1 centimeter apart may show an initial positivity next to an initial negativity. The dura, skull and scalp act as a filter, especially of the higher responses seen at the scalp level. Because of the complexity of the cortical generated wave forms it is interesting that the response seen with scalp electrodes is as reproducible as it is. The scalp amplitude of the p100 as measured from the peak of the p100 to the trough of the N145 and is only about 10 percent of that recorded directly from the brain. **(Fig. 5).**

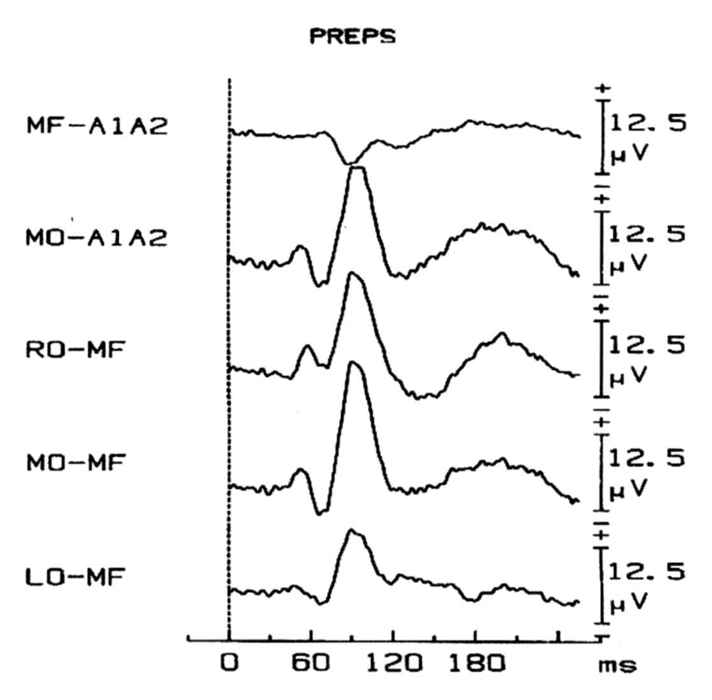

Figure 5. *In this scalp recording this patient has a well developed P55, followed by a well developed N75 then the P100. The N145 is not developed as well in MO-MF as in RO-MF.*

Interpretation of Results

In general, one will expect a series of waves including a N75, p100, and N145. Occasionally a p55 will be present. The only wave that is measured is the p100. The mean + 3 SD is used for identifying an abnormal response. This value is 110 milliseconds and depends upon normative data collected in each laboratory. In our laboratory, any p100 latency that exceeds 110 milliseconds is abnormal. Three standard deviations (SD) were selected to keep the false-positives at a minimum. At 3 SD, only about one normal individual out of 100 studied will be called abnormal. Clinically, it is much better to have a few false-negatives than false-positives. Calling a normal individual abnormal generally leads to the launch of a series of expensive diagnostic tests, some of which, such as the lumbar puncture are uncomfortable, and the results will be normal. The second measurement to consider for abnormalities is that of the intra-ocular difference. If the latency difference of the p100 from each eye exceeds 10 milliseconds, the longer latency is abnormal.

One additional normal variant needs to be considered here. That is the posteriorly located p135. Typically, the p135 is seen in the posterior temporal leads with hemi-field stimulation. In a few individuals this waveform is well formed and found more posteriorly. **(Fig. 6)..**

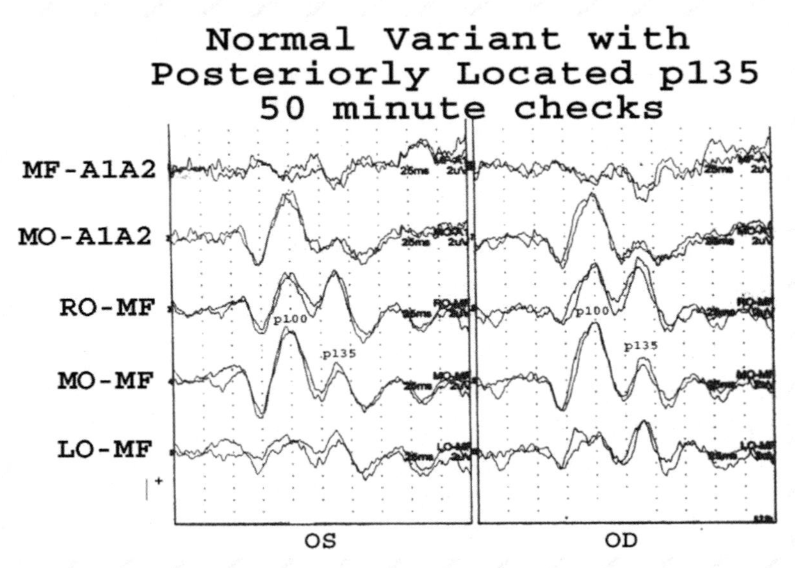

Figure 6 . *This normal study shows both well developed p100 and p135 over the posterior electrodes.*

When this type of study is identified one can confirm the p135 by doing a hemi-fields study. As a whole hemi-field studies are of little clinical value, but can be used to identify the N105 and P135. See Figure 7. The guidelines suggest that hemi-fields are more sensitive to detecting lesions of the chiasm and post-chiasmal areas. This has not proved to be true in our laboratory. Standard visual field perimetry is more likely to demonstrate an abnormality in lesions posterior to the chiasm than hemi-field PREPs. **(Fig. 7).**

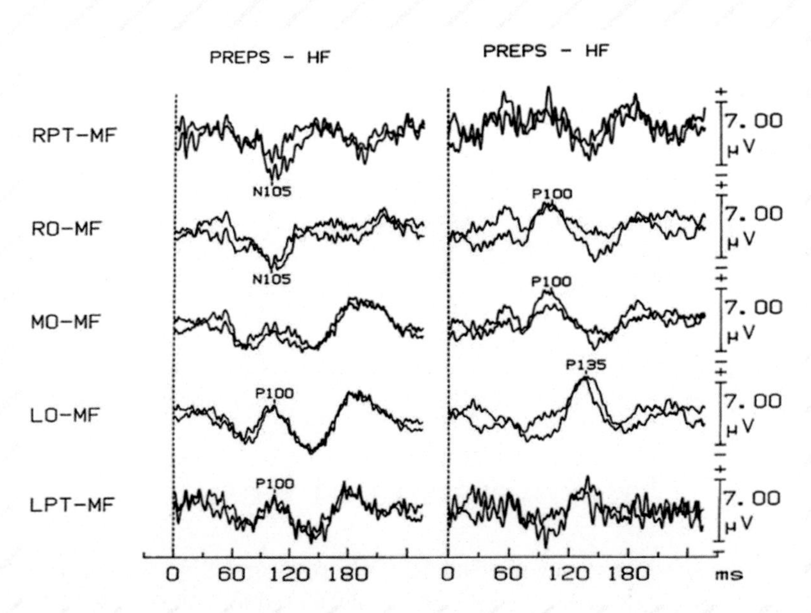

Figure 7 . *Hemi-field stimulation to show location of 2 wave forms not commonly seen on the standard studies. The N105 seen in RPT - MF and RO-MF in the left panel and the p135 seen in the LO-MF and LPT-MF electrodes in the right panel.*

The recognition of the posteriorly located p135 may be of clinical value if an individual is missing the p100 from an attack of optic neuritis and that person has a posteriorly located p135. It could be presumed that the normal p135 is a prolonged latency p100. A p100 of this latency would be abnormal, but that is not the correct diagnosis. The correct diagnosis is that the individual has an absent p100 and what is being seen posteriorly is a normal p135.

One final point in the interpretation of abnormalities in the PREP; certain hereditary retinal disorders can lead to a prolonged p100 and before assuming the delay is due to a demyelinative disorder, one has to make sure the patient does not have one of these retinal disorders. The most common retinal disorder to cause a prolonged p100 would be retinitis pigmentosa, but cone-rod dystrophy, and cone dystrophy can also prolong the p100 latency. Not all individuals with these retinal conditions will show a prolonged P100, but about 20 percent in each category are likely to do so.

Optic Neuritis and Multiple Sclerosis

Acute optic neuritis is a fairly common demyelinative disorder affecting the optic nerve. This condition can occur in isolation or as part of an attack of multiple sclerosis. Optic neuritis affects females more than males with a ratio of almost 2:1. The average age of onset is about 30 years. In general there is either acute blurring of vision or acute loss of vision in one eye. There may be a sharp pain that feels like it is behind the eye, especially when the eye is moved. The vast majority of cases resolve over 4 to 12 weeks. The optic disc frequently shows no abnormalities during the acute phase, but with time may develop optic atrophy**. (Fig. 8 and Fig. 9).**

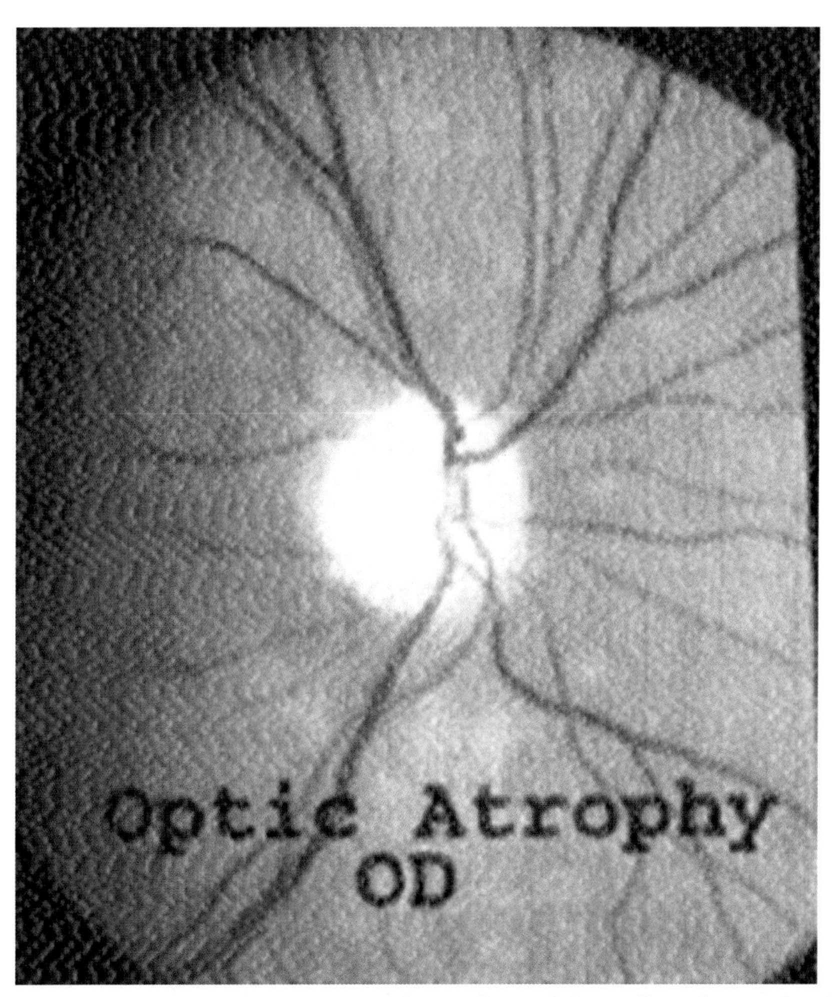

Figure 8. *Retinal photograph shows optic atrophy. The disc is white and the small vessels are reduced In number along the temporal margin.*

This is a typical pattern for retrobulbar neuritis. The demyelinative plaque is in the optic nerve and with time may progress to optic atrophy. However, a demyelinative lesion in the optic disc anteriorly shows a different pattern, that of acute papillitis.

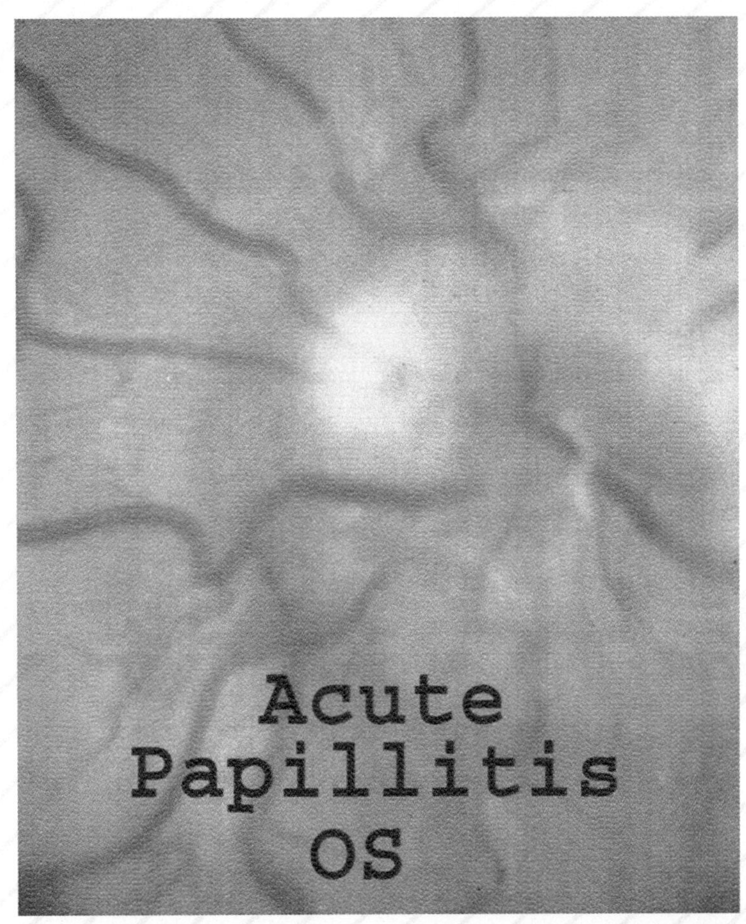

Figure 9. *Retinal photograph shows the swollen Optic nerve head with soft cotton exudates and congested retinal veins.*

Over the next 10 to 15 years approximately 40 percent of individuals with optic neuritis will develop additional symptoms of multiple sclerosis. Almost a third of patients who initially develop multiple sclerosis will develop optic neuritis as part of their syndrome over the next few years. About 55 percent of patients with multiple sclerosis will develop optic neuritis during the course of their life.

Pattern reversal evoked potentials are very valuable in the evaluation of a patient for optic neuritis. First, during an acute attack the p100 latency will prolong to abnormal values without any significant changes in the amplitude of the p100. This is a typical feature of the demyelinative lesion. However, with repeated attacks or extremely severe attacks, the axons can also be damaged by the process, and then there will be decreased amplitude of the p100. Patients with long standing chronic disease tend to show decreased amplitude of the p100. If a patient gives a history of optic neuritis years before, but has completely returned to normal in between, the PREP generally continues to have a prolonged latency for the p100. This finding may be very valuable in the establishment of a second lesion when trying to establish a diagnosis of multiple sclerosis. Third, subclinical episodes of optic neuritis occur and the PREP may demonstrate a prolonged latency for the p100 even though the individual has never had any eye complaints. Again, the presence of a subclinical optic neuritis may help establish a diagnosis of multiple sclerosis. Other diagnostic tests may be of value in establishing a diagnosis of multiple sclerosis, including magnetic resonance imaging (MRI) of the brain and spinal cord, examination of the cerebral spinal fluid for oligoclonal gamma globulin bands and myelin basic protein. Unfortunately, there is no single diagnostic test for multiple sclerosis; an appropriate clinical history with multiple lesions affecting the nervous system at different times remains a mainstay in establishing a diagnosis. Unilateral optic neuritis is more significant as a diagnostic finding for multiple sclerosis than bilateral optic neuritis which may occur in a number of different conditions. Bilateral optic neuritis does occur in multiple sclerosis frequently, but may be seen in a number of the hereditary spino-cerebellar degenerations. A series of examples of PREPs in Optic Neuritis and multiple sclerosis will now be follow. **(Fig. 10, Fig. 11, Fig. 12,Fig. 14, and Fig. 15)**

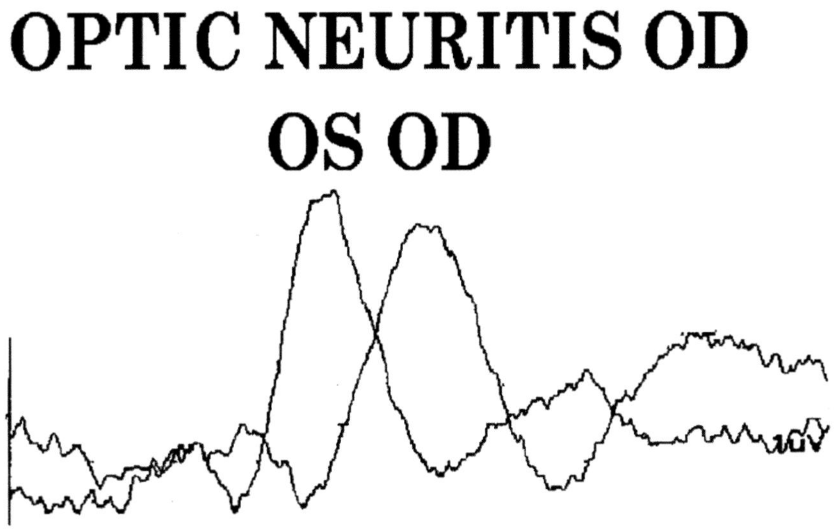

OPTIC NEURITIS OD
OS OD

10uV

Figure 10. *Superimposed p100 from normal left eye with a p100 latency of 97.0 ms and prolonged p100 right eye with a latency exceeding 120 ms. Note the nearly identical amplitudes of the waveforms.*

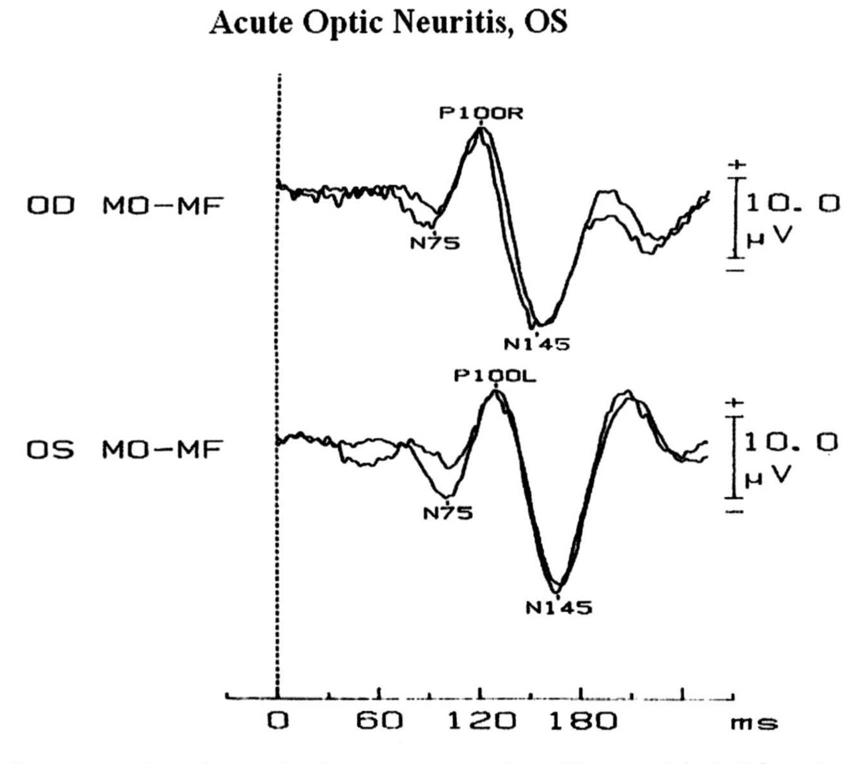

Figure 11. *In this example, the right eye p100 latency is prolonged at 118 msec, while the left eye is prolonged beyond 124 msec. Clinically, this individual had an earlier optic neuritis OD. Acute visual changes occurred OS.*

Full Field PREP
Old Optic Neuritis
with Axonal Loss

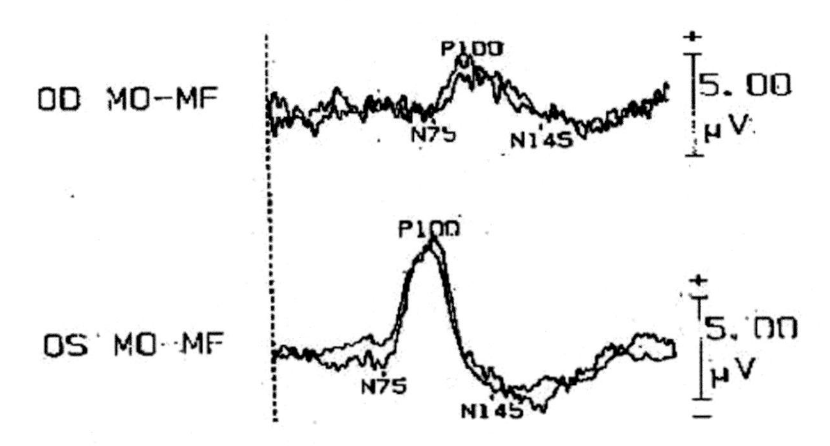

Figure 12. *In this example, the right eye not only has prolonged latency of the p100, but there is a dramatic reduction in the amplitude of the p100. This can be seen in old chronic lesions or in a person who has had recurrent attacks of disease..*

Severe Optic Neuritis, OD
Multiple Sclerosis

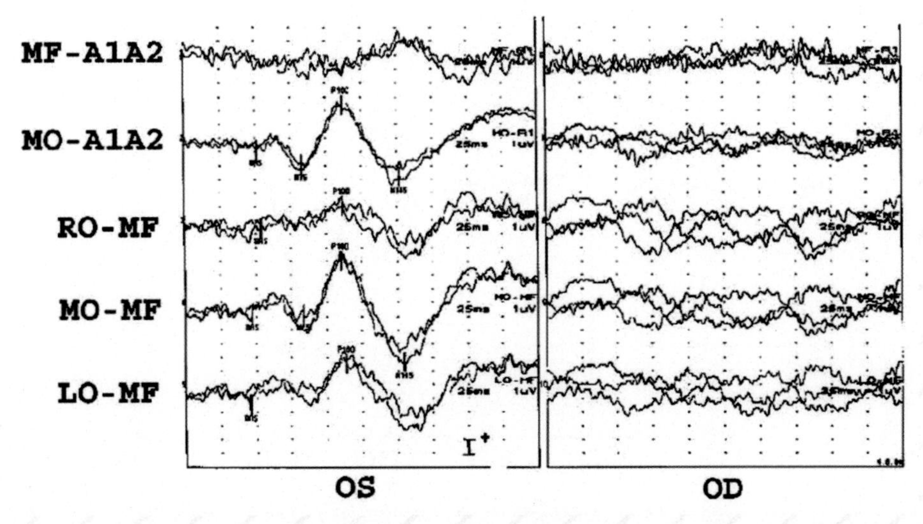

Figure 13. *In this example, there is no response in the right eye to the Checkerboard stimuli. This severe finding has a poor prognosis for visual recovery. This test was carried out with 50 minute Checks. To determine if any function is possible the person can then be tested with a very large check size, see Figure 10.*

1.5 Degree Checks, No Response

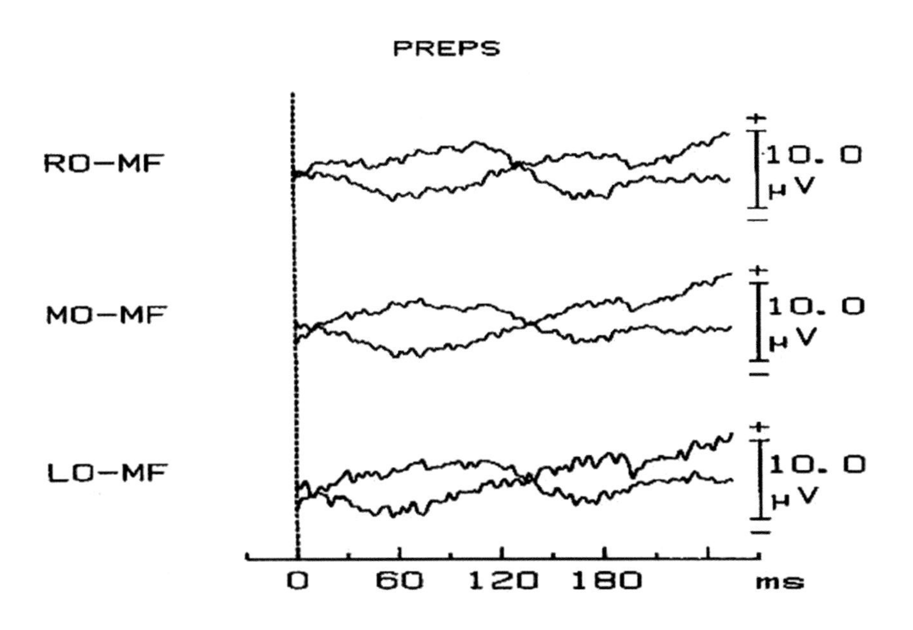

Figure 14. *No response to very large checkerboard pattern of 1.5 degrees of visual angle.*

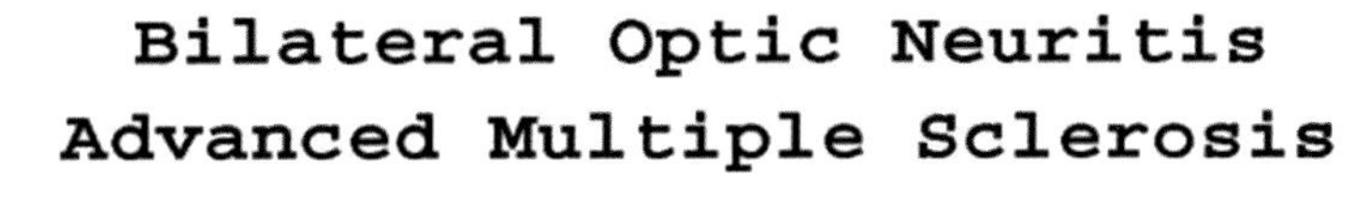

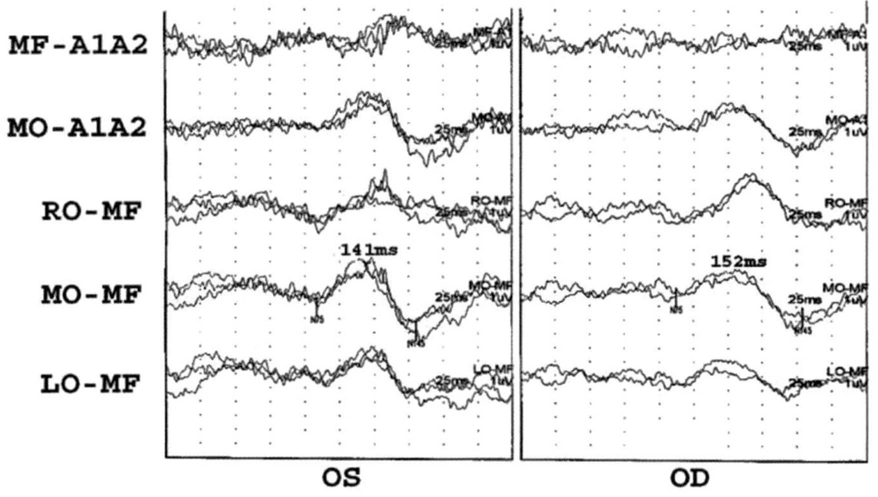

Figure 15. *Bilateral prolonged p100s in a patient with long standing MS. Patient had had multiple episodes of optic neuritis affecting both eyes at different times.*

Neuromyelitis Optica (Devic's Disease)

For many years this condition has had the distinction of being considered a severe form of multiple sclerosis or that it represents a specific demyelinating disease. Unfortunately, the combination of optic neuritis and spinal cord disease may be seen as part of multiple sclerosis, the primary difference being that the optic neuritis does not show the same type of clinical resolution seen with typical multiple sclerosis. The spinal cord lesions also tend to progress without the typical periods of improvement followed by additional symptoms at a later date. Typically, when multiple sclerosis causes this syndrome, other diagnostic tests found to be abnormal in a large percentage of multiple sclerosis patients should also be abnormal. This would include the peri-vertricular plaques seen on magnetic resonance imaging and the oligoclonal gamma globulin bands from the cerebral spinal fluid.

Neuromyelitis optica does not share these laboratory abnormalities. Recently, however, a specific antibody found in about 70 percent of cases has been discovered. This antibody named NMO-IgG should help to sort out the majority of these cases. Clinically, rapid blindness in one or both eyes occurs over a short period of time. Lesions affecting the spinal cord may develop weeks to even years after the eye symptoms.

One of my patients, a 32 year old Asian male with Neuromyelitis optica, first had sudden blindness in his left eye. Two months later that same thing happened to his right eye. Neither eye showed improvement with intravenous steroid therapy. This young man then went to the Washington State School for the Blind and learned Braille. For two years he was able to read utilizing fine tactile sensation. Then he developed a cervical demyelinative lesion leaving both hands devoid of meaningful sensation.

Pattern Reversal Evoked Potentials can document this disease and how it affects the optic nerve. Unfortunately, this test is not specific and a correct diagnosis requires additional testing as mentioned above.

For many years, neuromyelitis optica has been reported in the Japanese medical literature. In fact, in that country, reported multiple sclerosis is relatively rare and more cases of neuromyelitis optica are reported than multiple sclerosis. **(Fig. 16 and Fig. 17).**

No Responses to 1.4 Degree Checks in Devic's Disease

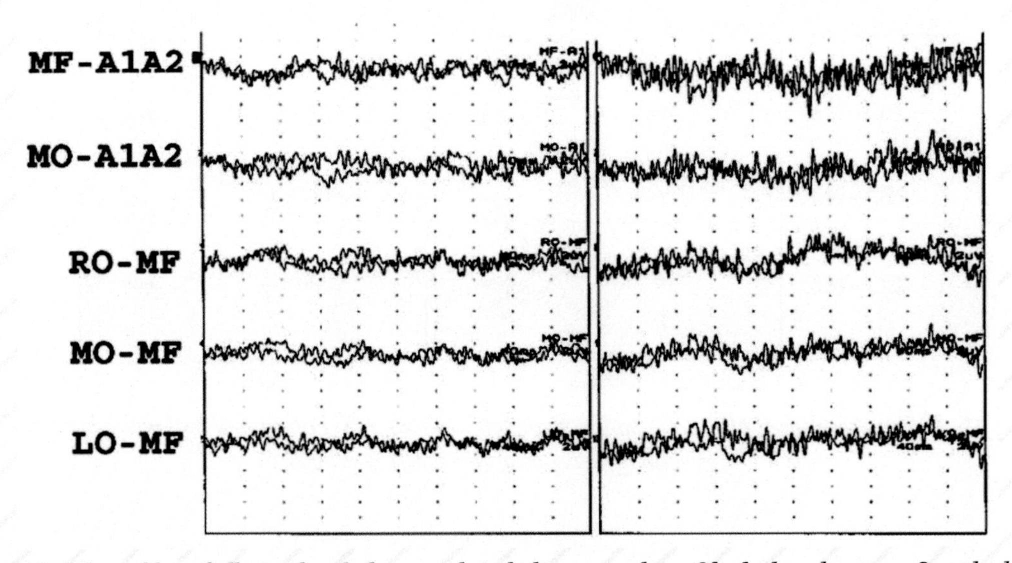

Figure 16. *Bilateral loss of all visual evoked potentials including a very large Checkerboard pattern. Completely absent responses would be very unusual in multiple sclerosis.*

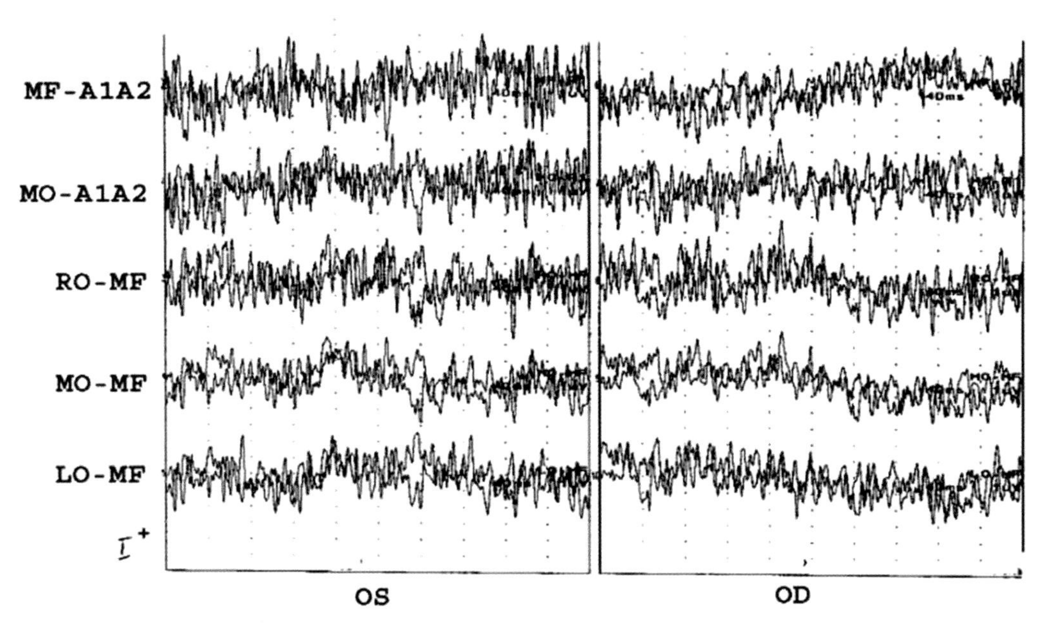

Figure 17. *Additional example of the findings in Neuromyelitis optica. Again, no responses are obtained with very large checks.*

Leber's Hereditary Optic Neuropathy

Leber's Hereditary Optic Neuropathy was first described in the 1870s. Only recently has the genetic basis of this condition been understood. It is a maternally inherited mitochondrial disorder in which the mother can pass on the mutation to her children, but fathers with the mutation cannot pass it on to any of his children. There are many more males than females affected. Severity of the disease varies with which point mutation has occurred. The most common mutation is the (G to A) mutation at nucleotide 16,569 this is followed by a (G to A) at nucleotide 3,460 and (A to G) at nucleotide 14,484. All of the point mutations affect Complex I, a NADH-ubiqinone oxidoreductase or one of its subunits. Complex I functions in the electron chain transport system of the mitochondria. Patients who have greater than 70 percent of the defective mitochondria will develop the severe (combination of number of abnormal mitochondria plus effects of X inactivation). Less severe forms had less than 30 percent of their mitochondria being abnormal.

Patients generally lose their vision during their adult years. There is progressive loss of vision symmetrically in both eyes. Until the genetics of this disorder was understood, this condition was frequently diagnosed as a toxic disorder, alcohol-tobacco ambylopia. At one time it was even considered to be caused by the excess cyanide produced in smokers. An example of the progressive nature of this disorder is shown in **(Fig.18).**

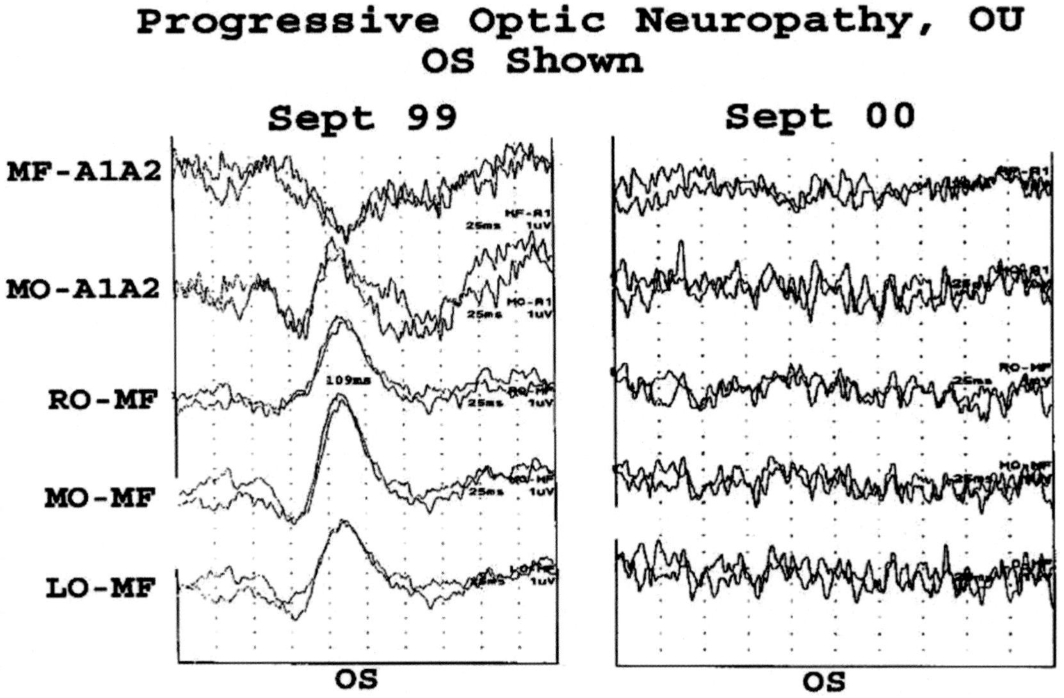

Progressive Optic Neuropathy, OU
OS Shown

Sept 99 | Sept 00

MF-A1A2
MO-A1A2
RO-MF
MO-MF
LO-MF

OS | OS

Figure 18. *These PREPs were taken a year apart. This male patient has Leber's Hereditary Optic atrophy. Note the normal response in 1999 when the patient was just beginning to have symptoms to the 2000 response when no responses could be obtained.*

Ischemic Optic Neuropathy

Ischemic optic neuropathy has two common causes, the first is atherosclerotic and the signs and symptoms are due to thrombo-embolism. Patients may have transient visual obscurations, amourosis fug ax preceding the infarction of the optic nerve and retina. The optic nerve has two sources of blood supply. The central artery is an end artery that arises directly from the carotid artery. This artery supplies the central core of the nerve while the secondary supply has a number of short penetrating vessels arising in the arachnoid supplying the outer rim of the nerve. The second major cause of ischemic optic neuritis is secondary to Temporal Arteritis. This is an inflammatory disorder of the elderly (7th and 8th decade) and leads to giant cell arteritis. Temporal arteritis may also occur as part of a systemic illness called myalgia rheumatica, characterized by low grade fevers, weight loss, and aches and pains around the shoulders and upper arms. Only about 5 percent of individuals develop ischemic optic neuritis, but this can be prevented with appropriate therapy. In Temporal arteritis the sedimentation rate is generally markedly elevated from 80 to greater than 100. Temporal artery biopsy is necessary to establish the diagnosis as patients with this condition are treated for 1 -2 years of oral steroids. Treatment can be started before the biopsy as the histologic changes remain present for long periods of time even after therapy has been started. Care must be taken in picking the site of biopsy. This is a segmental condition and the site of possible abnormality must be carefully selected, the affected artery is frequently nodular and one of the nodes needs to be included in the biopsy. **(Fig. 19, Fig. 20 and Fig. 21)**

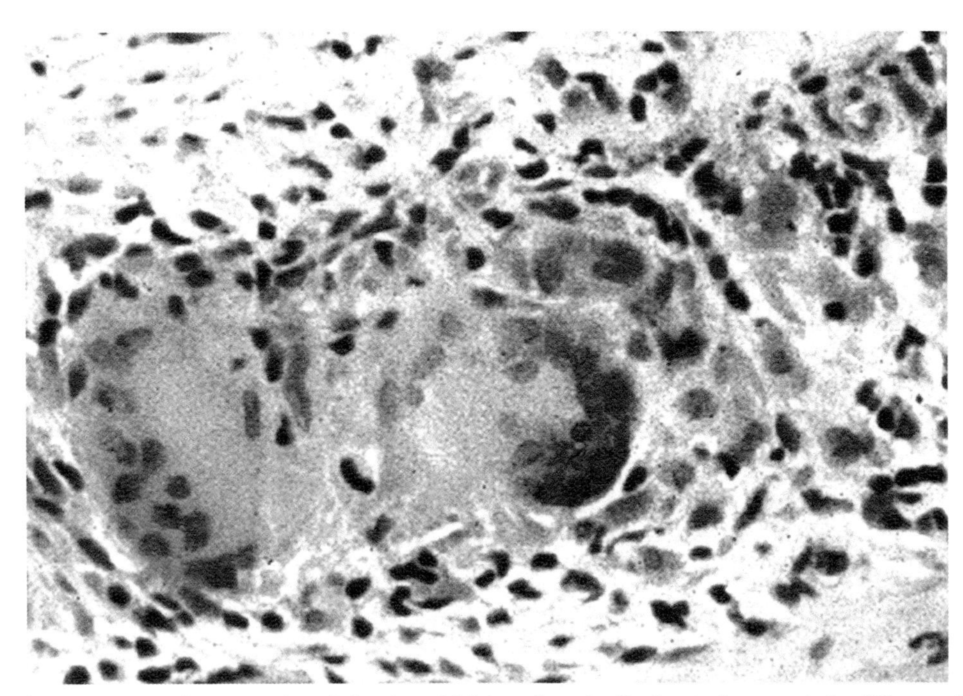

Figure 19. *Microscopic demonstration of the giant Multi-nucleated cells that is characteristic of Temporal Arteritis.*

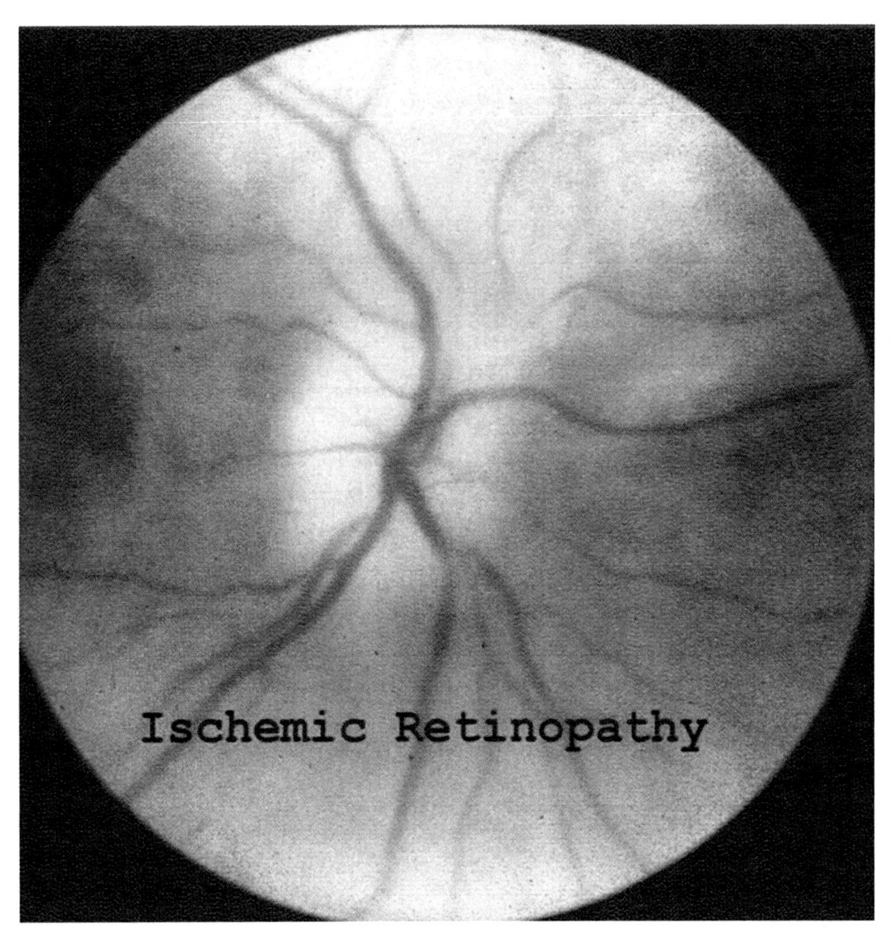

Figure 20. *Retinal photograph of Ischemic changes including a pale Retina, attenuated arteries and blurring of the optic nerve head.*

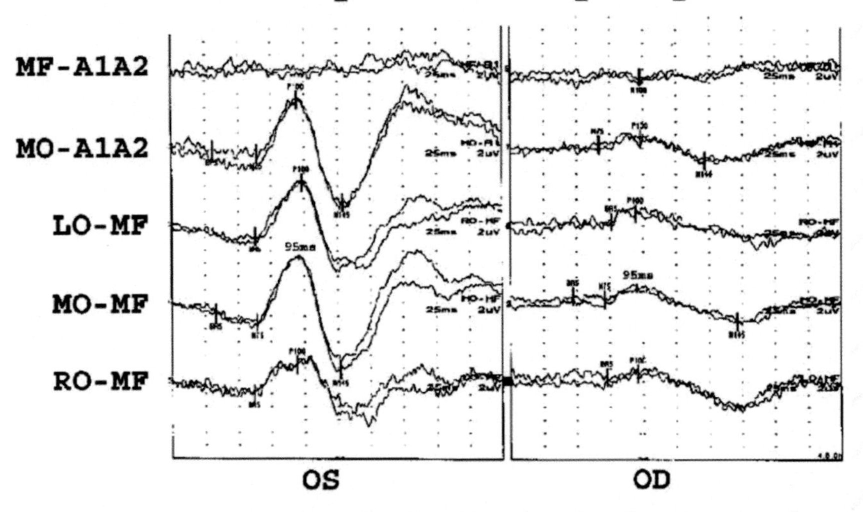

Figure 21. *Near complete loss of optic nerve function of the right eye.there remains only a trace of the p100, but continues to have a a normal latency. The left eye has normal responses.*

Subacute Combined Degeneration of the Spinal Cord (Vitamin B12 Deficiency)

Typical Subacute combined degeneration of the spinal cord is secondary to vitamin B12 deficiency related to the absence of intrinsic factor secretion by the stomach. B12 deficiency most commonly leads to megaloblastic anemia plus thoracic spinal cord degeneration. There are examples of the nervous system disorder developing in the absence of the anemia. Optic neuropathy is a rare complication of Subacute combined degeneration of the spinal cord and the patients develop a progressive loss of vision. This is a very important condition to recognize early as therapy may prevent the progression of the disorder or in some cases actually reverse the condition. The example shown below actually showed improvement in optic nerve function following injections of Vitamin B12. **(Fig. 22).**

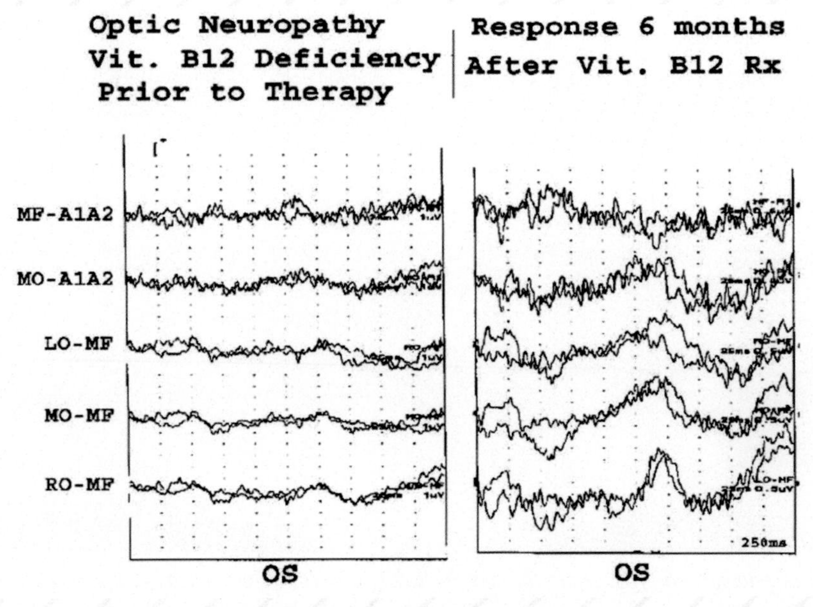

Figure 22. *Left sided panel shows the PREP in one eye prior to therapy. There is a trace of the p100 with a very prolonged latency. The right panel demonstrates the marked improvement in the amplitude of the p100, but the latency remains markedly prolonged.*

Hereditary Spino-cerebellar Ataxias

This is a very complex group of disorders that are generally inherited. The most common form is Friedreich's Ataxia and is inherited as an autosomal recessive disorder. This condition generally shows symptoms by late childhood. Severe ataxia progresses to the patient being wheelchair bound, scoliosis is a prominent manifestation as is a cardiomyopathy. Life expectancy is shortened and the majority of individuals with this condition die in their 20s. Bilateral optic nerve dysfunction is present in about 30 percent of affected individuals.

There are other groups of inherited ataxia that are generally inherited as an autosomal dominant condition. The classification of these disorders has undergone change from olivo-ponto-cerebellar degeneractions to Spino-cerebellar Atrophy (SCA) followed by a number to indicate the exact type. The most common of these is SCA 1. Currently the numbering system is up to SCA 17. The vast majority of these conditions are nucleotide repeat disorders where there are long repeats of certain DNA bases. Genetic testing for many of the repeat disorders is available. **(Fig. 23 and Fig. 24).**

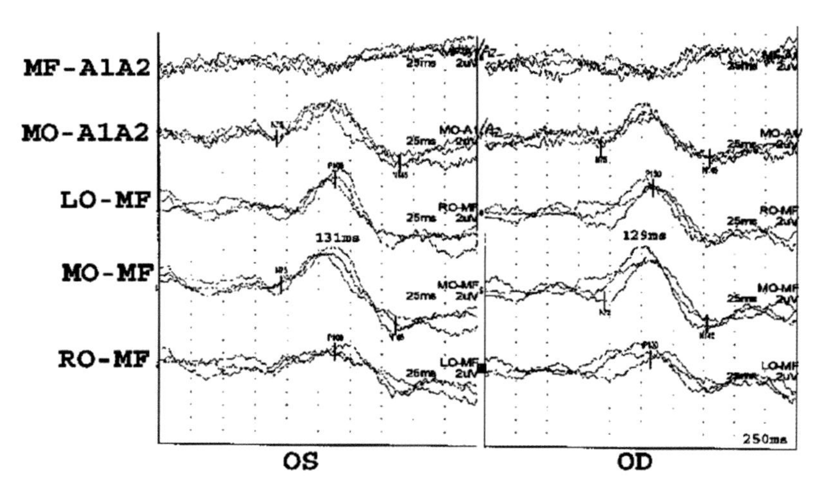

Figure 23. *Bilateral prolonged p100 in an individual from a family with SCA 7. The amplitude of the p100 is spared while the latency is markedly prolonged.*

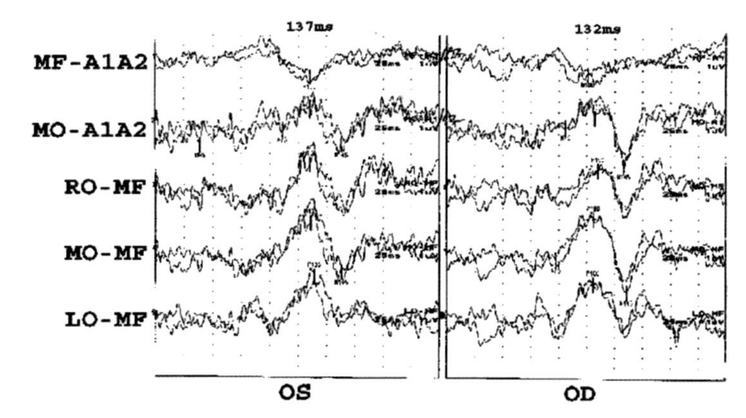

Figure 24. *This individual was adopted so there was no family history available. She was over 40 when her symptoms first began and it was impossible to determine whether she had a Spino-cerebellar degeneration or multiple sclerosis. Modern genetic diagnostic tests would help sort out cases like this.*

Mass Lesions

Tumors, especially meningiomas and pituitary tumors, as well as vascular aneurisms can compress the optic nerves and/or chiasm. The changes in PREP in these various conditions do not generally affect the latency, but are more likely to affect the amplitude of the p100. MRI scans are likely to identify these disorders early so that there can be no confusion about the cause of the physiologic abnormality.

Flash Evoked Potentials (FEP)

In general, flash evoked potentials are of little clinical value. Their primary use is in studies in which the subject cannot or will not fixate on a target. The ability to fixate on a target is necessary for PREPs.

Flash evoked potentials can be carried out in young children who are too young to be able to fixate. The bright flash is generally delivered with a Ganzfeld stimulator or LED goggles. An infra-orbital electrode is placed to record the electroretinogram as well as electrodes at MO, LO and RO to identify the occipital components. The reason for the limited value of this type of test comes from the fact that its response covers the entire occipital region and if there were partial lesions they would not be detected. For example, there are patients who have cortical blindness and the FEP remained normal. The one place in the adults where the FEP may be of value is in severe head trauma. If the optic nerve is completely severed from trauma, there will be no recorded response. If the optic nerve is partially damaged, the FEP may remain normal, but this could not be used to decide whether the individual will regain functional vision or not.

In certain disorders of infants or young children the FEP may be the only tool in which to examine the visual pathway. **(Fig. 25 and Fig. 26)**

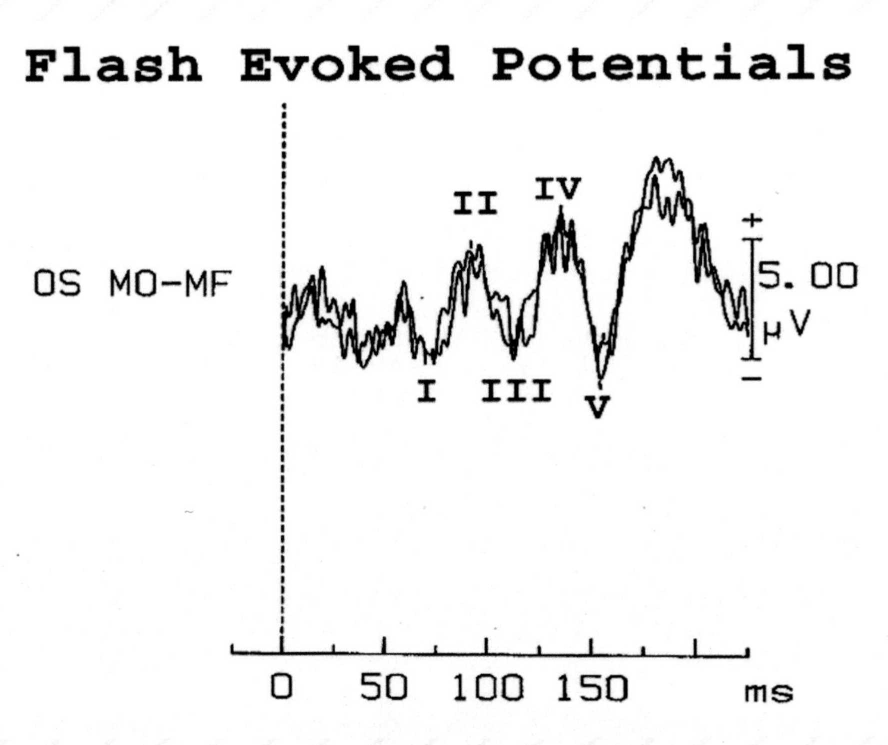

Figure 25. *The typical response to a flash is a series of Waveforms I to V. Note that wave II has the approximate latency of the p100.*

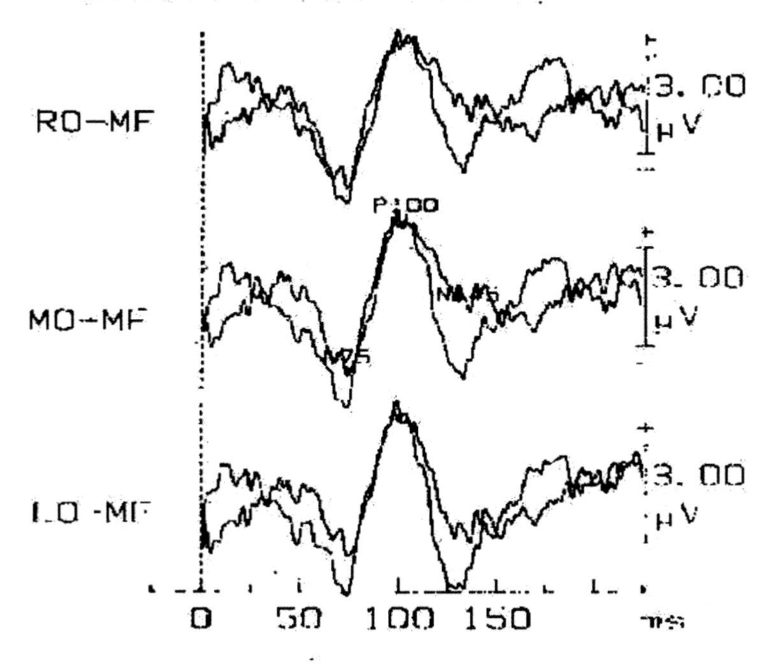

Figure 26. *At times the waveform may look more like a PREP without the following slowWaveforms.*

Leber's Congenital Amaurosis

Care must be taken not to confuse Leber's Congenital Amaurosis with Leber's Hereditary Optic Neuropathy. Leber's Congenital Amaurosis is present at birth and there is progressive loss of vision. Early on the infant shows eye movement abnormalities (pendular nystagmus, a finding that occurs with decreased visual acuity or blindness). Within a few months of birth there is complete blindness. This disorder is inherited as an autosomal recessive disorder and makes up about 10-20 percent of all congenital blindness. The gene causing this disorder in some individuals has been mapped to two different loci on Chromosome 17 and separate loci on Chromosome 19. **Fig. 27).**

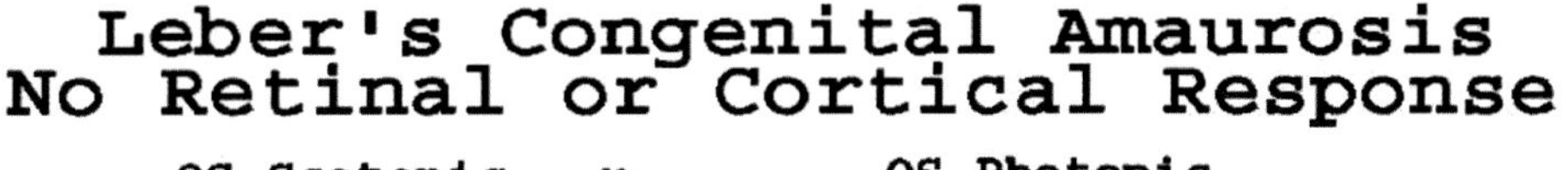

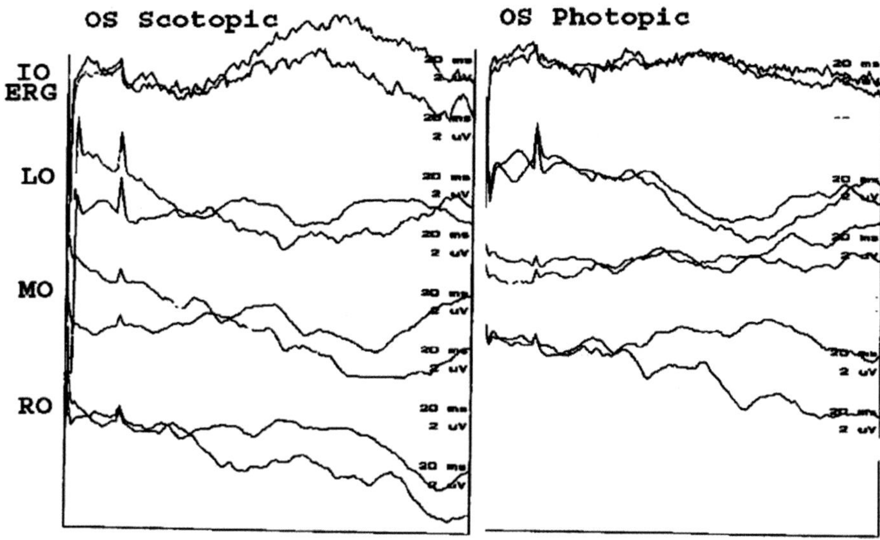

Figure 27. *There are no responses in the infra-orbital ERG Electrode. There are no responses cortically.*

Mitochondrial disorders

Genetic disorders affecting the mitochondria DNA are increasingly being recognized. Some of the more common include: Mitochondrial encephalomyopathy, lactic acidosis, and stroke-like episodes; Myoclonic Epilepsy with Ragged Red Fibers, and Leber's Hereditary optic neuropathy. There are a number of other conditions, but there are still many to be identified. It is probable that those so far discovered are caused by genetic mutations of the mitochondrial DNA, but others remain where nuclear DNA might provide only 1 subunit of an enzyme while the component from the mitochondrial DNA is normal. This combination would also lead to mitochondrial disorders.

Some of the mitochondrial disorders have defective Infra-orbital ERGs and absent FEP at the cortical level. **(Fig. 28)..**

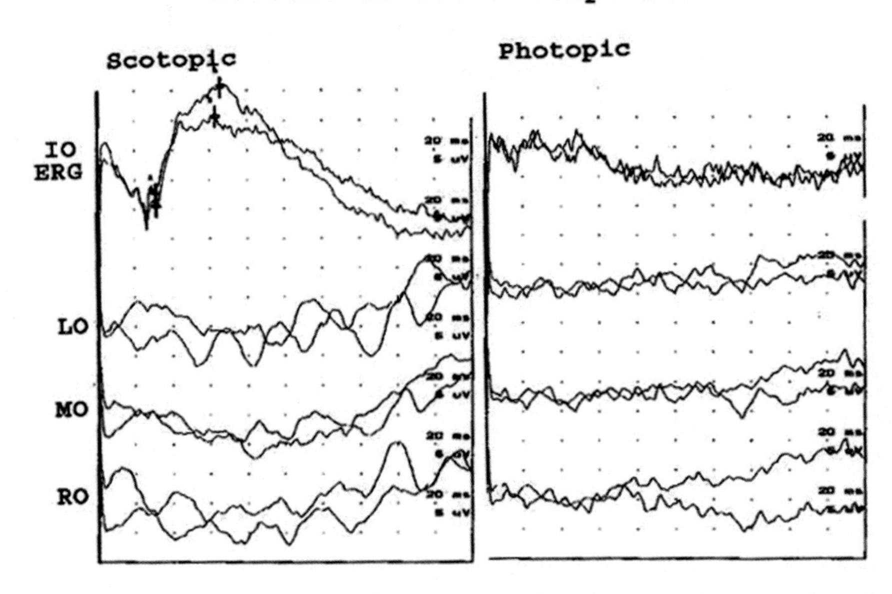

Figure 28. *This infant has a IO ERG response including an a- and b-wave I the dark adapted state, but no cortical response, suggesting the patient has functioning rods. In the light-adapted state there is no IO ERG response and no cortical response. These finding suggest loss of cone function and possibly optic nerve dysfunction.*

Functional Visual Loss

At times, clinicians suspect an individual to have functional vision loss. These individuals retain pupillary responses to light and may behave in such a way as to suggest that they can indeed see. However, they claim to have no vision. In general, they will not fixate on a target so are unable to undergo PREP testing which would be a better measure of central vision function.

The following example is such a case. While this test suggests the visual system to be intact, one cannot exclude the possibility of cortical blindness. Typically, cortical blindness occurs when both occipital poles are damaged. The patient is blind, but the pupillary reflexes are present. If the lesion is larger and involves the parietal lobe as well, cortical blindness with a denial of blindness (Anton's Syndrome) may occur. There are a number of examples in the literature of cortical blindness with normal FEPs. **(Fig. 29)**

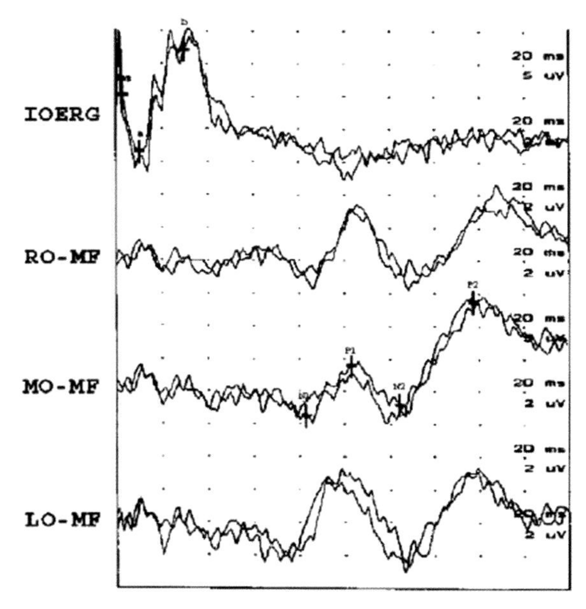

Figure 29. *This individual has a normal IO ERG and a well developed cortical response to Flash. Wave II is positive and has a latency equivalent to a p100. There is no history of trauma and no reason found for this individual to have cortical blindness.*

Behaviorally, this patient did not appear to be blind as she could maneuver around her environment without bumping into objects.

Selected Bibliography

Brindley, G.S. The variability of the human striate cortex. Proc Physiol Res (Oxford Meeting) 1972; 1P-3P

Stensaas, S.S., Eddington, D.K., Dobelle, W.H. The topography and variability of the primary visual cortex in man. J Neurosurg 1974; 40: 747-755

Cracco, R.Q., Cracco, J.B. Visual evoked potentials in man: early oscillatory potentials. Electroencephalogr Clin Neurophsiol 1978;45: 731-739

Whittaker,, S.G., Siegfried, J.B. Origin of wavelets in the visual evoked potential. Electroencephalogr Clin Neurophysiol 1983; 55: 91-101

Halliday, A.M., McDonald, W.I., Mushin, J. Delayed visual evoked responses in optic neuritis. Lancet 1972; 2: 982-984

Halliday, A.M., McDonald, W.I., Mushin, J. Visual evoked potentials in diagnosis of multiple sclerosis. BMJ; 1973; 4: 661-664

Halliday, A.M., McDonald, W.I., Mushin, J. Delayed pattern-evoked responses in optic neuritis in relation to visual acuity. Trans Ophthalmol Soc UK 1973; 93; 315-324

Farrell, D.F., Leeman, S., Ojemann, G.A. Study of the Human Visual Cortex: Direct Cortical Evoked Potentials and Stimulation. J. Clinical Neurophysiol 2007; 24: 1-10

Guideline 9B: Guidelines on Visual Evoked Potentials. J. Clin. Neurophysiol. 2006: 23: 138-1

CHAPTER 2
Brainstem Auditory Evoked Potentials

Introduction

BRAINSTEM AUDITORY EVOKED Potentials (BAEPs) were probably the first of the short-latency evoked potentials to be introduced as a clinical diagnostic tool. Prior to either CAT or MRI scans it was at times difficult to separate the various causes of coma. Structural lesions, especially in the Pons, including pontine hemorrhage or infarction tended to obliterate the central responses of the BAEPs. Drug overdoses that cause coma tend to spare the BAEPS responses based on their resistance to even deep levels of anesthesia. The advent of imaging studies, especially MRI, reduced the need for this type of physiologic study. For a while the BAEPs was used as an adjunct test in the evaluation of patients suspected of having multiple sclerosis. In recent years, the American Academy of Neurology has recommended against the use of BAEPs in diagnosing multiple sclerosis because of the very low yield of abnormalities.

Currently, BAEPs are most useful in the diagnosis of Vestibular Schwannomas and other tumors of the Cerebello-pontine angle. It also remains most useful as an intra-operative test that is continuously repeated to monitor auditory function during the course of the skull-base surgery.

Sound Intensity Measurements

There are three ways in which sound intensity can be measured. One is a physical measurement and two are biological measurements. The first and the measure that is recommended is Db pe SPL (decibels peak equivalent Sound Pressure Level. This is a physical measure and can be reproduced by any laboratory in the world. It is accomplished by hooking up a sound generator to a storage oscilloscope and recording the wave generated by 120 Db. The earphones that provide the stimulus are then hooked up to the storage oscilloscope and the test equipment stimulus is increased until the waveform has the same amplitude as created by the sound generator. An example, our most recent diagnostic equipment had to be set at 126 Db to reproduce the waveform matched to deliver the desired 120 Db.

The first of the biologic methods of sound intensity is called Db HL (Decibels Hearing Level). This is the value most often used by audiologists in the evaluation of the hearing system. This value is obtained by establishing a hearing threshold for 20 "normal young individuals". The mean of the threshold is then set to zero and in their studies the stimulus is set at 80Db above threshold. For example these studies will be done at 80 Db HL. This type of quantification is only useful for comparing individuals within a given laboratory. The values cannot be used to compare individuals from different laboratories or with individuals reported in the literature.

Finally the last method and not recommended is Db SL. In this instance a threshold is established for a single individual and one ear is then compared to the other. Again, the threshold is set to zero and 80 Db added to the hearing threshold. Example: 80 Db SL.

Polarity of the Sound Stimulus

Ear phones can deliver the sound to the ear either when the diaphragm is moving away from the ear (Rarefaction) or when the diaphragm is moving towards the ear (Condensation). In many laboratories all studies are done with rarefaction. This is good for about 80 percent of the population. Rarefaction stimulation gives the best resolved waveforms I -V, both in amplitude and latency. Rarefaction stimulates the receptor hair cells near the basal turn of the cochlea. This leaves about 20 percent where the waves are not as well resolved. Condensation stimulation is appropriate for this smaller population. Condensation stimulation tends to activate the receptor hair cells in the apical turn of the cochlea. The latency of the AP may differ slightly for the two stimulation modes. In each instance of testing the cochlea, one set of data is collected in rarefaction and a second set in condensation. The two waveforms can then be added for the best overall results. Alternating the rarefaction and condensation responses simultaneously and adding the results immediately may cause a cancellation of the waveforms and is not recommended. After the rarefaction and condensation responses are collected separately the waveforms can safely be added without cancellation.

Electrocochleogram

The electrocochleogram is obtained as a direct recording of the cochlea using an electrode placed deep in the external auditory canal close to the tympanic membrane. This electrode is frequently a small silver ball electrode held by folding wings that hold the electrode in place. The electrode in the external auditory canal is the active electrode and is referred to the ipsilateral ear. The electrocochleogram is made up of a number of different responses including cochlear microphonics, a summation potential (SP), a non- traveling wave generated by the hair cells and finally the action potential (AP) a traveling response that occurs in the auditory nerve while it is still contained within the cochlea. This AP wave is the same as Wave I of the brainstem auditory evoked potential. The simultaneous running of the Electrocochleogram and BAEPs allows one to always identify Wave I when it is present. To see all of the components described above, mathmatetical manipulations of the waves are necessary. **(Fig. 1).**

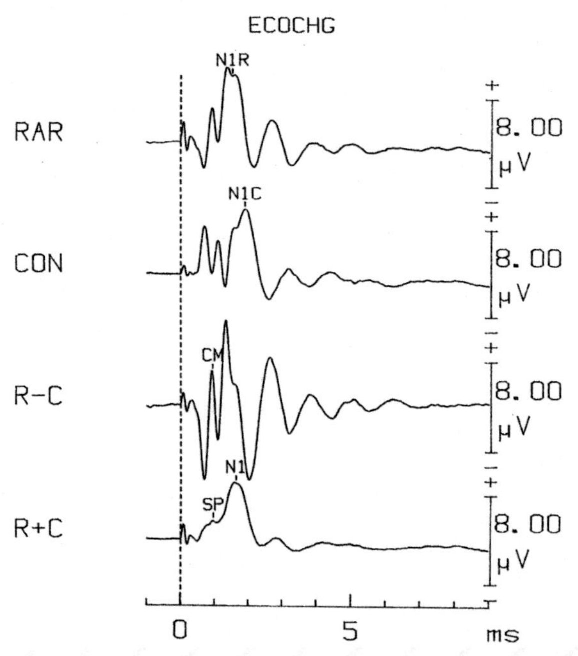

Figure 1. *First channel is collected with rarefaction stimulation at 120Db pe SPL.Note the cochlear microphonics. The second channel is the same except that the data is collected with condensation stimulation. The resulting cochlear microphonics is 180 degrees out of phase to those collected with rarefaction. When R-C is plotted (third line), the cochlear microphonics add and they become much larger. When R+C is plotted (forth line) the cochlear microphonics cancel and the summation potential (SP) and N1 or (AP) can be clearly identified.*

A second sample of the subtraction and addition of the rarefaction and condensation responses is shown in **(Fig. 2)**..

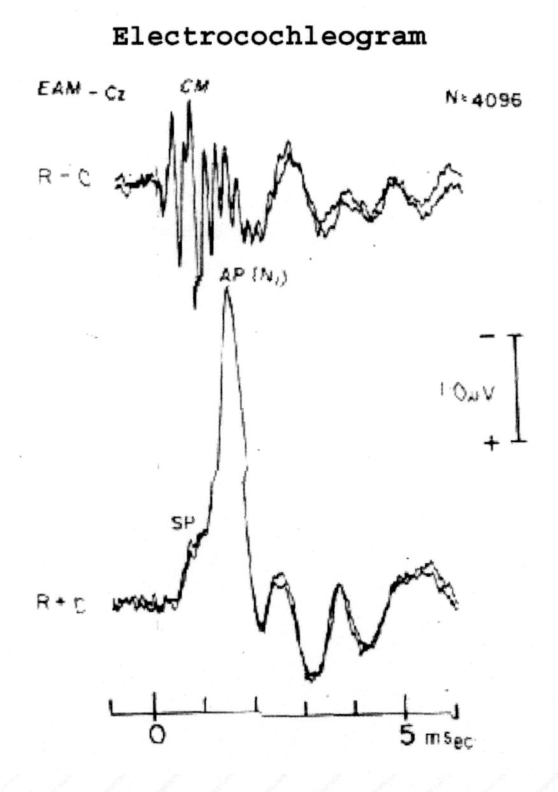

Figure 2. *Line 1, R-C shows better developed cochlear microphonics. The second line shows the small SP on the leading edge of the AP (N1) response.*

The effect of the stimulus intensity on the cochlear responses is shown in Figure 3. As the intensity of the stimulus is dropped by 20 Db intervals the latency of the response lengthens and the amplitude drops. This **(Fig. 3)** shows such a latency-intensity series.

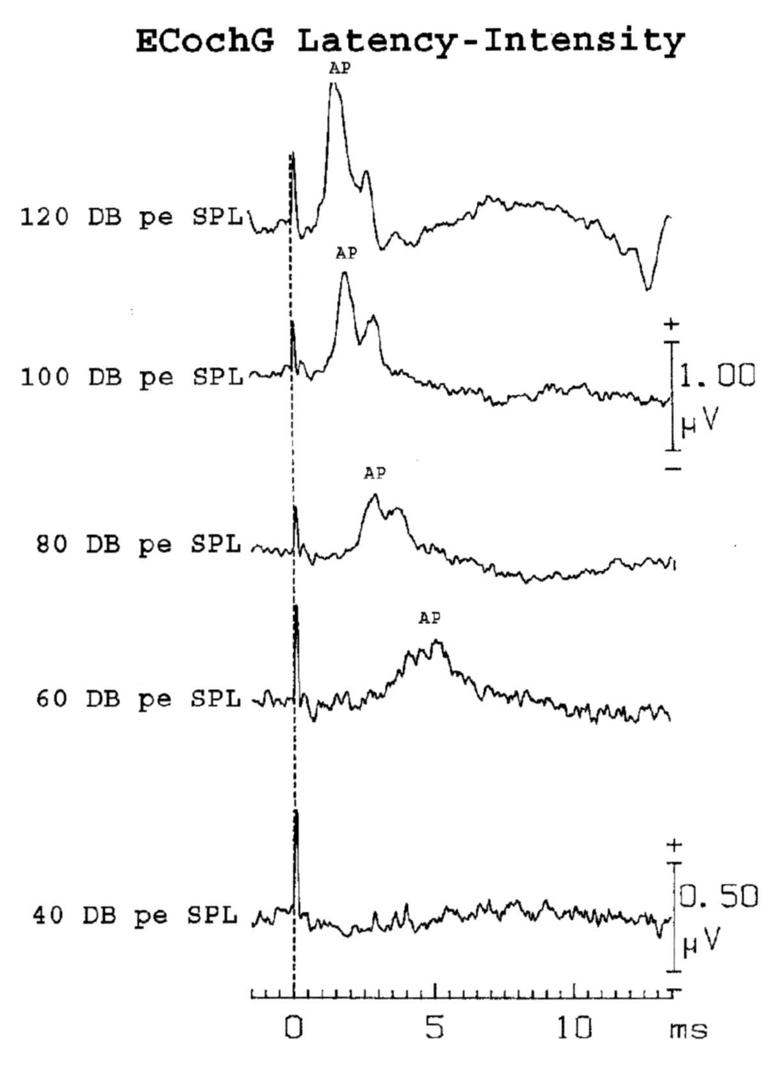

Figure 3. *Latency-intensity series as the intensity of the stimulus delivered to the ear is reduced by 20 Db pe SPL for each channel, the latency of the AP lengthens and the amplitude drops. The final channel at 40 Db pe SPL is below this persons hearing threshold.*

In addition to the identification of Wave I of the BAEPS, the ECochG may be of some value in the diagnosis of Meniere's disease. Meniere's disease or hydrops of the labyrinth is a disorder that causes very intense bouts of vertigo. The vertigo is severe enough that an individual may literally be thrown to the ground. The paroxysms of vertigo are frequently associated with nausea and in some individuals vomiting. This disorder is very difficult for the sufferer, as the attacks come in very unpredictable periods of time and the symptoms may last for hours. These patients frequently have tinnitus and develop a low tone hearing loss. The ECochG shows a characteristic change in that the summation potential (SP) increases in amplitude and exceeds 50 percent of the amplitude of the action potential (AP)**. (Fig. 4)..**

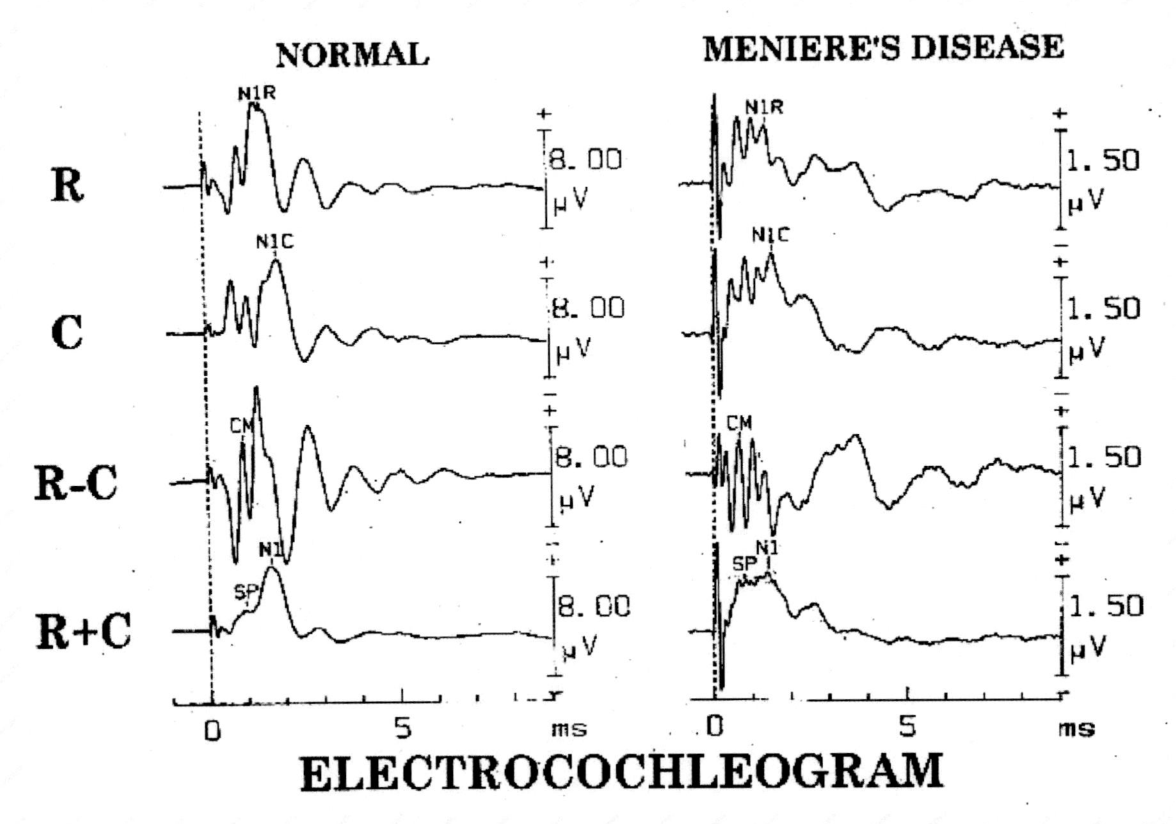

Figure 4. In the left panel in the R+C channel note the SP wave makes up about 1/3 of the amplitude of the N1 (AP). This is the normal ear. In the right panel the R+C channel shows the marked increase in the amplitude of the SP in relation to the N1 (AP), This is the typical finding in Meniere's disease.

BRAINSTEM AUDITORY EVOKED POTENTIALS - GUIDELINES

Recommended Guidelines for BAEPs. The stimulus is delivered by ear phones in the laboratory and by external auditory ear inserts in the operating room.

The stimulus is a broad band click with a mixed frequency from 4000 to 8000 Hz. The pulse is a rectangular and lasts 100 µSecond. The click generator of the stimulator should be capable of producing rarefaction only and condensation only, and alternating rarefaction and condensation clicks. The stimulation rate should be from 8 to 10 Hz, again the choice being one in which 60 is not divisible. My laboratory routinely uses 8.3 Hz. If the Wave V is difficult to identify (very rare) the stimulation rate can be increased up to 70 Hz which will reduce Waves I-III in amplitude, but enhance the amplitude of Wave V.

Stimulation intensities should be between 40 and 120 dB pe SPL in the tested ear while contra-lateral masking uses 60 dB pe SPL white noise. In the vast majority of individuals this level of masking prevents the stimulus from activating the contra-lateral cochlea.

Recording of the BAEPs in our lab uses a system band pass of 30 to 3000 HZ. This filter setting allows a long duration slow wave to be recorded and the BAEPs will ride on this slow wave. To eliminate the slow wave, the low frequency filter can be raised to either 100 Hz or 300 Hz. These higher low frequency filters demonstrate BAEPS waveforms that appear similar to what seen in the literature. It also allows one an additional means of identifying Wave V as it is the last of the series of waves that on its down slope crosses the baseline. The low frequency filter has a roll off of< 12dB/octave and the high frequency filter a roll off of<24 dB/octave.

Analysis time is either 10 or 15 milliseconds, and the number of trials varies with the individual from 1000 to 4000.

The scalp electrode should be placed at Cz of the standard 10-20 system. This is the active electrode for Waves II through V. Additional electrodes should be placed on each ear, A1 and A2. When doing an electrocochleogram the reference electrodes for the electrocochleogram are A1 or A2 and the active electrode is the external auditory meatus electrode (EAM). The montages for the BAEPs include: Cz to Ipsi-lateral ear (A1 in this case) and Cz to Contra-lateral ear (A2). With input 1 being the positive amplifier input and input 2 being the negative amplifier input. With this combination two channels of BAEPs are recorded. The stimulated ear shows all five Waves of the BAEPs. It is not uncommon for Waves IV and V to be fused on the ipsi-side, while the contra-lateral channel which does not show a wave I, but more consistently separates Wave IV and V. **(Fig. 5).**

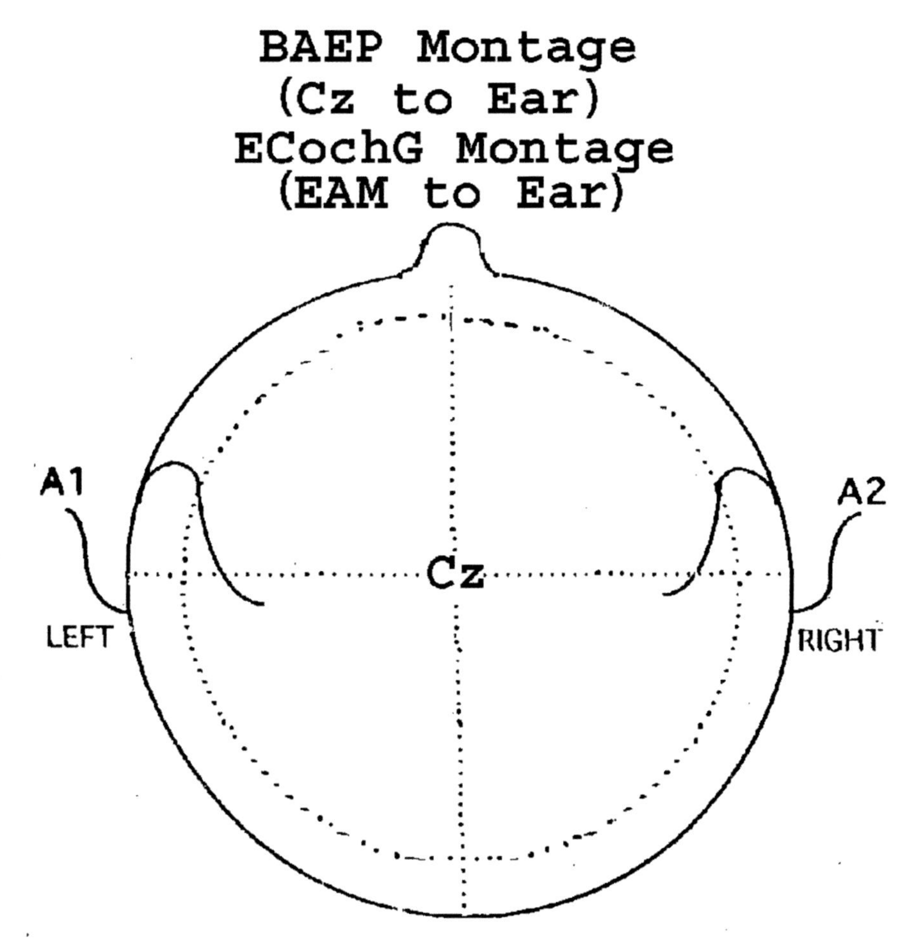

Figure 5. *Diagram shows the location of the electrodes used to record the BAEPs, contra-BAEPs, Electro-cochelogram and contra-ECochG.*

The following samples show the details of the BAEPS. The first is a 4 channel BAEPS demonstrating the usefulness of the ECochG in the identification of Wave I. **(Fig. 6).**

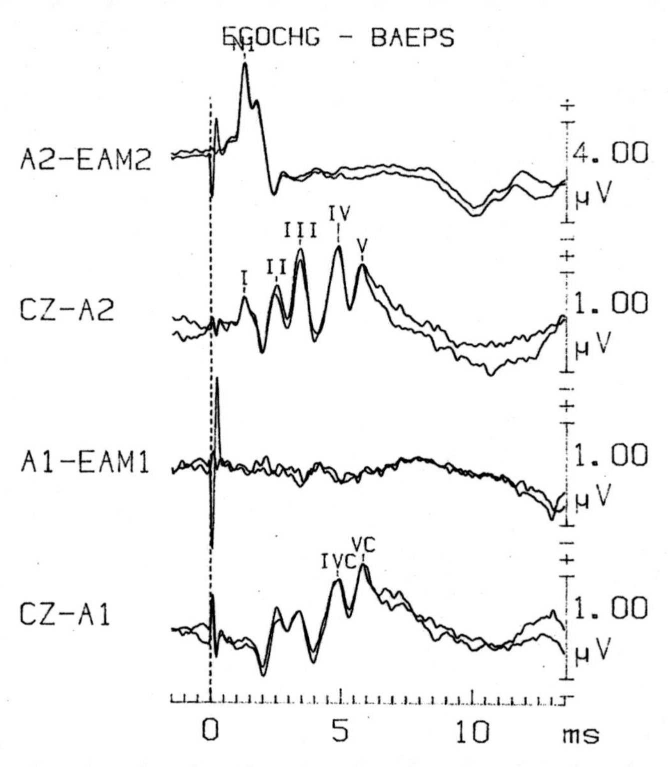

Figure 6. *This 4 channel BAEPs shows the ECochG –R+C derivative in channel 1. The second channel shows the Ipsi-lateral BAEPs with all five major Waves (I-V) clearly defined. Channel 3 is the contra-lateral ECochG showing no contamination from the Ipsi-lateral side. The 4th channel is the contra-lateral BAEPs showing the absence of Wave I, but well developed Waves II-V. In both instances the Wave IV and V is well separated.*

The next figure is a 2 channel BAEPS showing only the Ipsi-lateral and Contra-lateral BAEP. **(Fig. 7).**

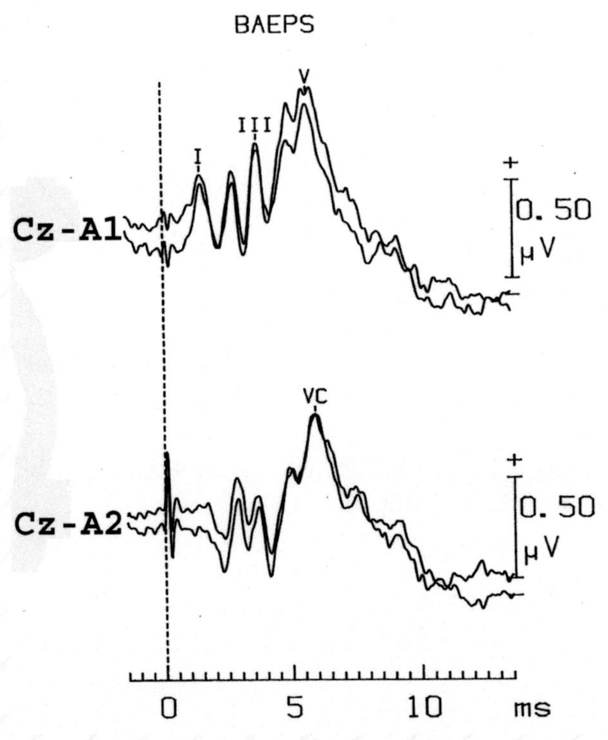

Figure 7. *This 2 channel BAEPS shows the typical responses recorded from both the Ipsi-lateral and Contra-lateral side. Note the separation of Waves IV and V is better seen from the contra-lateral ear.*

A plot of the normative data used in my lab is shown in the next Figure. This includes a Latency-Intensity Series for the major Waves I, III, and V as well as the more important Wave intervals, I - III, III - V and I – V. **(Fig. 8).**

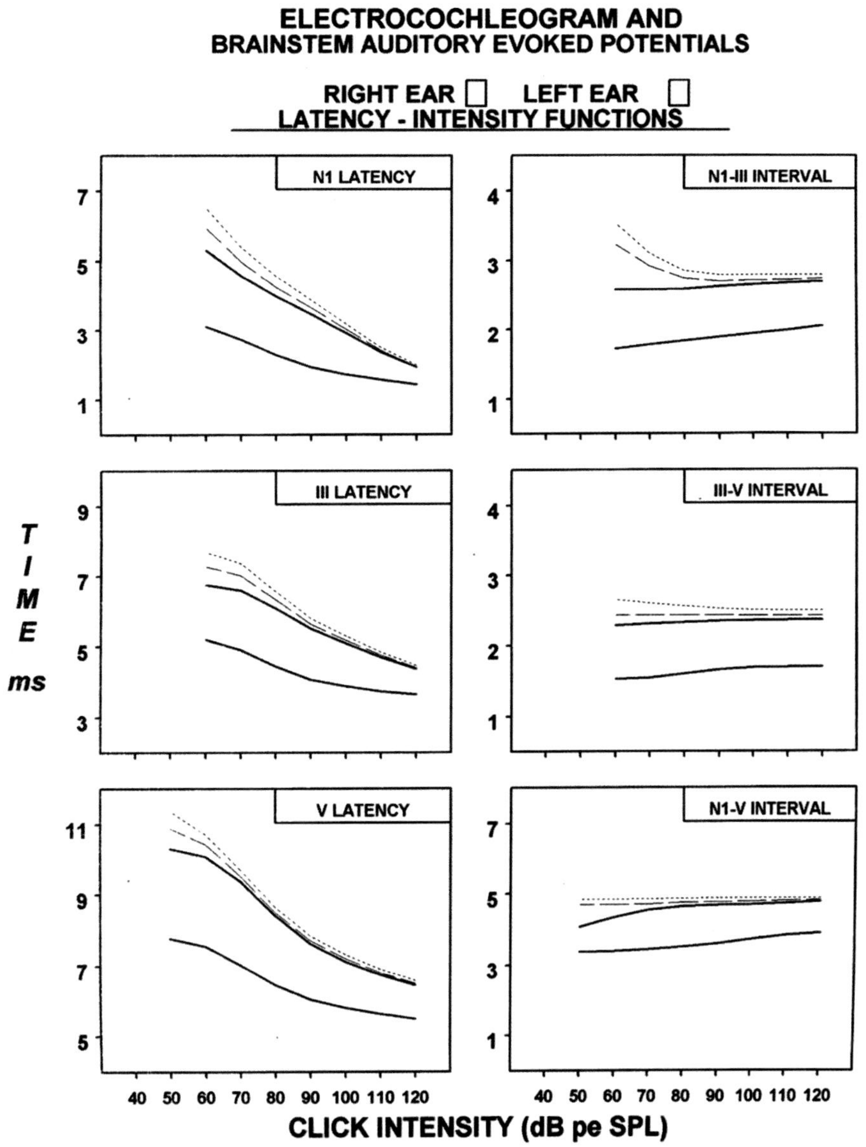

Figure 8. *Latency-Intensity series. The 3 blocks on the left side of the form shows the normal latencies for Waves I, III, and V. Note, there is a fairly sharp slope with increasing latencies as the intensity of the stimulus is reduced. The 3 blocks on the right show the inter-peak intervals which remain nearly constant as the stimulus intensity is reduced. The solid lines represent the mean +/- 3 SD. The dotted line represents a correction based on the individual's audiogram. This form is corrected for any hearing loss that is greater than 20 Db at 4000 Hz.*

Generators of the various Waves that make up the BAEPS

Wave I is formed in the Acoustic nerve before the nerve exits the Cochlea. The Wave is actually negative in polarity with the ear (A1 or A2) being the active electrode and the vertex (Cz) the reference electrode. All the other Waves are positive as the Vertex (Cz) is the active electrode and the ear (A1 or A2) the reference electrodes. Wave II is generated in the Acoustic nerve, most likely as the nerve leaves the boney canal and enters the subarachnoid space. This change in environment probably accounts for this standing wave. Direct Auditory nerve recordings have shown the action potential to have a latency approximate to Wave II. At no time were there any direct nerve latencies long enough to be Wave III. This included placing an electrode at the junction of the Auditory nerve and the pons (See Figure 9). Wave III has two possible generators, the first is the synaptic connections in the auditory nucleus the second is the first relay station after the auditory nucleus, the superior olivary nucleus. After synapsing in the auditory nucleus complex, some of the auditory fibers ascend to the superior olivary nucleus and synapse. Others fibers originating in the auditory nucleus descend then cross and synapse in the mid-line trapezoid nucleus. The ipsilateral fibers after ascending to the superior olivary nucleus where the fibers synapse, then ascend to the level of the inferior colliculus where the fibers make a near right angle before synapsing in both inferior colliculi.

After synapsing in the Trapezoid nucleus, the contralateral fibers continue laterally then ascend to the level of the inferior colliculi and synapse in both colliculi. This complex route up the brainstem and the further bilateral distribution to the medial geniculate nucleus and finally to both superior temporal lobes, the reason why hearing loss from central lesions is rare except in bilateral cortical disease. Wave IV is thought to result from the sharp change in direction as the fibers reach the level of the inferior colliculi and Wave V from synaptic activity occurring in that nucleus. **(Fig. 9).**

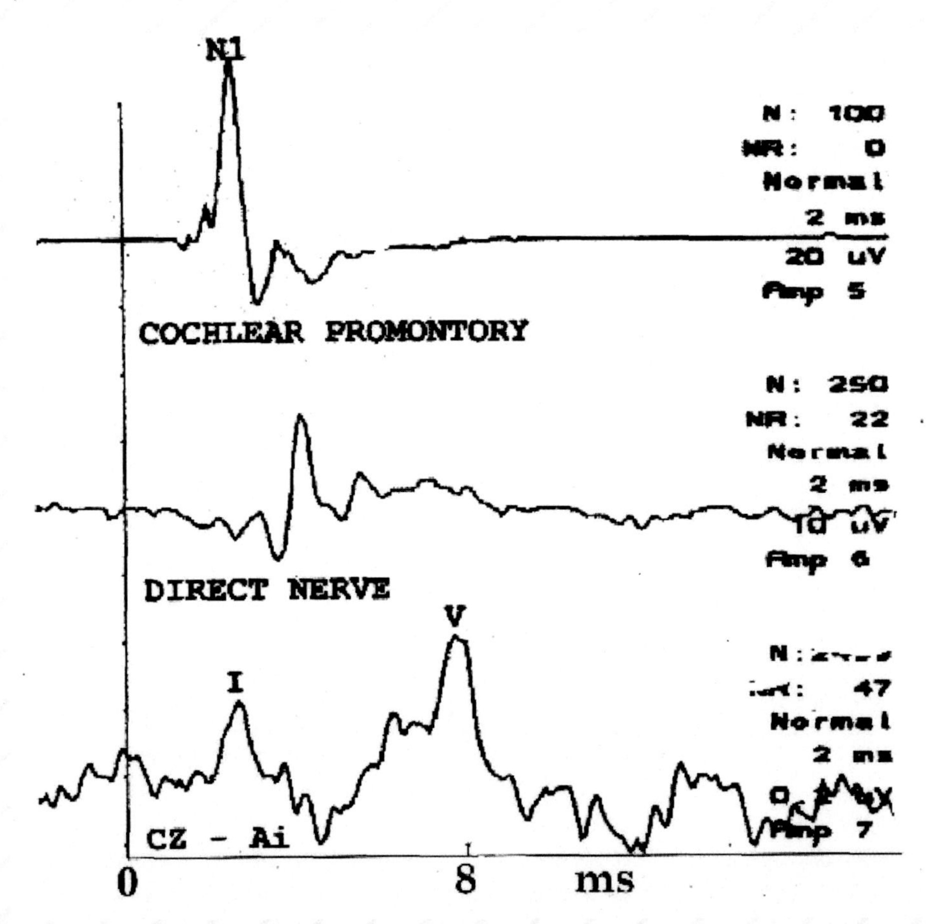

Figure 9. *Channel 1 shows the AP (N1) of the Electrocochleogram and its relationship to Wave I of the BAEPs. The second channel is the direct recording of the action potential from the Auditory Nerve . The initial downward wave has a positive polarity, like all action potentials recorded from the external surface of a nerve. This direct response approximates Wave II of the BAEPs. Channel III shows Waves I and V in a patient undergoing surgery for a large Vestibular Schwannoma. Wave III is not marked, but the direct response occurs with a shorter latency than Wave III.*

From a neuro-diagnostic point of view, absolute latencies of the various Waves, I to V are less important than the intervals between the waves. Audiologic disorders such as conductive hearing loss secondary to middle ear disease or even sensory neural-hearing loss affect the absolute latencies, but do not generally affect the inter-wave intervals. A side to side latency difference of greater than 0.5 milliseconds of the wave V is abnormal. On the other hand, disorders of interest for the neurologist or neurosurgeon, namely tumors, affect the absolute latency but they also affect the inter-wave intervals.

Vestibular Schwannomas

Vestibular Schwannomas are estimated to occur approximately once in every 100,000 people. They may occur at any age, but tend to occur in persons over 40 years of age. This tumor has in the past been mistakenly called an Acoustic neuroma. It is a slow growing benign tumor that begins on the vestibular nerve in the boney canal leading from the labyrinth to the porous acousticus. The auditory nerve lies adjacent to the vestibular nerve; the plane between these two nerves is frequently marked by a small artery. The Facial nerve (VII) runs in proximity to the vestibular-acoustic nerve complex. These tumors may be present for many years before symptoms develop. A small percentage of small tumors that start deep in the auditory canal and cannot decompress themselves by growing into the subarachnoid space may lead to sudden complete hearing loss. These small tumors are thought to compromise the blood supply to the nerve or cochlea. More commonly, the tumor starts closer to the subarachnoid space and as the tumor grows it erodes the boney opening and expands into the subarachnoid space. By far the vast majority of vestibular schwannomas are sporadic in nature and occur only on one side of the head. About 5 percent of these tumors are bilateral and occur as part of a genetic condition, Neurofibromatosis II. This condition is so different from the sporadic form of the disorder that it will be discussed in a later section. The earliest symptoms from vestibular schwannomas include mild tinnitus and hearing loss. The nature of the hearing loss is fairly specific for vestibular schwannomas. The spoken word is affected earliest so that the patient may have trouble understanding spoken language when using a telephone. This is especially true when the tumor occurs in the ear used for the phone. If this tumor is not diagnosed at this stage, it will continue to grow and eventually get large enough that it distorts the brainstem and the affected patient develops an unsteady gait. The diagnosis of vestibular schwannomas is based on the history, especially the development of a unilateral hearing loss. An audiogram at this stage typically shows a mild high tone hearing loss with a great many errors in the spoken language portion of the test. One hallmark thought to be very specific is the presence of "rollover". In this case the patient will have an increase in the error rate to spoken language as the intensity of the stimulus is increased. BAEPs provide a useful and sensitive diagnostic test for vestibular schwannoma. A number of different patterns can be seen. A very small percentage of cases with a very small tumor will have a normal BAEPS. The most common specific change is a prolongation of the Wave I - III interval. **(Fig. 10 and Fig. 11)..**

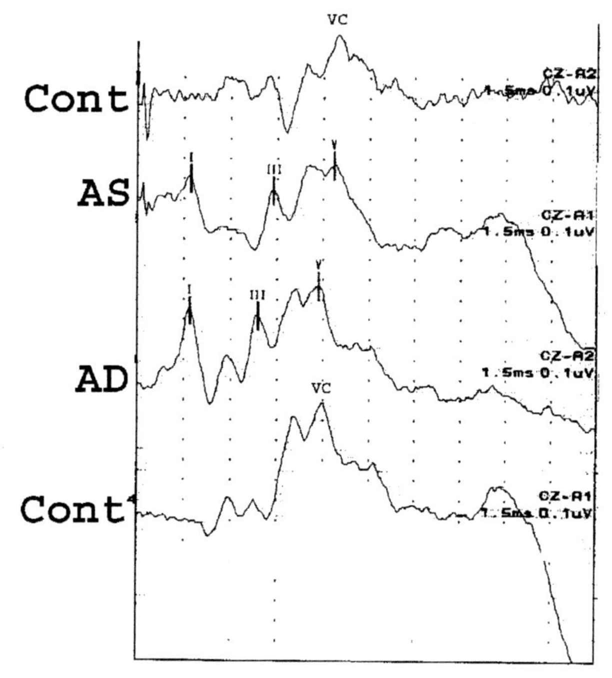

Figure 10. *Individual has a small tumor on the left. The right BAEPS remains normal. On the affected side, the Wave I – III interval is prolonged while III-V interval remains normal. The left BAEPS is diagnostic for a vestibular schwannoma.*

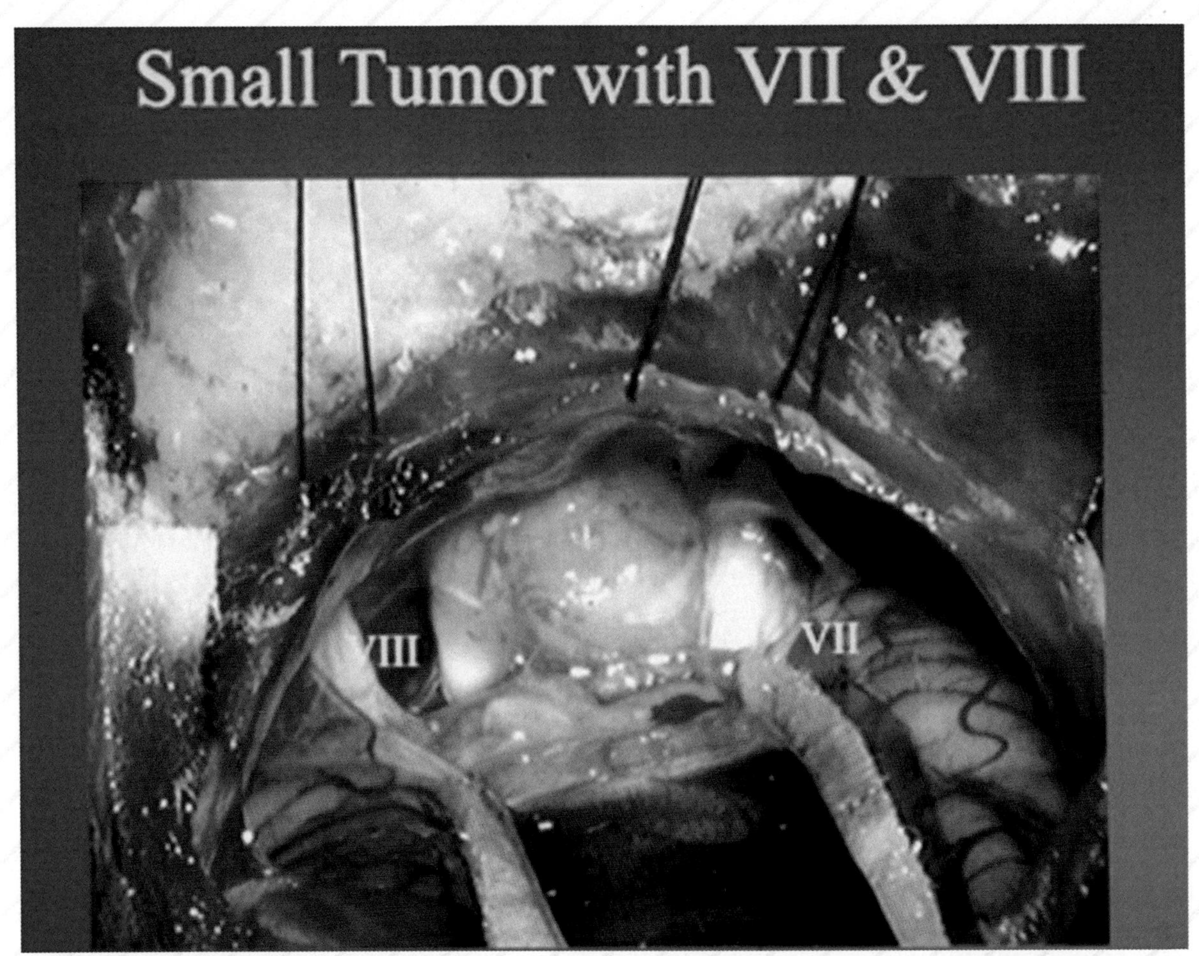

Figure 11. *Small vestibular schwannomas which separates the VIIIth Nerve on the left and the VIIIth Nerve on the right. This is a very unusual relationship as it is more common to see VIII under the tumor and VII medial to that.*

Probably the second most common abnormality seen in the BAEPS is when the patient has lost Wave III and only Waves I and V remain. In this instance there is a prolonged I – V interval. This finding is common, but not as specific as when the Wave III is present.

When the tumor enlarges enough to compress the brainstem additional changes in the BAEPS occur. If Wave III remain, then both the Wave I - III interval and the Wave III - V interval become prolonged. At times the III - V interval lengthening is also seen in the contra-BAEPs.**(Fug. 12 and Fig. 13).**

Prolonged I-III,III-V,I-V
Large Vestibular Schwannoma

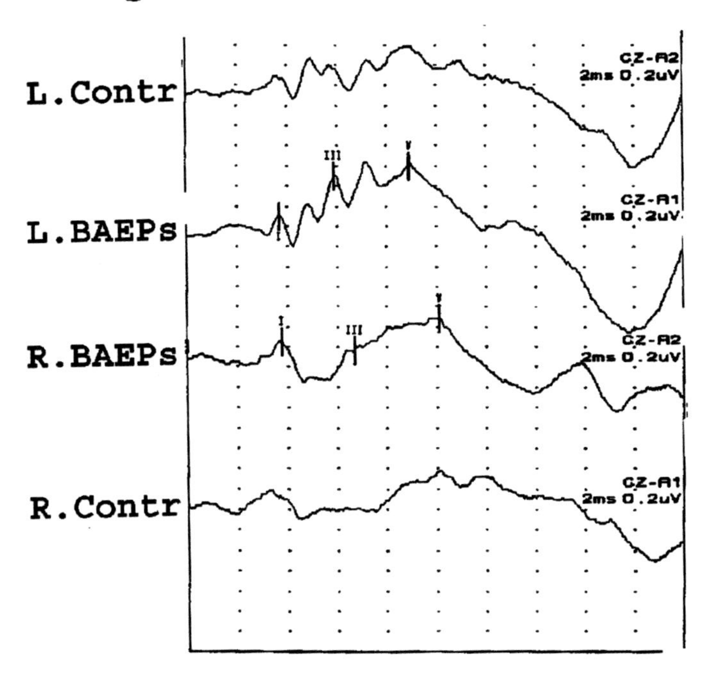

Figure 12. *The affected ear is the right. Note that The Wave I – III and the Wave III – V intervals are both prolonged.*

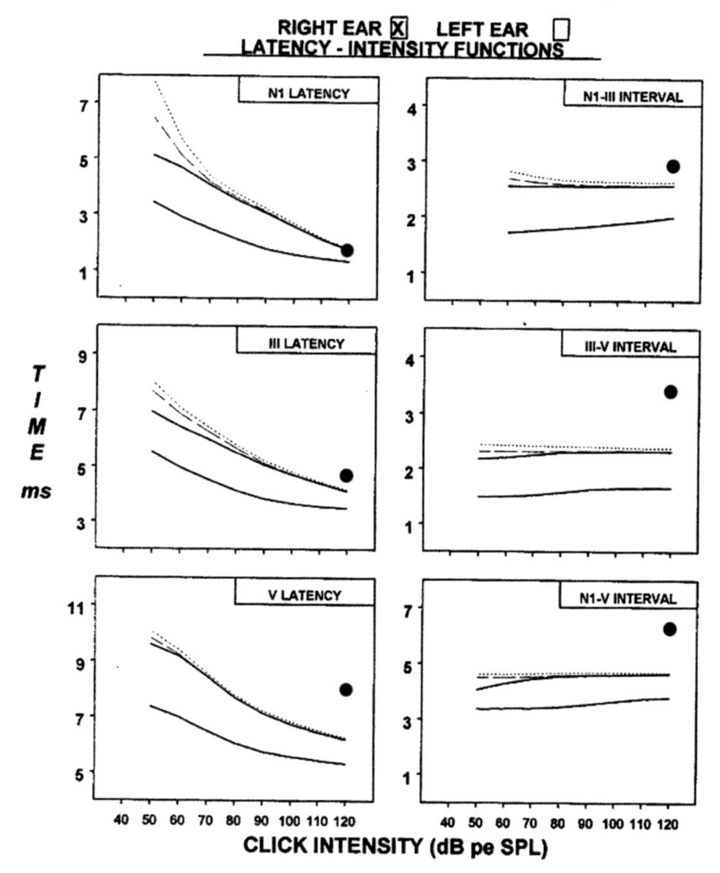

Figure 13. *Form shows the marked prolongation in the intervals I – III, III – V and I – V.*

The next two figures are MRI scans of both a large Vestibular Schwannoma and a Giant Vestibular Schwannoma. The MRI is very sensitive in demonstrating Vestibular Schwannomas. Tumors as small as 2 millimeters can be seen in the internal auditory canal. **(Fig. 14 and Fig. 15).**

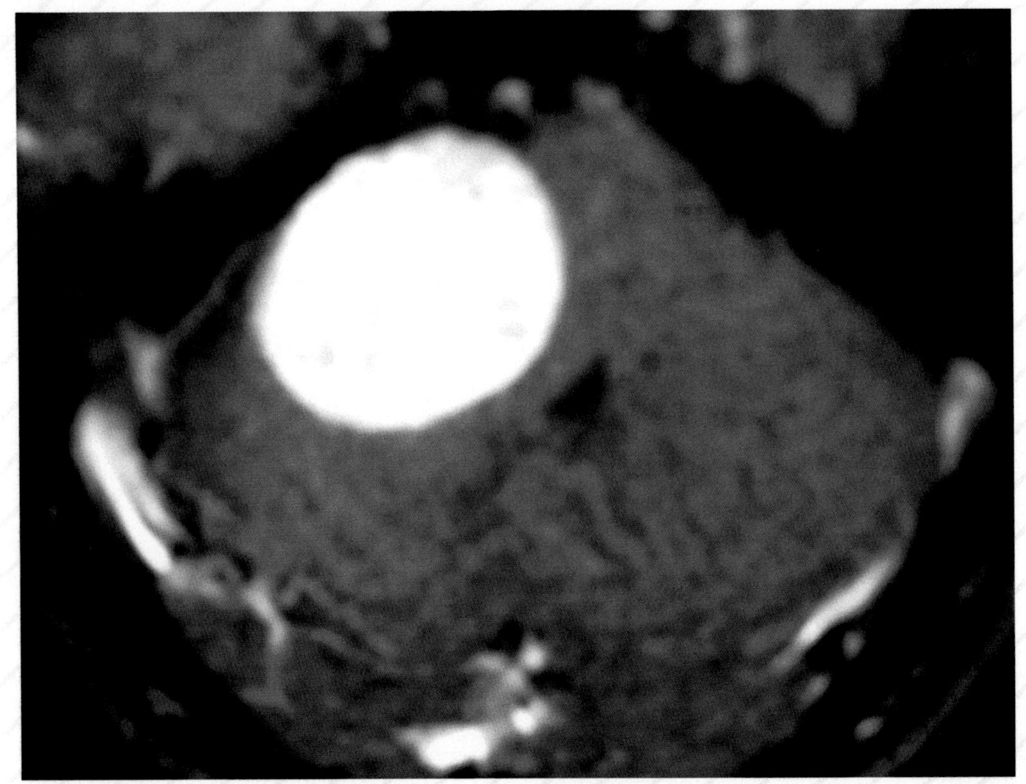

Figure 14. *MRI scan of a large vestibular schwannomas compressing the Pons and distorting the IVth Ventricle.*

Giant Vestibular Schwannoma

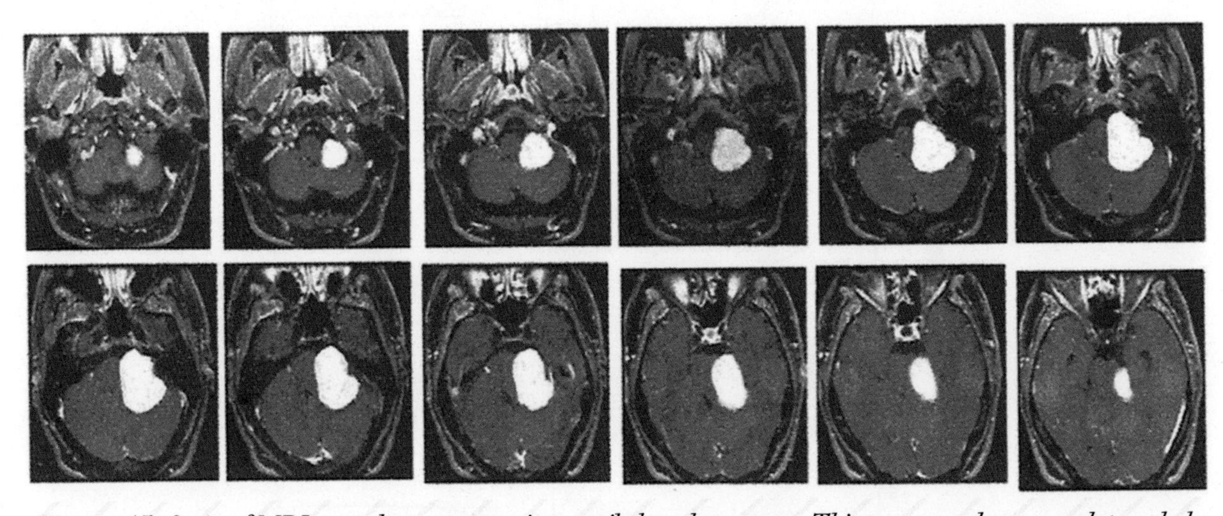

Figure 15. *Series of MRI scans demonstrate a giant vestibular schwannoma. This tumor was large enough to occlude the IVth ventricle and compress the cerebellum. Note the temporal horns are not enlarged.*

Further discussion of vestibular schwannomas will continue in the Chapter on Skull-base surgery.

NEUROFIBROMATOSIS, TYPE II

Neurofibromatosis, type II is a rare disorder generally inherited as an autosomal dominant disease. It has been estimated to occur at a rate of 1 in 60,000 persons. About 50 percent of cases are inherited, while another 50 percent are thought to result from new mutations. The defective gene product has been called "Merlin" or "Schwannomin". The defective gene is located on chromosome 22 band q11-13.1. The protein is thought to have a tumor-suppression function. The defective protein allows for a lack of contact-mediated cell suppression and

results in uncontrolled tumor growth. A number of different mutations are known to occur and there is no predominant form. These different mutations probably account for the differences in tumor growth. The pathologic characteristics of this disorder are the presence of bilateral vestibular schwannomas. In addition, patients with this condition may also have meningiomas and ependymomas. There appear to be two phenotypic forms of this disorder. The first form shows symptoms under the age of 20 and the tumors show rapid growth and re-growth after debulking. The second form occurs after age 20 and the vestibular schwannomas are slower growing, and additional solitary tumors such as meningiomas are frequently present.

Schwannomas may also affect almost any cranial nerve. The spinal cord can also be affected by schwannomas, meningiomas and ependymomas.

Hearing loss is an early symptom of the Neurofibromatosis II and is the most frequent complaint. Tinnitus is next in frequency followed by unsteadiness of gait. Additional symptoms and signs may result from tumors in other locations. Tumors may vary in size from side to side or may be large and relatively symmetric as demonstrated in the next 2 Figures. **(Fig. 16 and Fig. 17).**

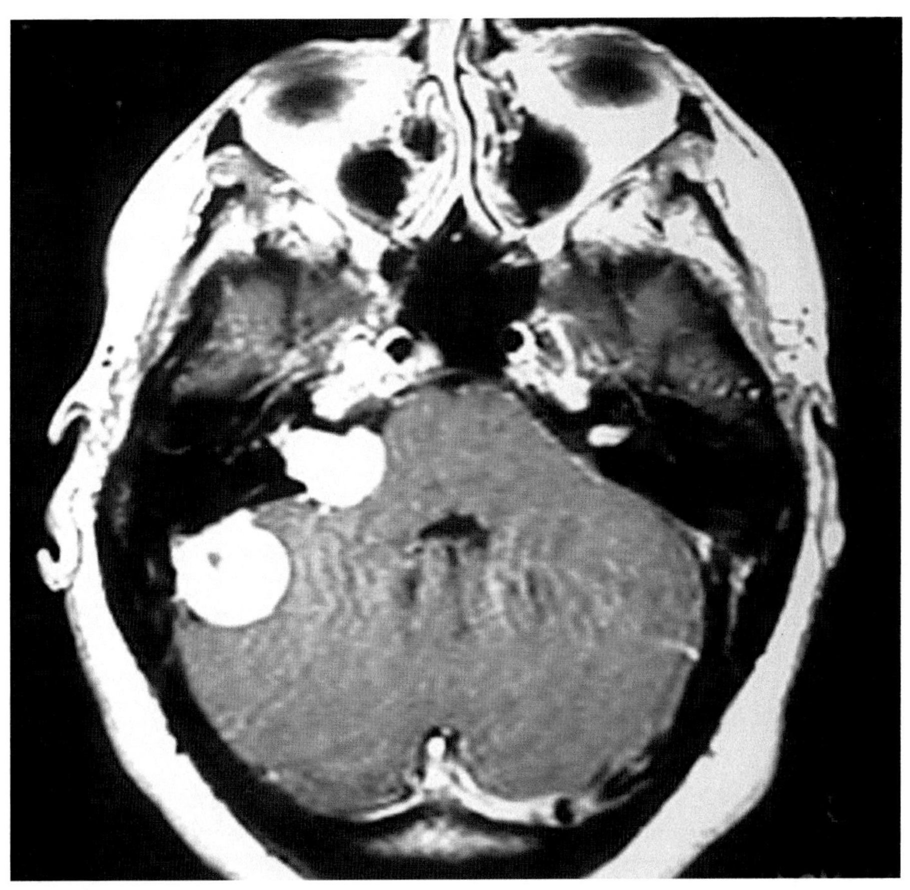

Figure 16. *MRI Scan demonstrating bilateral vestibular schwannomas. The smaller of the two is entirely within the auditory canal while the larger has eroded out of the canal and is compressing the pons. Note, this patient also has a meningioma compressing the cerebellum.*

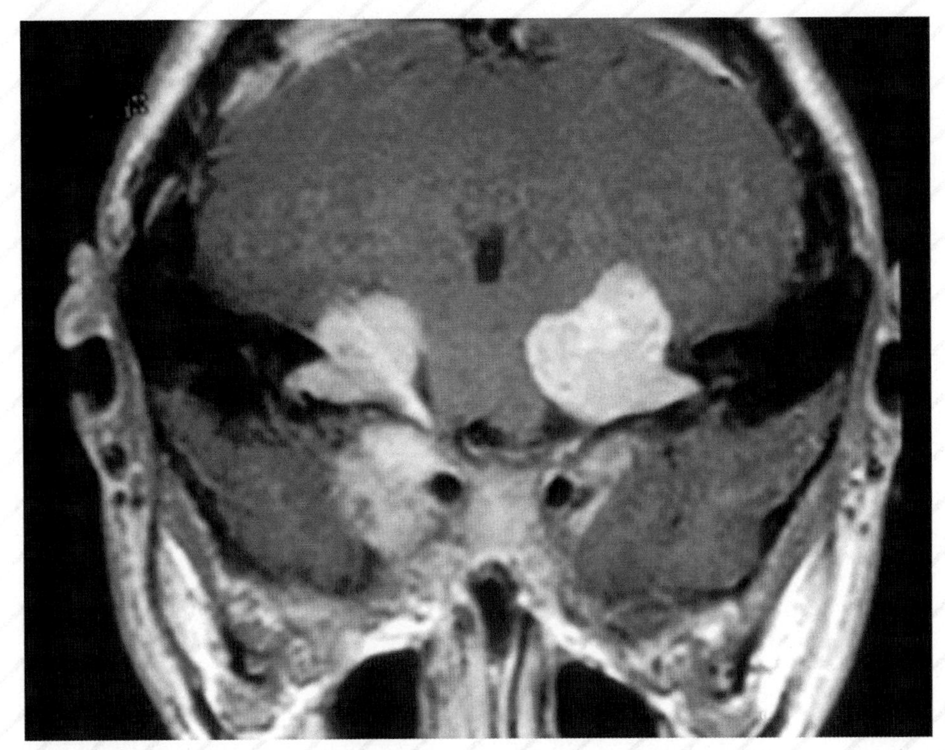

Figure 17. *This patient has bilateral vestibular schwannomas that have eroded out of the Internal acoustic canals and are both compressing the Pons and to a lesser extent the cerebellum.*

BAEPs in individuals with NF II are almost always abnormal on both sides as shown in the next figure. **(Fig. 18 and Fig. 18).**

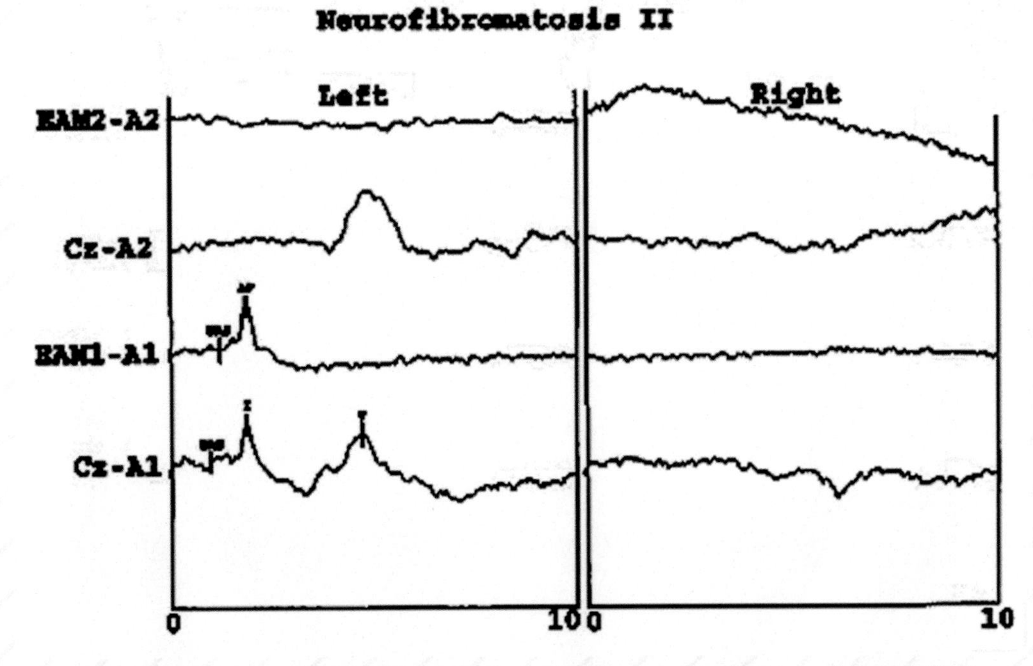

Figure 18. *The BAEPs are abnormal on both sides. The right side shows No function while the left side continues to have a Wave I which has a normal latency and an I-V interval that is prolonged.*

Neurofibromatosis, type II is a complicated disorder and the surgical management is difficult. The goal is to preserve hearing and facial function as long as possible. Further discussion of the surgical management of this disorder will be found in the intra-operative section of this book.

Developmental BAEPs

BAEPs have been used for a number of years to evaluate hearing in premature infants and high risk newborn infants. Pre-mature infants with a birth weight less than 1500 grams (<34 Weeks gestation) are at a high risk for hearing loss and the development of cerebral palsy. Certain full-term infants are also at high risk. This includes infants with hyperbilirubinemia, congenital perinatal infections, craniofacial abnormalities, use of ototoxic antibiotics, and a family history of congenital hearing loss.

Since all premature infants are at risk for hearing impairment, there is no definitive control BAEPS data. Abnormalities are generally the absence of all responses in both ears. There is control data for the normal full-term infants and the latencies of Waves I, III, and V are prolonged compared to the adult values. Wave I latency is about 0.3 milliseconds longer than the adult value. Wave III about 1.0 milliseconds longer than the adult values and Wave V latency is 1.3 milliseconds longer. Over the first two years of life there is a gradual reduction in the various latencies and adult values are obtained by 2 years of age.

Unlike adult studies, the intensity of the broad band click uses 80 dB HL. In general in the normal neonate and or full-term infants at high risk, Waves I, III and V are present. At times only the Wave V is obtained and as long as the latency is reasonable, the response is considered normal. These studies are usually carried out using two levels of stimulation, 40 dB HL and 80 dB HL. Zero dB HL is threshold under these conditions so that at the two levels being tested the latencies of the waves should be shorter at the higher levels of stimulation. Rarely one finds that as the intensity of the stimulus is increased the latency of Waves III and V prolongs. This finding is an artifact, and is caused by the collapse of the external auditory canal leading to a severe conduction block. To correct this artifact, ear inserts need to be used to replace the ear phones and the studies repeated. Any infant found to have abnormal BAEPs in the screening period should have the studies repeated at 6 month of age. It is not uncommon for the infant to have fluid in the middle ear and that may be the cause of an abnormal study. One must also remember that the standard BAEPS only measures the hearing pathway up to the level of the inferior colliculus. It is possible to have a normal BAEPS and the infant still have a significant hearing loss.

It is extremely important for hearing loss in premature and at high-risk full term infants to be identified as early as possible, so that special educational techniques for language development can be started early.

CORTICAL AUDITORY EVOKED POTENTIALS

It is possible to record direct auditory evoked potential from the primary auditory cortex. This area is present in the posterior superior temporal cortex and is organized in a sound frequency orientation with the lower frequencies occurring laterally and the higher frequencies occurring medially.

Testing conditions must be changed from those used to generate BAEPs. Pure tone sounds (PIPs) are used as the stimulus and the lower frequency PIPs (500 Hz) yield a better response than the higher frequency PIPs (2000 Hz). The rate of stimulation is reduced from about 10 Hz to 1 Hz to obtain these responses. Unlike short-latency BAEPs, the patient must remain alert in order to record these long-latency auditory cortical potentials. **(Fig. 19).**.

Auditory Evoked Potentials

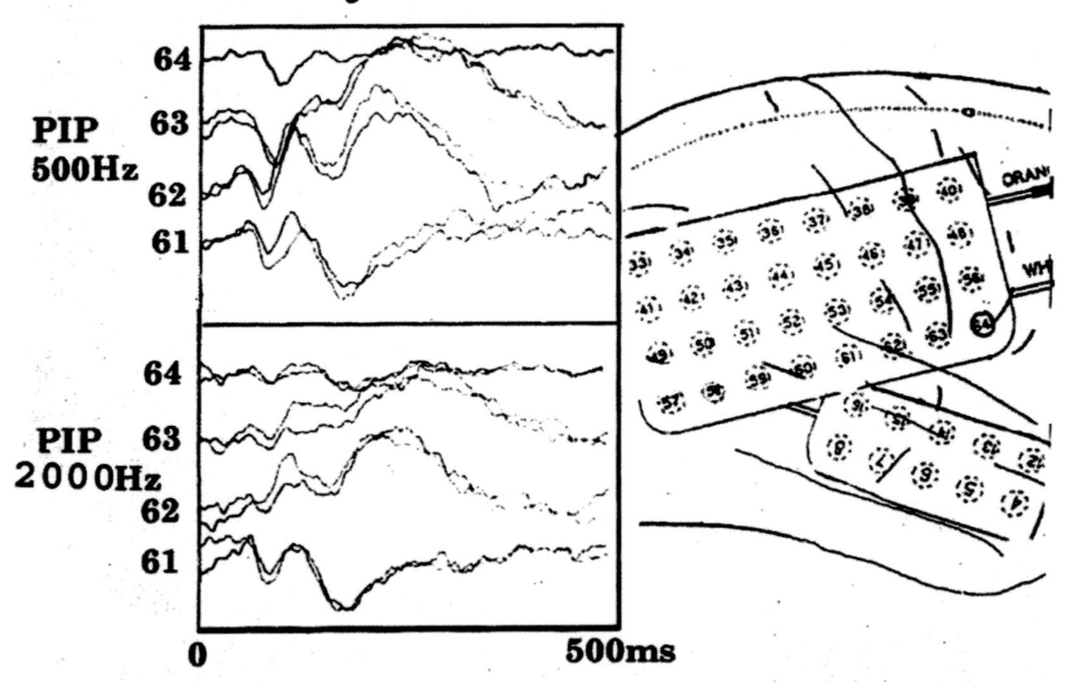

Figure 19. *Direct recording from the auditory cortex. The response to a 500 Hz PIP are much better defined than those which derive from deeper Locations in the temporal cortex (2000 Hz PIP).*

Unfortunately, the cortical auditory responses have little to no practical use as this is a bilateral system and unilateral lesions are unlikely to cause hearing abnormalities. Cortical hearing loss is almost always due to bitemporal strokes involving Heschel's gyrus.

SELECTED BIBLIOGRAPHY

Chatrian, G.E., Wirch, A.L.,Lettich, E.,, et. al. Click-Evoked Human Electrocochleogram. Noninvasive Recording Method, Origin and Physiologic Significance. Am J EEG Technol 1982; 22: 151-174

Chatrian, G.E., Wirch, A.L., Edwards, K.H.,, et. al. Cochlear Summating Potential recorded from the External Auditory Meatus of Normal Humans. Amplitude-Intensity Functions and relationships to Auditory Nerve Compound Action Potential. EEG and Clin Neurophysiol. 1984;59: 396-410

Cann, J. and Knott, J. Polarity of acoustic click stimuli for eliciting brainstem auditory evoked responses: a proposed standard. Am J EEG Technol. 1979, 19:125-132

Coats, A.C. Human auditory nerve action potentials and brainstem evoked responses: Latency-intensity functions in detection of cochlear and retrocochlear pathology. Arch. Otolaryngol. 1978;104: 709-717

Coats, A.C. The summating potential and Meniere disease. I. Summating potential amplitude in Meniere and non-Meriere ears. Arch Otolargynol .1981; 107: 199-206

Goin, D.W., Staller, S.J., Asher, D.L., et. al. Laryngoscope 1982; 92: 1383-1389

Ojemann, R.G., Levine, R.A. Montgomery, W.M., et. al.. Use of intraoperative evoked potentials to preserve hearing in unilateral acoustic removal. J Neurosurg. 1984; 61: 938-948

Starr. A., Picton, T.W., Sininger, Y., et. al.. Auditory Neuropathy. Brain 1996: 119: 741-753

Parker, S.W., Chiappa, K.H., and Brooks, E.B. Brainstem auditory evoked responses in patients with acoustic neuromas and cerebello-pontine angle meningiomas. Neurology ; 30: 413-414

Maurer, K. Strumpel, D. and Wende, S. Acoustic tumour detection with early auditory evoked potential and neurological methods. J. Neurol. 1982: 227: 177-185

Eggemont, J.J., Don, M. and Brackman, D.E. Electrocochleography and auditory brainstem electric responses in patients with pontine angle tumors. Ann. Otol. Rhinol. Laryngol. 1980: 89: Suppl. 75

House, J.W., amd Brackman, D.E. Brainstem auditory potentials in neurologic diagnosis. J. Otolaryngol. 1979: 105: 305-309

Collet, L., Delorme, C., Chanal, J. M., et.al. Effect of stimulus intensity variation on brain-stem auditory evoked potentials: Comparison between neonates and adults. Electroencephalogr. Clin. Neurophysiol. 1987: 68: 231-233

Gafni, M. Sohmer, H., Gross, S., et. al. Analysis of auditory brain-stem responses (ABR) in neonates and very young infants. Arch. Ototrhinolaryngol. 1980: 229: 167-174

Krumholtz, A., Felix, J.K., Goldstein, P.J., et. al. . Maturation of the brain-stem auditory evoked potential in premature infants. Electroencephalogr. Clin Neurophysiol. 1985: 62: 124-134

Schulman-Galambos, C. and Galambos, R. Brain stem evoked response audiometry in newborn screening. Arch. Otolaryngol. 1979: 105: 86-90

Bahls, F.H., Chatrian, G.E.. Mesher, R.A., et. al. A case of persistent cortical deafness: Clinical neurophysiologic and neuropathologic observations. Neurology 1988: 38: 1490-1493.

Guideline 9C: Guidelines on Short-Latency Auditory Evoked Potentials. J. Clin. Neurophysiol. 2006: 23: 157-167

CHAPTER 3
Short-Latency Somatosensory Evoked Potentials

SHORT-LATENCY SOMATOSENSORY EVOKED potentials (SSEPs) are created when one of the major peripheral nerves are stimulated. Typically in the upper extremity the median nerve is the nerve of choice, however, the ulnar nerve can also be used without changing the recording sites or the latencies of the various waves. In the lower extremity, the posterior tibial nerve at the ankle is the nerve of choice, but occasionally that nerve cannot be used (for example in a person with a below the knee amputation) then the common peroneal nerve at the knee can be used. Again, the recording sites for the various responses are the same as posterior tibial nerve responses, but 10 milliseconds must be subtracted from the posterior tibial nerve latencies.

SSEPs generate a series of waves along the neuraxis and therefore are of great value in localizing conduction delays in peripheral nerve, spinal cord, brainstem and cerebral cortex. Combinations of median nerve SSEPs and posterior tibial nerve SSEPs may help localize lesions to the spinal cord, especially between the mid-cervical region and the lower thoracic region.

Because of the amount of the nervous system traversed the posterior tibial nerve response has the highest yield for identifying plaques in multiple sclerosis. This test along with PREPs, provides a combination of tests that will yield the highest number of abnormalities in patients with multiple sclerosis.

Prolonged inter-wave intervals are very valuable is localizing a "lesion"; however, these findings are not etiologically specific and require additional diagnostic tests such as MRI scans. One needs to recall that the evoked potentials provide useful physiologic data and that the MRI scans provides anatomic data. The tests are complementary and should be used together on many occasions.

Median nerve SSEPs have been found to be very valuable in determining both diffuse cortical damage or cortical and brainstem damage as a result of cerebral anoxia. It is also a valuable tool in the study of brachial plexus injuries and may guide the surgical repair of this vital collection of nerve roots. For the purpose of clarity median nerve SSEPs will be discussed first followed by posterior tibial nerve SSEPs.

Median Nerve Somatosensory Evoked Potentials

Guidelines for median nerve SSEPs testing follows and include placement of the stimulating electrode over the median nerve at the wrist. The Cathode is placed between the tendons of the Palmaris longus and the flexor carpi radialis 2 centimeters proximal to the wrist crease. The anode is place 2 to 3 centimeters distal to the cathode or on the back of the hand. The ground is placed between the stimulating electrodes and the first recording electrode. **(Fig. 1).**

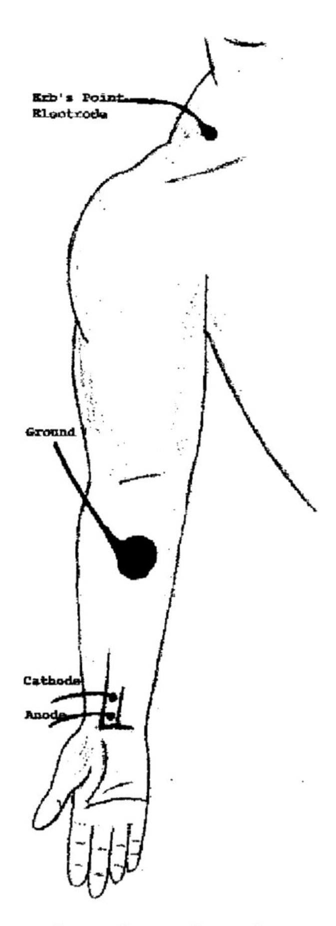

Figure 1 *Schematic demonstrating location of stimulating electrodes over the median nerve at the wrist, recording electrode at Erb's point and the ground on the forearm.*

Stimulation rate is 3-5 Hz with duration of 100-300 µsecond. The recording montages includes 4 recording channels (minimal number), however it is possible to carry out an 8 channel study when appropriate. Samples of both 4 and 8 channel recordings will be demonstrated throughout this chapter.

A 4 channel montage consists of the following 4 channels.

Channel 4 consists of electrodes placed at CPc – CPi (modified 10-10 system). CPc is the scalp electrode that is contra-lateral to side of the stimulated nerve. CPi is the electrode that is ipsi-lateral to the stimulated nerve. This combination of electrodes identifies the initial cortical response which is negative in polarity and has a latency of about 20 milliseconds therefore, known as the N20 or cortical response. The N20 is generally followed by a distinct wave that is positive in polarity and has a latency of about 27 milliseconds.

Channel 3 consists of a scalp electrode referred to an electrode off the scalp. This combination CPi – EPc (Erb's point contra-lateral) shows a series of subcortical responses (far field) that are generated in the medulla. The first of these responses is the P14 or a positive wave with a latency of about 14 milliseconds. This wave is generated in the low medulla and thought to result from the crossing sensory fibers making a sharp turn and ascending in the sensory fasciculus. The second of the 2 far field potentials is a long duration negative wave with a latency of 18 milliseconds (N18). This long acting wave is thought to be generated at the level of the medulla by local synaptic activity. At one time the N18 was thought to arise higher up in the neuraxis at the level of the upper midbrain or even thalamus, but a series of patients with pontine hemorrhages and thalamic hemorrhages did not obliterate the N18.

Channel 2: This channel is to record the cervical response generated by collateral synapses at the C 5 level of the spinal cord. This wave has a latency of about 13 milliseconds and has been called the N13. The active electrode is placed over the C5 spine (C5S) and the reference over the anterior neck (AC) or another inactive reference. The cervical response is a stationary wave generated locally in the cervical cord. The maximum amplitude of this wave is at C5, but may be seen a couple of segments above and below this level... The latency for the response is the same at all locations where the wave is identified.

Channel 1 provides the most peripheral response, the Erb's response (EP). This is a peripheral response with the typical waveform of an initial positive wave (P9) followed by a negative wave (N10) [EP]. The numbers provide the approximate latency for the response. The montage is generated with the active electrode being Epi (ipsi-lateral) and the reference electrode being Epc (contra-lateral). The expected responses (obligate waveforms) from peripheral to central include: 1 Erb's response (N10/EP); 2. Cervical response (N13); 3. Sub-cortical P14 and N 18; 4. Cortical response (N20).

Stimulus Intensity is selected first by starting with very low stimulus intensity then gradually increasing the intensity until the patient identifies the stimulation. This value identifies the sensory threshold. The stimulus intensity is then increased until a motor response is obtained in the muscles innervated by the median nerve causing abduction of the thumb. If no motor response is obtained, correct stimulator placement has been verified and appropriate troubleshooting techniques have proved unsuccessful, then the stimulus intensity is set at 2 and one-half to three times the sensory threshold. Both of these stimulation intensities guarantee a maximal sensory volley. The stimulus is delivered at a rate of 5.4 Hz. **(Fig. 2).**

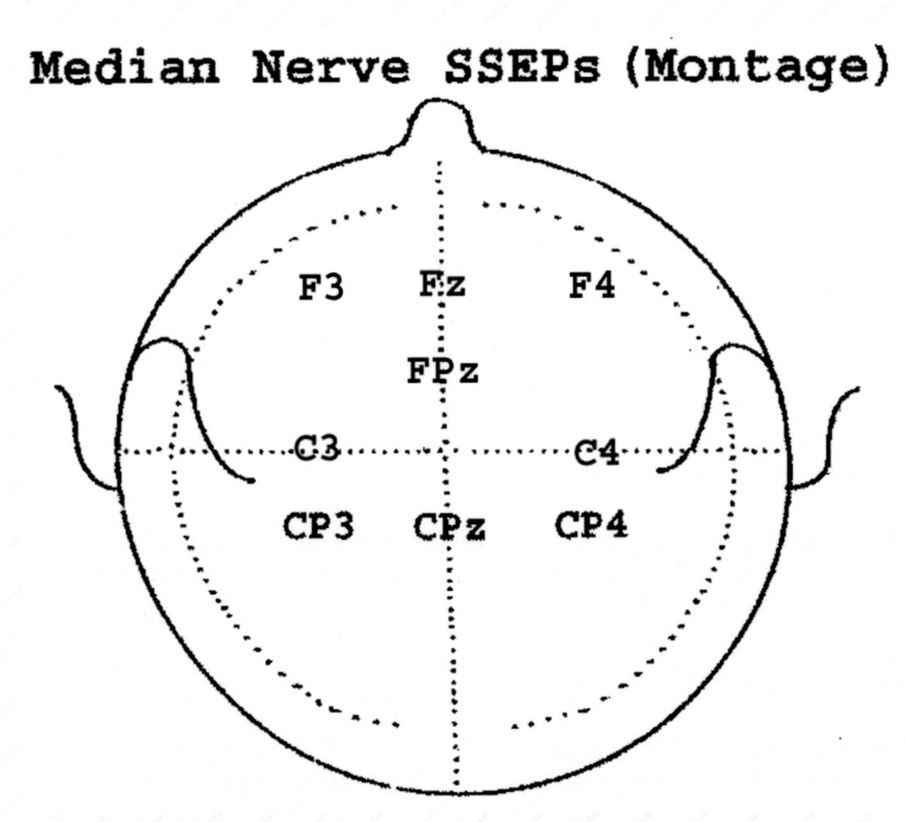

Figure 2. *Diagram of the electrode placements for median nerve SSEPs.*

An eight channel median SSEPs is demonstrated to show the wide spread sub-cortical responses from different scalp locations.

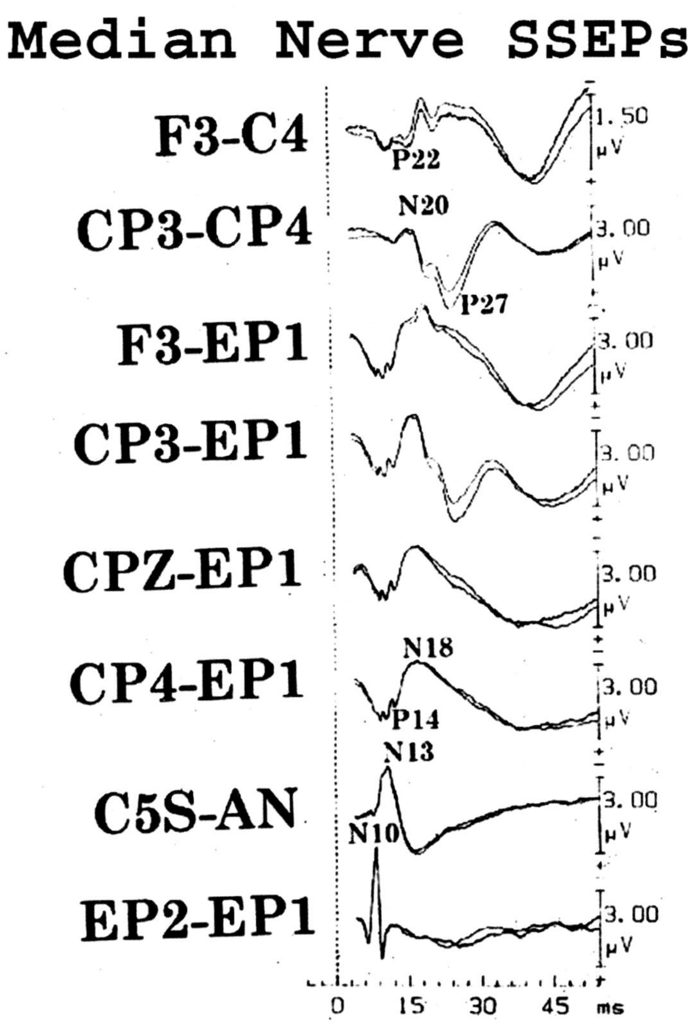

Figure 3. *This is an 8 channel MNSSEPs to demonstrate the wide distribution of the Sub-cortical responses. The lowest channel is the Erb's response (N10). The second channel is shows the Cervical response (N13). The next 4 channels, CP4, CPZ, CP3, and F3 all are referred to an electrode off the head and show the well develop P14 and N18. Note there is an Extra negativity superimposed on the N18 in CP3-EP1, this is the N20 (cortical response). The near field cortical response is shown in CP3-CP4 and consists of an N20 (Cortical response) and it's trailing P27. The F3-C4 shows the P22 an independent response generated over the motor strip.*

The sub-cortical P14 and N18, responses generated in the medulla, are seen in many different montages as shown in Figure 4.

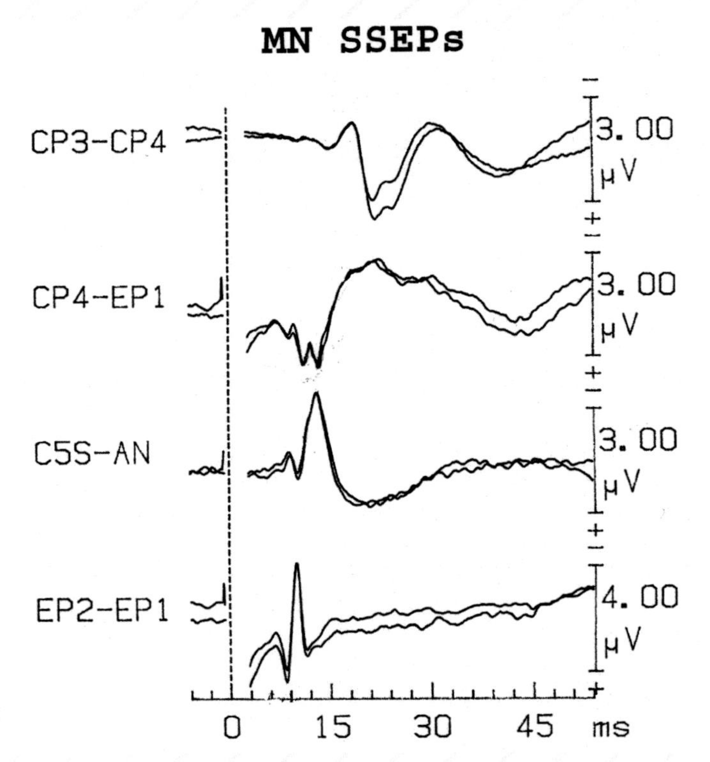

Figure 4. *This sample shows the wide spread distribution that demonstrates the far-field Potentials P14 and N18. The two positive waves before the P14 are the inverted N10 and N13 seen from even lower levels of the nervous system. Note the near identical nature of the Sub-cortical waveforms in spite of the different electrode locations on the scalp. This is characteristic of far-field potentials.*

The standard 4 channel median nerve SSEPs is shown in Figure 5.

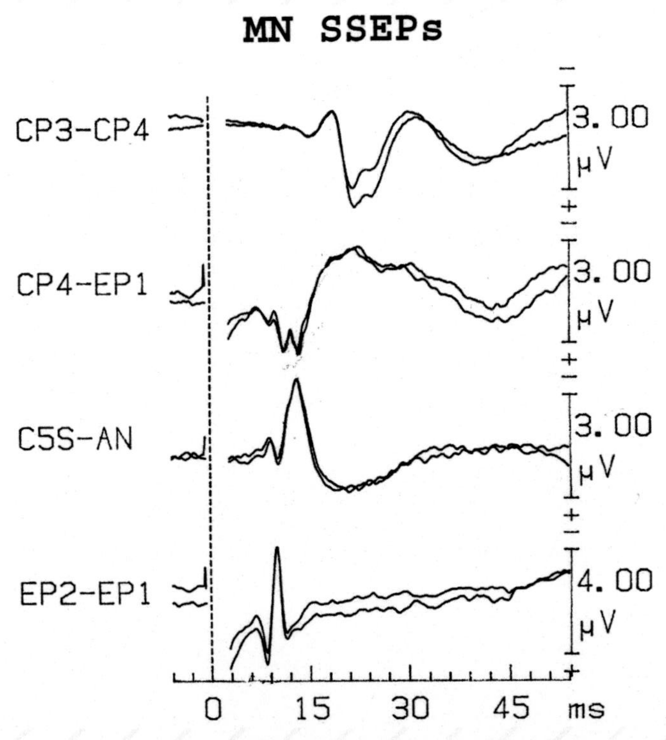

Figure 5. *Standard 4 channel median nerve SSEPs showing the peripheral Erb's response, cervical response, sub-cortical responses and cortical response. All evoked potential studies should be done in duplicate to demonstrate the reliability and reproducibility of the responses.*

The peripheral Erb's response (N10/EP), cervical response (N13) and the sub-cortical responses (P14 and N18) are quite resistant to sedative drug influences; however the cortical response (N20) can be suppressed with sedative drugs. See **(Table 1)** for range of normal median nerve SSEPS latencies.

TABLE 1

Median Nerve SSEPs Latencies and Intervals (N=92)

(Milliseconds)

	Minimum	Maximum	Mean	+3SD
ERB's (N10)	7.93	12.01	9.95	12.80
CERVICAL (N13)	10.70	15.64	13.13	16.61
SUBCORTICAL P14	11.80	16.71	14.20	17.65
SUBCORTICAL N18	14.85	20.20	17.50	21.10
CORTICAL(N20)	16.40	22.20	19.30	23.35
INTERVALS				
ERB's(N10) – CERVICAL(N13)	2.70	3.60	3.20	3.83
CERVICAL(N13)-CORTEX(N20)	5.70	6.60	6.15	6.75
SUBCORTICAL P14-CORTEX(N20)	4.50	5.98	5.23	6.43

Median Nerve *SSEPs in Brachial Plexus Injuries*

Brachial plexus injuries are relatively common injuries in young adults. The vast majority of individuals who suffer injury to the brachial plexus are motorcycle riders, but I have seen similar injuries in roll-over accidents involving a short wheel-based jeep and dune buggies. Following such an injury the question arises if there is any chance for recovery or should the useless denervated arm be removed? This severe form of surgical therapy should be reserved for those individuals who will have no innervation to the limb and are not potential candidates for surgical intervention to reconstruct the plexus. If all the nerve roots are avulsed from the spinal cord then there are no surgical measures to correct this terrible injury. The next question that comes up is where in the plexus is the lesion and in this situation, is the lesion proximal (avulsion) or distal in the various nervous components that make up the brachial plexus. The median nerve SSEPs along with high resolution MRI can help localize the location in the plexus. In this situation, the lesion in the distal parts of the plexus interrupts the generation of the Erb's response and supports the distal nature of the lesion. **See Figure 6.**

Left Brachial Plexus Injury

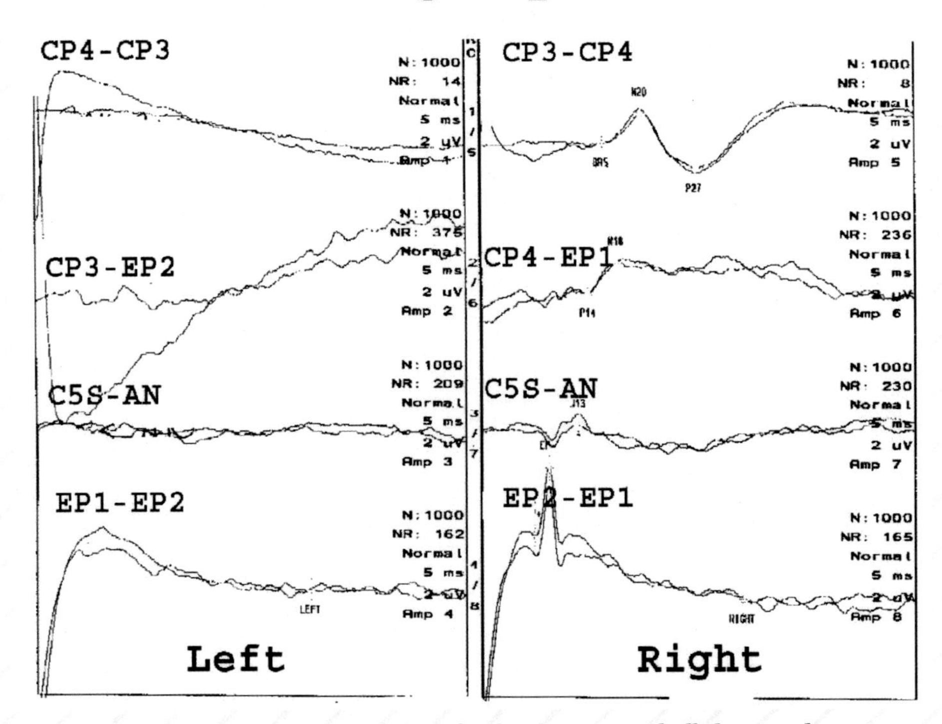

Figure 6. *This sample demonstrates the complete loss of Erb's Response and all the central response on the left side. The right side has normal responses and latencies. This pattern supports that the lesion is in the distal brachial plexus and should be surgically treatable.*

Ulnar Nerve SSEPs

Brachial Plexus Injury - Lower Root Avulsion

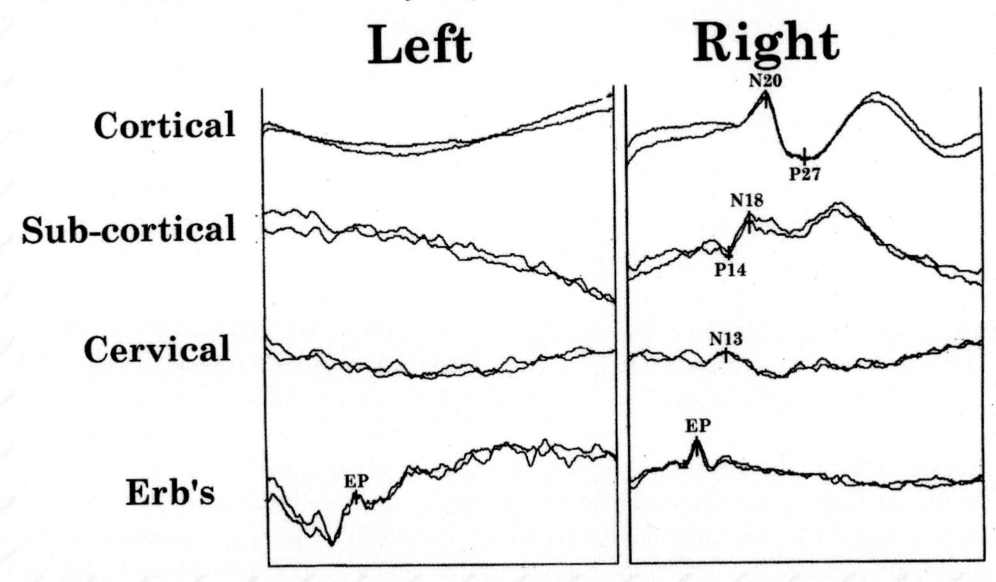

Figure 7. *This example shows the Erb's response to be present and not Central waves being formed. This type of response is highly suggestive that the roots on the left side have been avulsed and that surgical treatment will not help. (Along with MN SSEPS showing the same pattern)*

Additional physiologic studies may help sort out a complex case. One has to remember that each complex peripheral nerve has multiple roots.

Dermatomal SEPs show detailed sensory function for each cervical root**. See Figure 8.** Dermatomal SEPs will be discussed more fully in Chapter 4.

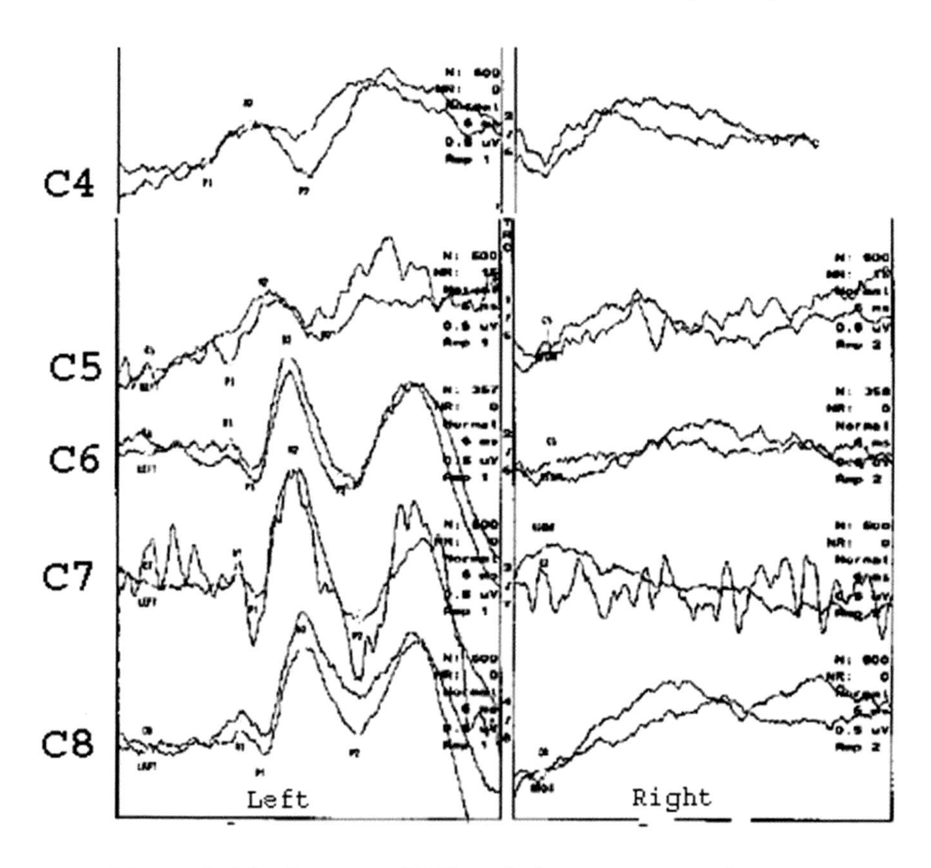

Figure 8. *This Dermatomal SEP study demonstrates normal response*

From C4 *to C8 in the left brachial plexus. The right side shows no response at any of the levels from C4-C8. Dermatomal SEPs will not distinguish whether the lesions are distal or proximal in the plexus. They just indicate that n physiologic volley is getting through the roots. The responses are recorded at the cortical level for each stimulated root.*

Electromyographic studies and high resolution MRI scans may be of great value in the evaluation of traumatic injuries to the brachial plexus. An accurate diagnosis is critical if optimum therapy is to be provided to these unfortunate individuals.

Median Nerve SSEPs in the Diagnosis of Anoxic Brain Damage

Anoxic damage to the central nervous system is a common condition and results most commonly from cardio-pulmonary arrest or a ventricular arrhythmia. For many years attempts have been made to provide prognostic information in those individuals who became comatose following cardiac arrest. The electroencephalogram patterns can be of help and include 1) a non-responsive isoelectric EEG, 2) burst-suppression patterns and 3) alpha coma. When these patterns are present and result from anoxia they all have very poor prognosis for full neurologic recovery. For those individuals who do not show any improvement within 24 hours of the event have a poor prognosis for full neurologic recovery. Marked elevations in cerebrospinal fluid CK, bb bands at 72 hours have a very poor prognosis for full neurologic recovery and support the EEG findings. The electroencephalogram is sometimes difficult to record in the intensive care unit. The EEG is generally accomplished with the most sensitive settings for the equipment. Because of this, there are many artifacts that can be introduced into the recording. Electrodes are generally placed at double distance to maximize the chance of identifying any brain activity. Cerebral spinal fluid CK, bb band elevation is usually done at 72 hours because this is a time where most studies of this test have been accomplished. The clinician in charge of the patient would like to make decisions much earlier. Median nerve somatosensory evoked potentials are more readily accomplished in the environment of an intensive care unit. It is more rapid to carry out than the EEG and does not require that the patient be at 72 hours after the event. As with EEG, it is essential to document any sedative medications as well as the core body temperature at the time of testing. Three major patterns are seen in comatose patients who have suffered anoxic brain damage. The first is the absence of both cortical and subcortical responses. This is seen in unresponsive comatose patients who do not have any brainstem reflexes present. **See Figure 9 and Figure 10.**

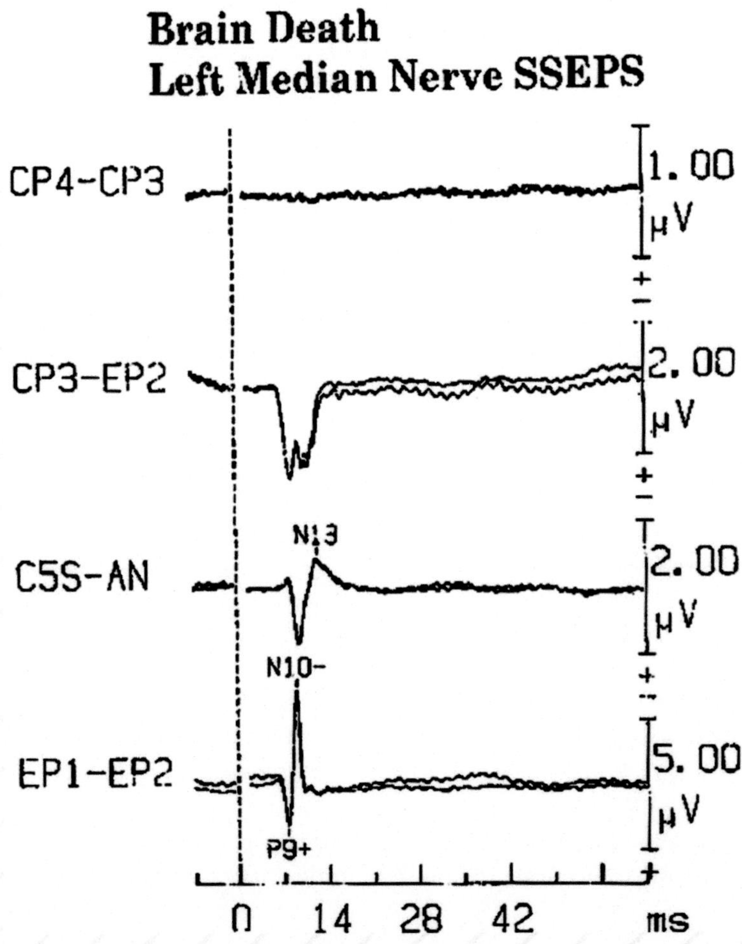

Figure 9. *The peripheral Erb's response and cervical response remain present. The sub-cortical responses from the medulla are absent, only the cervical response and Erb's response remain in the CP3-EP2 channel, the P14 and N18 are absent. the cortical channel shows no evidence of activity. This study would qualify for brain death.*

The second major pattern is that in which the brainstem components of the median nerve SSEPs remain and the cortical responses are absent.

Anoxic Brain Damage

Subcortical Responses Present

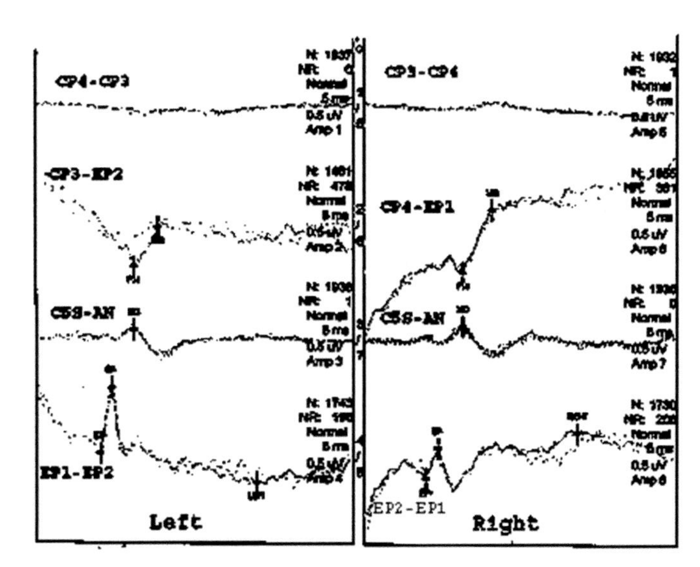

Figure 10. *This example shows an absent cortical response with the Sub-cortical P14 and N18 remaining. The peripheral and cervical responses are present.*

This pattern shows that the patient has suffered from a significant anoxic event, but does not meet requirements for brain death. The prognosis for full recovery is poor and outcome.

Rupture of aneurisms involving major arteries such as the aorta may be extensive enough to cause major cerebral damage as shown **in Figure 11.**

Anoxia after Thoracic Artery Aneurism Rupture

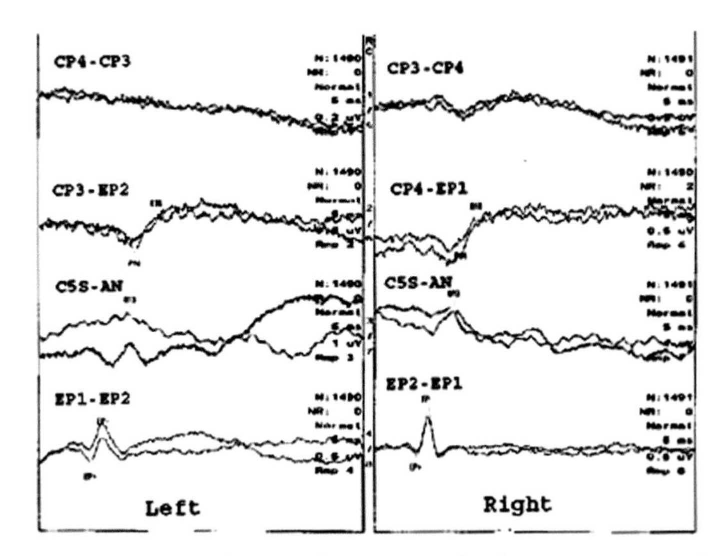

Figure 11. *Patient suffered major CNS damage after rupture of a thoracic aorta aneurism. Note the asymmetry with the left side having no cortical response while the right Cortical response is too early to be an N20 and probably represents an inverted sub-cortical response. The sub-cortical responses remain normal. Prognosis is poor for significant neurologic recovery.*

Figure 12 demonstrates responses in a patient with traumatic transection of the upper cervical spinal cord. The absence of all responses above N13 confirms the location of the lesion being above the C5 level. If the lesion is at the C5 level the cervical response (N13) will be lost as well. In that case only the peripheral Erb's response will be present. **See Figure 12.**

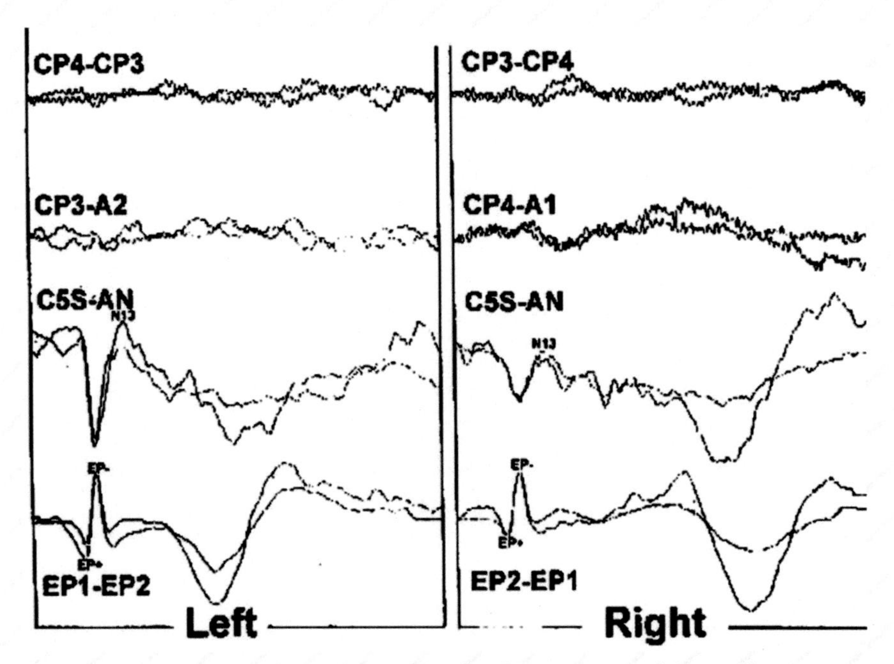

Figure 12. *This median nerve SSEPs shows the peripheral Erb's response and Cervical response generated at C5 to be intact. The sub-cortical response generated in the medulla and the cortical responses are absent. This localizes the lesion to be between C5 and the medulla.*

Central pontine myelinolysis is rare condition most frequently caused rapid correction of severe hyponatremia. The example shown below in **Figure 13** had sodium of 114 meq/liter.

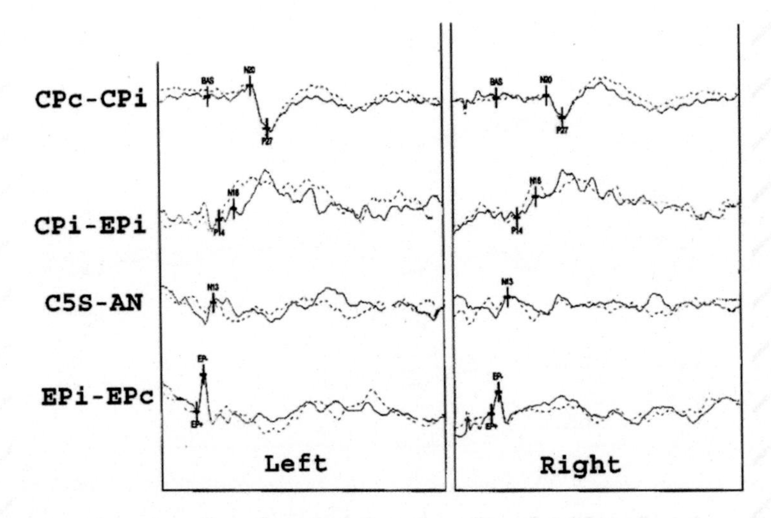

Figure 13. *All waves are present at the various locations in the peripheral and central nervous systems. The cortical response is delayed making the Sub-cortical to cortical interval prolonged. This finding was caused by the CPM.*

Extrinsic and Intrinsic tumors such as vestibular schwannomas, meningiomas, pontine gliomas, ependymomas, etc. may all cause changes in the median nerve SSEPs. The SSEPs cannot distinguish between intrinsic and extrinsic tumors. **Figure 14** is an example where two large vestibular schwannomas compress the ponto-medullary junction.

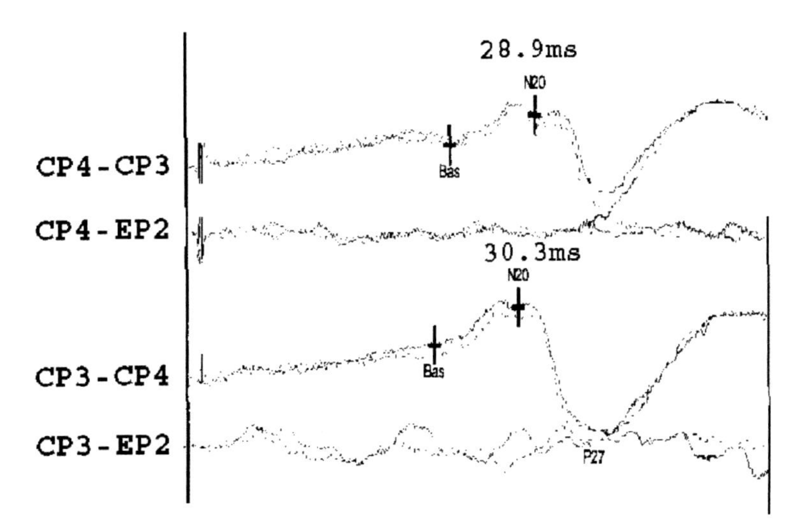

Figure 14. *This sample shows the absence of sub-cortical responses bilaterally and markedly prolong cortical response latencies. This patient suffered from Neurofibromatosis II. Other types of tumors in this location would give similar results.*

Finally, the median nerve SSEPs may be of value in certain forms of acute and chronic inflammatory neuropathy. This study is particularly helpful in those cases that have proximal demyelination and relatively normal peripheral nerve conduction velocities, as shown in **Figure 15.**

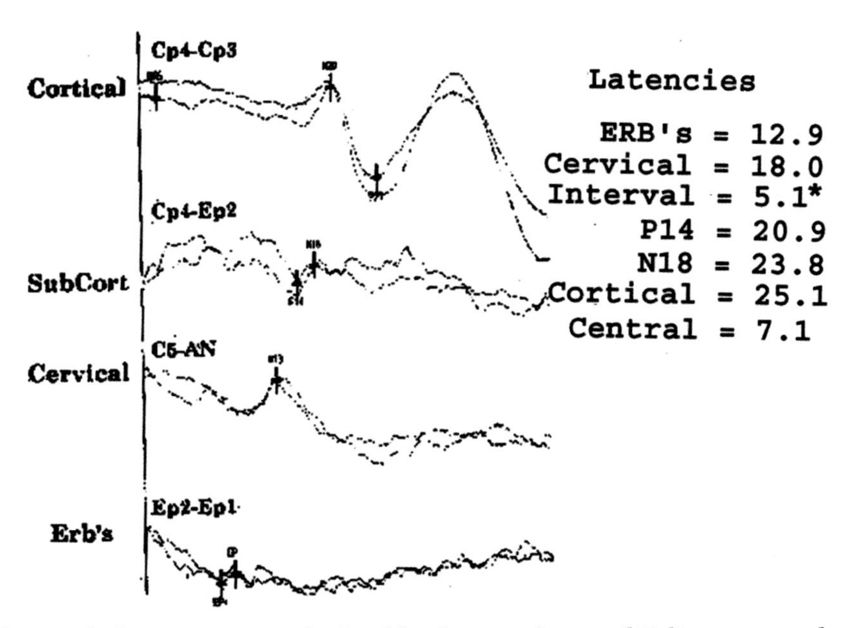

Figure 15. *This sample demonstrates a conduction delay between the normal Erb's response and a delayed cervical response. Central conduction is the interval between the cervical response and cortical response and in this instance is normal. These findings are consistent with a delay at the radicular level resulting from an inflammatory polyradiculopathy.*

Direct Cortical Responses to Median Nerve Stimulation

Direct recordings from the motor-sensory cortex are valuable to identify eloquent cortex during long-term monitoring for epilepsy surgery, at the time of electrocorticography removal of tumors that approximate these important structures and for epilepsy surgery. When the median nerve is stimulate and the response recorded directly from cortex, the responses are a bit different than seen at the scalp level. 1. The initial cortical sensory response has a short latency negativity that is close to the central fissure and a much larger response posterior. These two responses blend to give the scalp cortical response. The cortical field size varies greatly from individual to individual; the size may vary up to 100 percent **(Figure 16).**

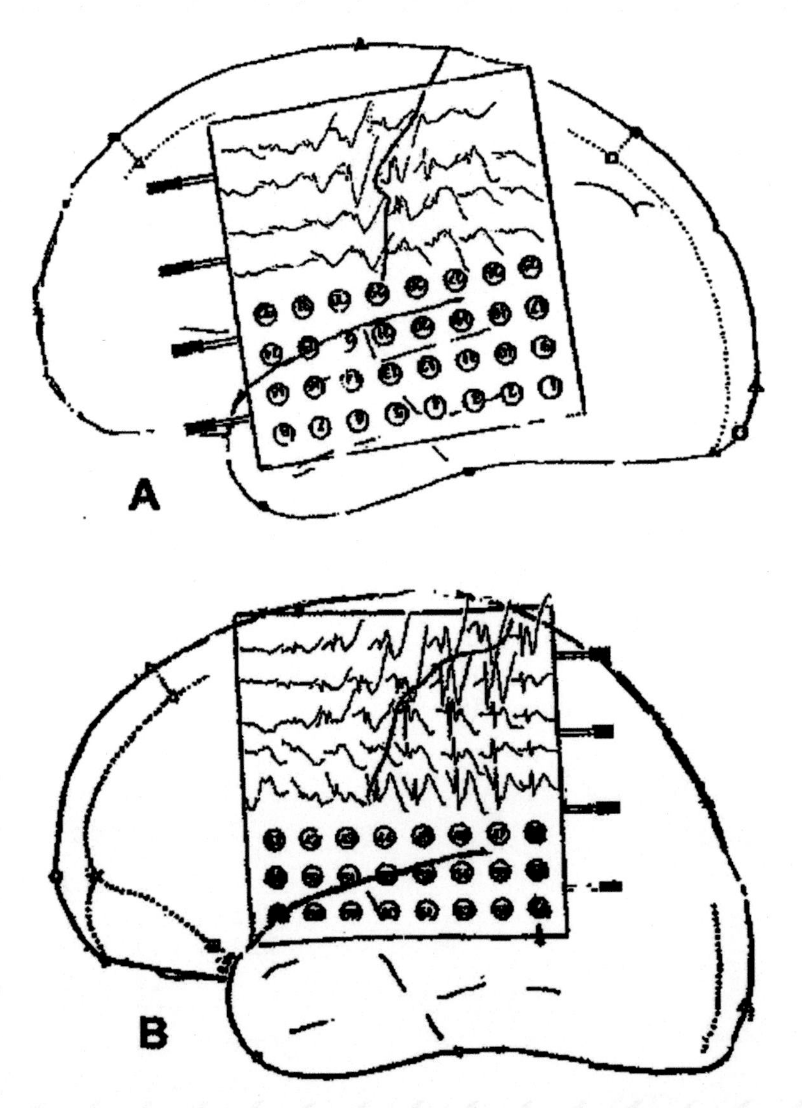

Figure 16. *The upper map (A) shows the small field for the median Nerve responses while (B) the lower map shows an individual with a large median nerve field.*

The exact location of the response may also vary greatly, centering in an area from 2 centimeters to 8 centimeters from the central fissure. Over the motor strip there is a large independent response, the positive (P22). This positivity was at one time thought to represent one end of a dipole with the N20 being the other end. Over a number of years, this was proven not to be the case, but that the P22 was independent of the N20. The central sulcus which separates the motor strip from the sensory strip is identified by these waves as shown in **Figure 17.**

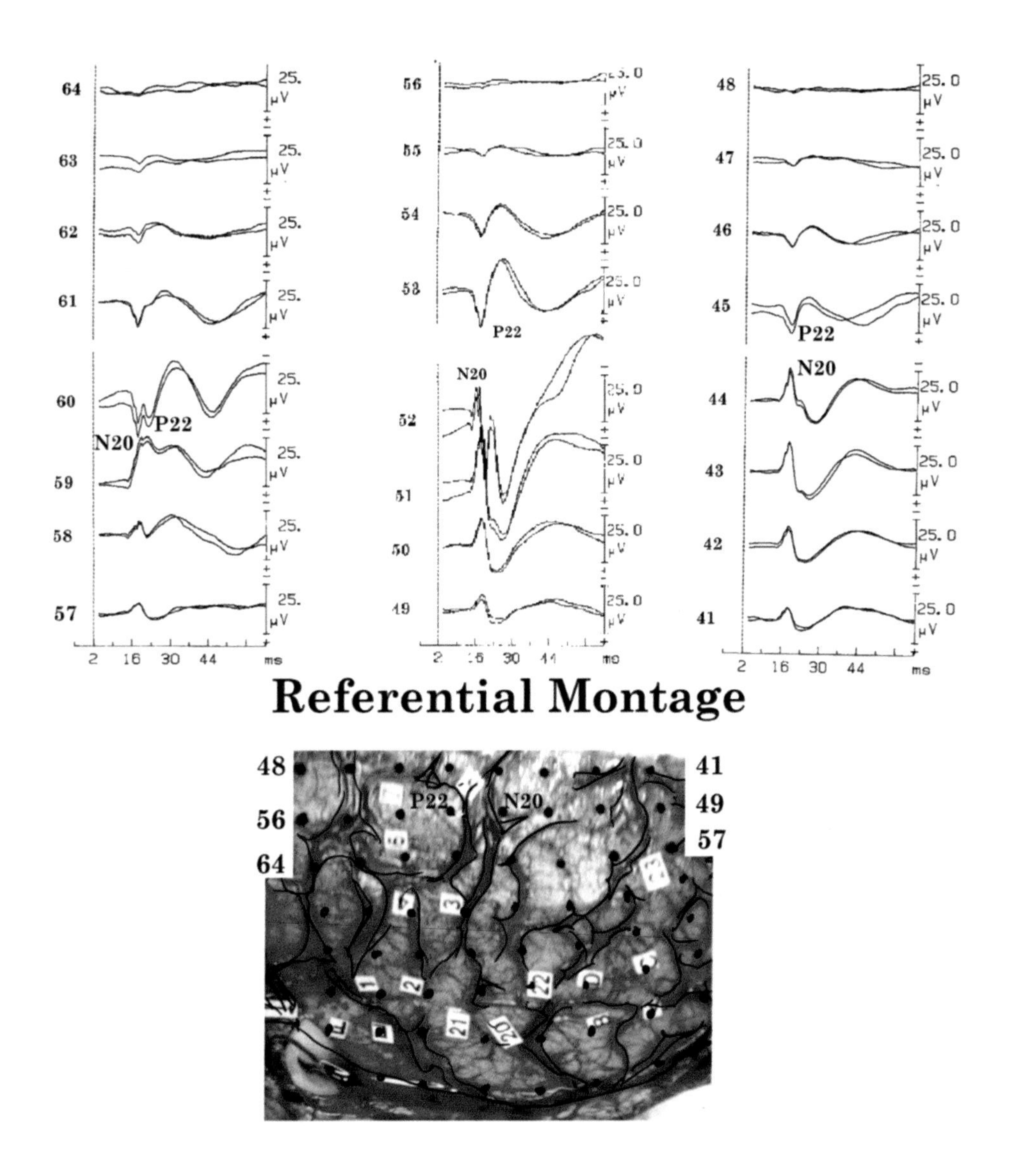

Figure 17. *This sample shows a series of electrodes overlying the Motor-sensory cortex. The central sulcus is identified by the phase reversal between the motor (anterior) P22 and the sensory (N20). In the middle panel note the different latencies for the N20. The shortest latency response is just across from the P22 and the highest amplitudes N20 is 1 centimeter posterior.*

Additional discussion of the direct cortical responses will be presented in the Section on Intraoperative Monitoring.

Posterior Tibial Somatosensory Evoked Potentials (PTN SSEPS)

The PTN SSEPs are of great value in the evaluation of spinal cord function. Unlike the median nerve SSEPs, the PTN SSEPs cover the entire spinal cord as well as the sub-cortical responses and the cortical responses.

The minimum guidelines recommend that a 4 channel montage be used for this recording. The most peripheral response is recorded in the popliteal fossa with the following electrodes, Pfd is placed 2 centimeters above the popliteal crease and the reference electrode (Pfp) is placed 5 centimeters above the popliteal crease. An alternative electrode combination is Pfd referred to the MK (medial knee). The wave form is typical for a peripheral nerve response, positive wave followed by a negative wave then an additional positive wave. Only the negative wave is used for latency measurements and has a latency of about 8 milliseconds (N8 response). The next channel is to record the Lumbar response (N22). This is a standing response similar to the cervical response of the median nerve SSEPs. The largest amplitude of the Lumbar response occurs over the T12 spine, so the active electrode is placed at T12S and referred to the Iliac Crest (IC). This response has an approximate latency of 22 milliseconds and is a negative wave (N22).

This wave can be identified above and below the T12 location, but will have lower amplitude, but the latency will be identical. The sub-cortical responses probably are generated in the same locations as with the median nerve SSEPs. In the PTN SSEPs there is a positive wave with a latency of 31 milliseconds (P31) followed by a negative wave with a latency of 34 milliseconds (N34). Electrodes are placed at Fpz (Modified 10-10 system of electrode placement) and referred to C5S on the back of the neck. Finally, the cortical response is recorded with the active electrode over Cpi (ipsilateral) referred to CPz (midline electrode). The initial cortical response when stimulating the posterior tibial nerve is a positive response with a latency of about 37 milliseconds followed by a negative response with a latency of 45 milliseconds (N45).

Unlike, the median nerve cortical response which is contra-lateral to the stimulated side, the posterior tibial nerve cortical response is best seen over the ipsilateral cortex or over the midline.

The PTN stimulation electrodes are placed as follows. The Cathode is placed one half way between the Achilles tendon and the medial malleolus. The Anode is placed 3 centimeters distal to the cathode. A ground electrode is placed over the Tibia (shin). **(Figure 18).**

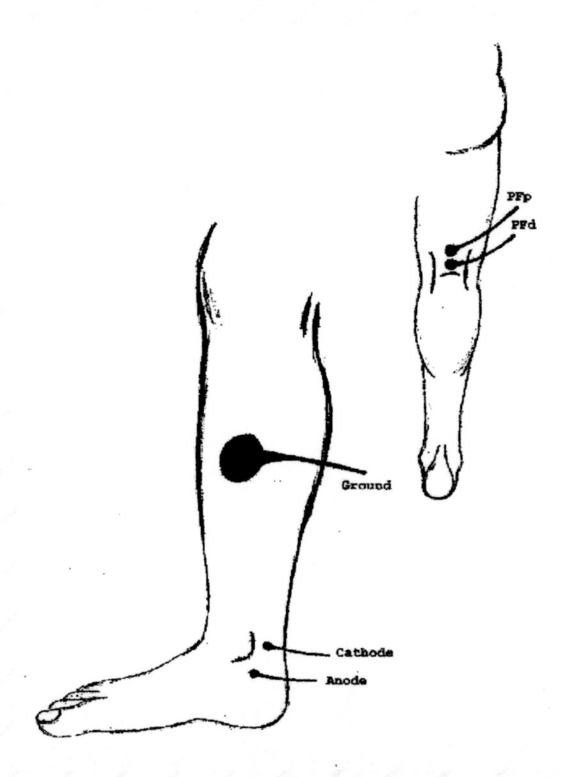

Figure 18 *Schematic of leg demonstrating the location of the stimulating electrodes over the posterior tibial nerve at the ankle. Rocording electrodes are located over the upper extent of the Popliteal fossa and the ground over the shin.*

Stimulus intensity is obtained in an identical way as for the median nerve studies. First the intensity is increased to determine the sensory threshold, current at which the patient can first identify the shock and then increase until a motor response occurs in the toes. If threshold is used. Well developed, reproducible wave forms are obtained with between 500 and 2000 repetitions. In general, 2 replications are carried out for each study, but when there is a great deal of EMG artifact present the number of replications may have to be increased. This is particularly true when the patient has spasticity in the lower extremities. Each electric shock may increase the muscle tone and results in increased EMG artifact. **(Figure 19).**

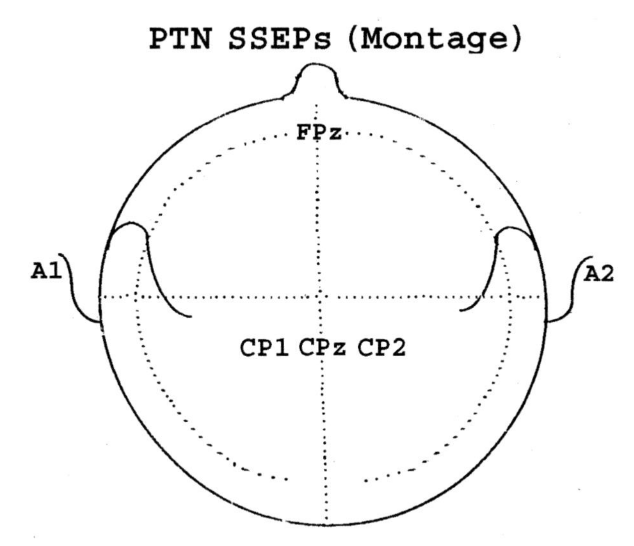

Figure 19. *Scalp montage for recording the PTN SSEPs Cortical response. This response is known as the P37 and is almost always followed by an N45. Because of the location of the generators in the intra-hemispheric fissure the Best responses are seen on the ipsilateral or midline electrode rather than over the actual side where the response is generated. This has been called a "Paradoxical response".*

The standard is to record 4 channels of data for the PTN however, an 8 channel recording is shown here to make the point of the cortical response being seen best over the midline or ipsilateral side. **(Figure 20).**

Eight Channel PTN SSEPs

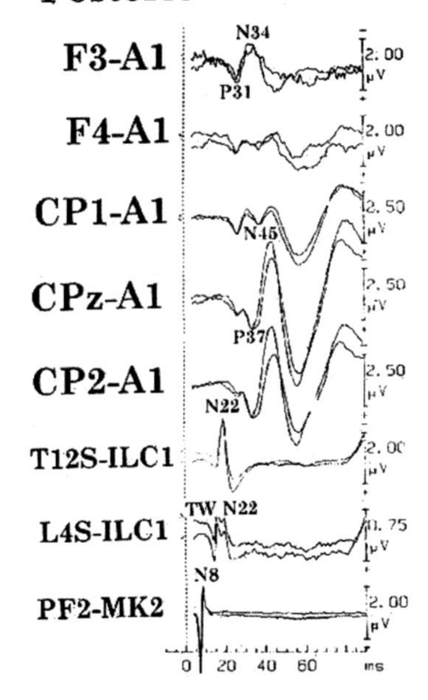

Figure 20. *This 8 channel PTN SSEPs shows the popliteal response (N8) in the first channel. The second channel shows a bifid response at L4S, the first response is the traveling wave, The second wave the Lumbar response (N22); this volume conducted response has the same latency as seen in T12S, but is of lower amplitude. The third channel shows the Lumbar response which has a maximum amplitude at T12S. The next three channels show the cortical responses, note the largest response is found in the ipsilateral electrode, the next largest at the midline and the smallest over the contra-lateral electrode. The top response (F3-A1) shows the sub-cortical responses, P31, N34. The F4-A1 derivative is similar to thatseen in CP1-A1 though smaller in amplitude. This is a good example of the paradoxical response seen in PTN SSEPs.*

Typically, PTN SSEPs are accomplished with 4 channels, but the next **Figure 21** demonstrates the cortical localization of the P37 and the N45.

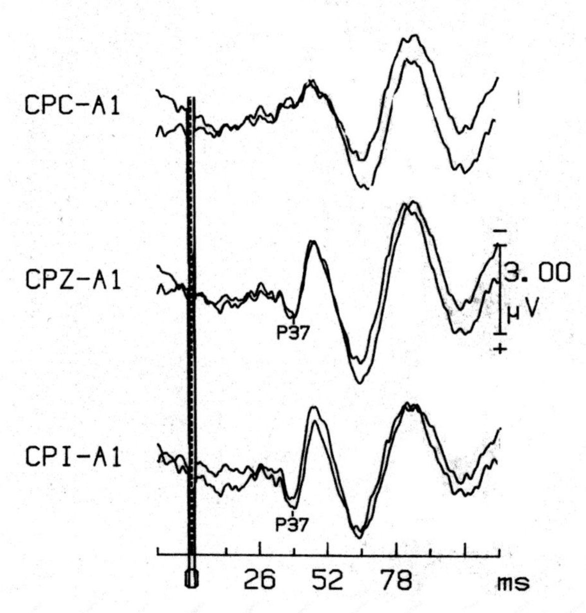

Figure 21. *This figure shows the highest amplitude P37 in CP1-A1 when the left PTN was stimulated. The second largest response in over the midline while the P37 is not seen over the contra-lateral hemisphere.*

A five channel PTN SSEPs is shown in the next figure. Typically, a 4 channel PTN SSEPs is carried out in clinical studies. The L4 channel is left out as it does not provide any additional useful information.

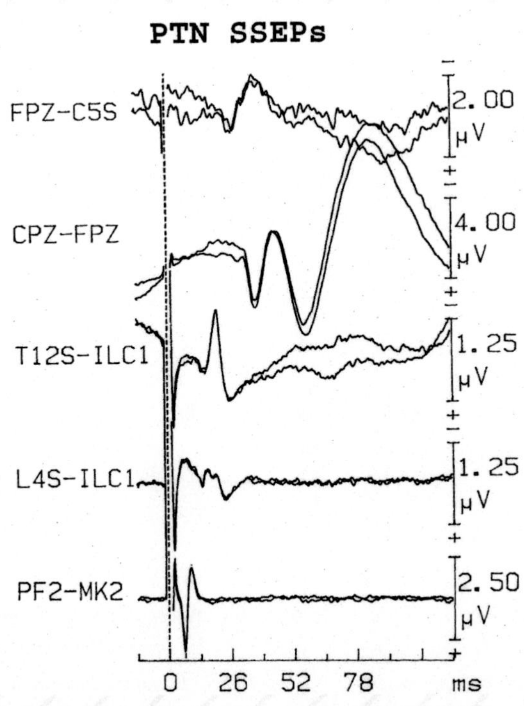

Figure 22. *This PTN SSEPs shows the popliteal response (N8). The L4S derivative shows the bifid response which includes a traveling wave followed by the Lumbar response (N22). The 3rd channel shows the Lumbar response (N22) at its maximum amplitude. The cortical response P37 is shown at the midline electrode and the subcortical responses are shown in the midline frontal to C5S reference.*

The traveling wave is typically seen in the lower lumbar or sacral electrodes. To record this traveling wave the patient must be young and very slim. The measurement of the traveling wave is of no clinical value, but in rare occasions can be measured.

PTN SSEPs
Traveling Wave

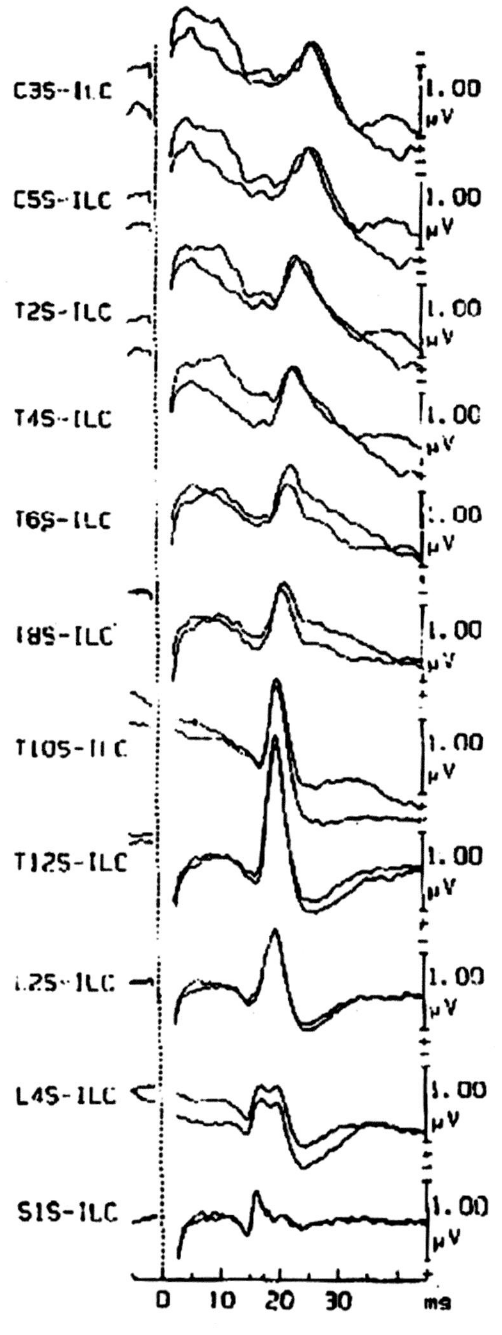

Figure 23. *This figure demonstrates the traveling wave resulting from the electrical volley at various levels of the spinal cord. The lowest level at S1S shows only a peripheral response, at the L4S level the peripheral traveling wave precedes the lumbar response. At the L2S level the traveling wave is a shoulder on the front of the lumbar response. At T12S the traveling wave is not seen as it is behind the large lumbar response. At the T10S level the lumbar response has disappeared and only the traveling wave is present. Electrodes from T10S to T3S show the traveling wave to have a gradually increasing latency.*

Control Latencies and Intervals are shown in **Table II** (Mean + 3SD) are used as the outer limits of Normal. Extremes of height, either very short or tall must be taken into account in the interpretation of an abnormality.

POSTERIOR TIBIAL SSEPs - Latencies and Intervals

	Mean	SD	3SD
POPLITEAL (n8)	9.91	1.27	3.81 (13.72)
LUMBAR (L4)	22.56	2.09	6.27 (28.83)
LUMBAR (T12)	23.12	2.17	6.51 (29.63)
SUBCORTICAL (p31)	31.02	3.03	9.09 (40.11)
SUBCORTICAL (n34)	34.13	2.53	7.59 (41.72)
CORTICAL (P37)	39.84	2.82	8.46 (48.3)
INTERVALS			
Popliteal to Lumbar (T12)	12.91	1.27	3.81 (16.72)
Lumbar to Subcort. (p31)	8.39	0.65	1.95 (8.54)
Subcort. (p31) to Cortical	8.54	0.0012	0.0036 (8.54)
Central Conduction (T12 - P37)	16.86	0.58	1.74 (18.6)

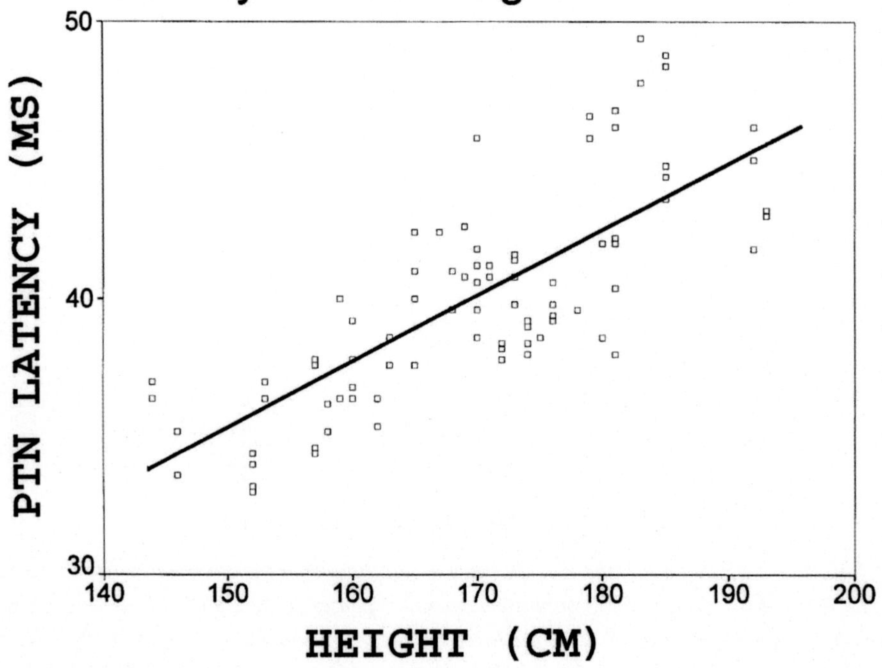

Figure 24 *This figure demonstrates the effect of height on the latency of the cortical p37. Extremes of height, both short and tall must be taken into account in interpreting whether a given result is normal of abnormal.*

The effects of height are shown on the latency of the cortical P37 **(Figure 24).** The Solid line represents the mean for height. When Mean + 2 SD are plotted, at least 3 individuals fall into the abnormal range. When the Mean + 3SD are used none of these control individuals fall into an abnormal range. The reduce the influence of height on this control data, a plot of the Interval between the Lumbar response and the Cortical P37 are much more linear. **(Figure 25).**

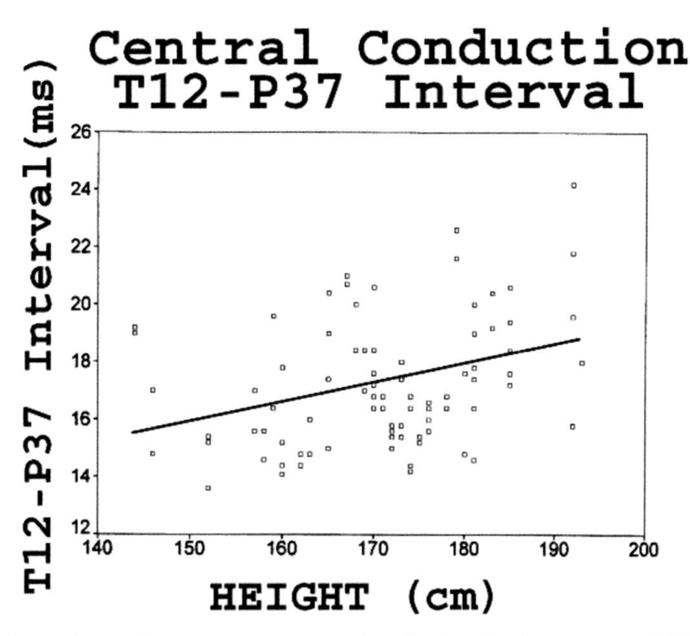

Figure 25 *This figure shows the results of measuring central conduction, lumbar response (T12) to the cortical response (p37). This measurement effectively reduces much of the height variation due to leg length. Effectively, it measures sitting height. Sitting height is much more consistent from individual to individual than standing height. Notice the slope of the line is much reduced when compared to the standing height.*

The difference in height is do to differences in leg length, whereas sitting height is much more similar among individuals. The PTN SSEPs is of diagnostic value in many different conditions affecting the spinal cord and brain.

MULTIPLE SCLEROSIS

In the northern climes, multiple sclerosis is a relatively common disorder in both Europe and North America. As one moves to more southern climes, the prevalence of this disorder declines. Migration of large populations from northern and central Europe to the Mid-east following World War II allowed epidemiologists to identify the age at which an individual would pick up the reduced risk for the development of this condition. If an individual was less than 15 years of age when the move to a warmer climate was accomplished, they would pick up the lower risk for development of Multiple sclerosis. Above 15 years, the individual would keep the higher risk from the place of origin. These studies have been confirmed by studying aerospace workers and their families moving between the states of Washington and California.

Certain HLA haplotypes have been identified with an increased risk for the development of multiple sclerosis, especially among familial form of this disorder. This data confirms that there is likely a genetic predisposition, but environmental factors also play a role. A number of viruses have been thought to be the cause of MS, but in each instance it turned out to be a contaminant from the laboratory or the results could not be confirmed by other laboratories.

No single diagnostic test has been developed to diagnose multiple sclerosis. Clinical criteria before MRI scans required that an individual have multiple lesions in the nervous system separated in time, i.e., the relapsing form of the illness. These criteria have been modified since the advent of the MRI scan. In many individuals, especially during their first attack or in those older individuals (above 40 years) that start off with the chronic progressive form of the disease other diagnostic tests have been of great clinical value. Cerebral spinal fluid oligoclonal bands, while not specific for MS provide helpful information as does the measurement of myelin basic protein in that same fluid. Posterior tibial SSEPs have the highest abnormality rate of any of the evoked potential studies in patients with MS. This is followed in frequency by the PREP visual examination. In both the optic nerve and spinal cord the MRI scan may not resolve plaques that can be diagnosed by functional evoked potential studies. **Figure 26** shows the conduction delay seen in patients with multiple sclerosis.

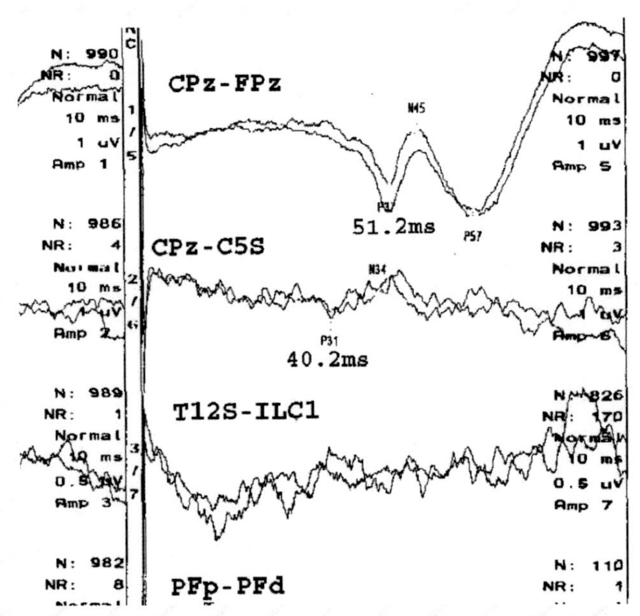

Figure 26. *This figure is an example of a patient with the spinal cord Form of multiple sclerosis. Both the sub-cortical P31, N34 and cortical Responses are delayed significantly. This is diagnostic of conduction Delay in the spinal cord. The intervals above the level of the medulla are Normal. The failure to resolve the lumbar response occurs in about 20 percent of a control population.*

Transverse myelitis is a monophasic demyelinative illness in which no additional symptoms develop over many years of follow-up. Unfortunately, a first attack of multiple sclerosis may mimic this condition. If the case is to turn out to be multiple sclerosis a new event is likely to occur within a year of the initial event. After MS is eliminated as a diagnostic possibility, the outcome of transverse myelitis is divided into thirds. One-third do not improve, one-third show some improvement and the final third have significant improvement. **(Figure 27).**

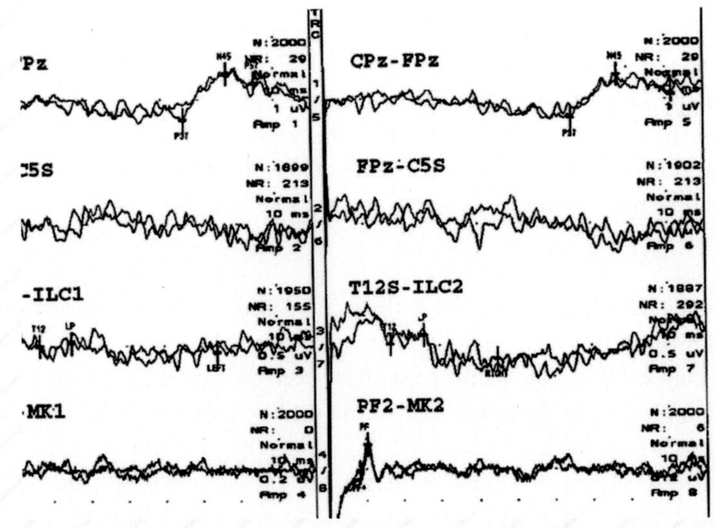

Figure 27. *This example shows the effects of transverse myelitis on the PTN SSEPs. The peripheral response is normal. The lumbar and sub-cortical responses are poorly resolved. The cortical response is much delayed and of low amplitude.*

TUMORS OF THE SPINAL CORD AND BRAINSTEM

Mass lesions of the spinal cord are relatively common entities. These consist of intrinsic and extrinsic masses. The most common are the extrinsic mass lesions including herniated discs, schwannomas, meningiomas, dermoid cysts, and lipomas. Intrinsic tumors include ependymomas and gliomas. PTN SSEPs does not distinguish the histologic nature of the tumor, but is helpful in demonstrating a conduction delay or block at the level of the lesion. Samples of some of these lesions are shown in the next few figures. **(Fig. 28, Fig. 29 and Fig. 30).**

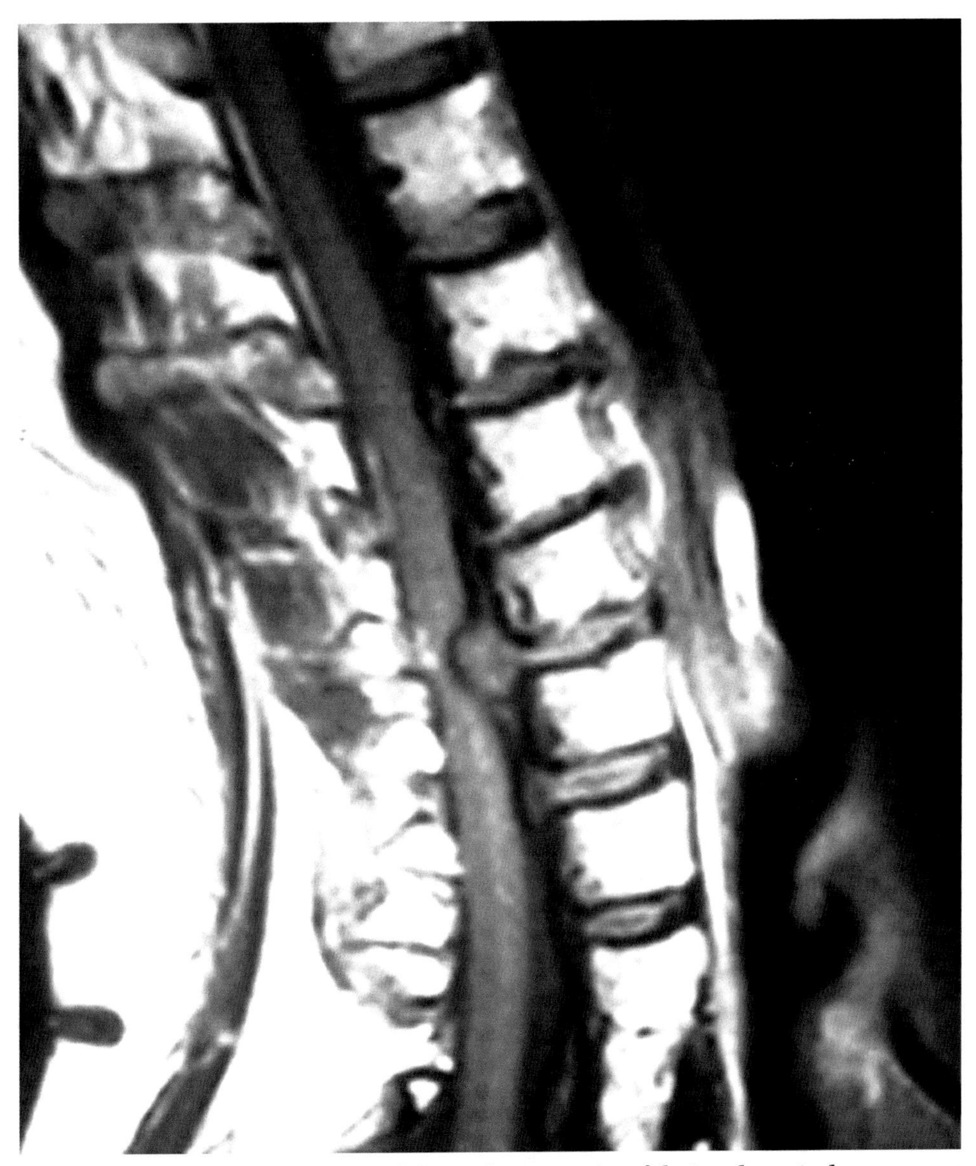

Figure 28. *Central herniated disc with compression of the Lumbar spinal roots.*

Acute herniation of a cervical disc may cause either spinal cord symptoms or with lateral herniations lateralized radicular symptoms.

Whereas, lumbo-sacral discs generally affect only the roots as the protruded discs are below the level of the spinal cord.

Lesions such as the one shown in Figure 28 must be treated surgically. Small lateral herniations frequently respond to conservative therapy and should not be operated on until there has been a failure of conservative therapy. Any significant motor loss from an acute herniated disc is an indication for early surgery.

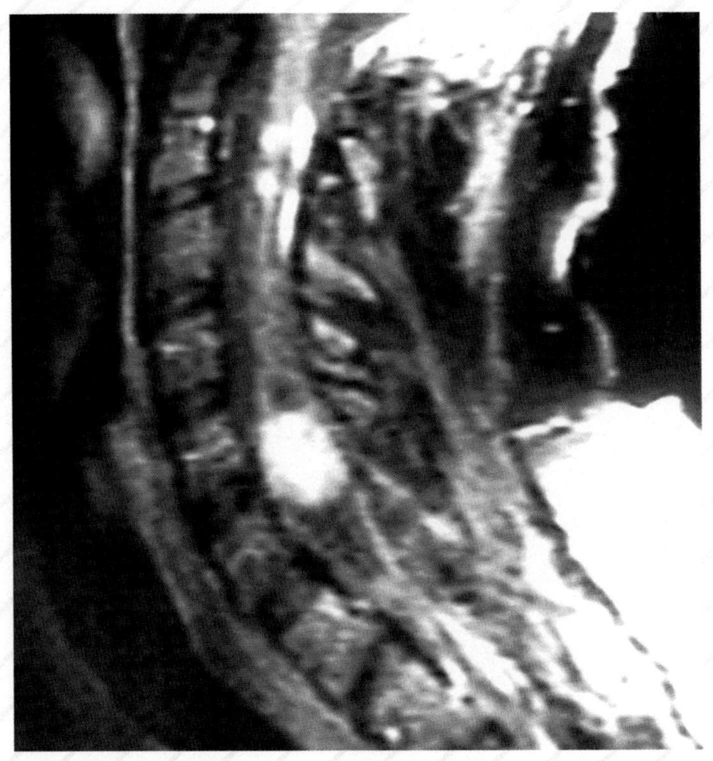

Figure 29. *Spinal cord ependymoma. This mass lesion occurs in the central portion of the spinal cord. A syrinx- like structure is seen around the tumor.*

Ependymomas are tumors derived from the ependymal lining of the ventricular system. These tumors can frequently be removed, but have a propensity to recur. Drop tumors may also occur with this histologic type of tumor.

The PTN SSEPs may be utilized to demonstrate delayed responses that are generated above the level of the mass lesion. See **Figure 30.**

Delayed Cortical P37
Spinal Cord Tumor

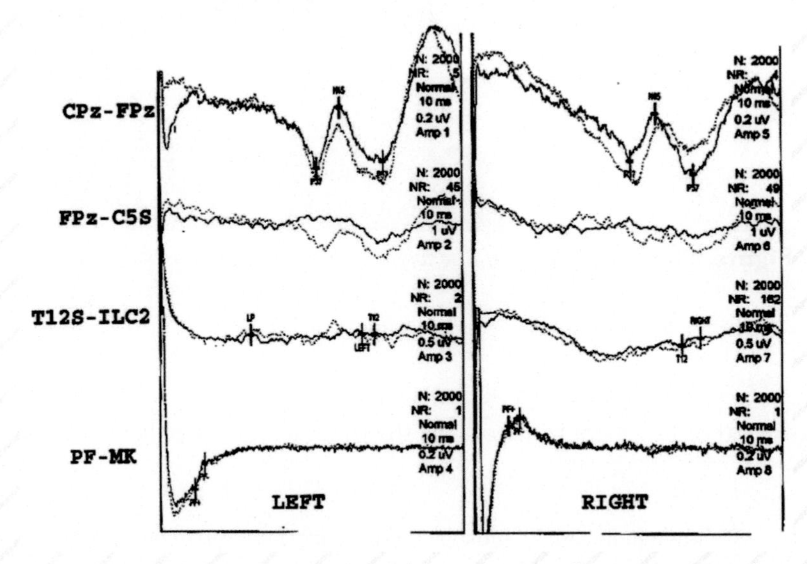

Figure 30. *This PTN SSEPs demonstrates a marked delay in the cortical responses. Note the sub-cortical responses are not resolved and the lumbar response is only present on the left side.*

A number of other clinical disorders may cause the PTN SSEPs to develop prolonged latencies. These conditions may be peripheral such as chronic inflammatory neuropathy, compressive such as in the mucopolysaccharidosis, or central such as found in syringomyelia.

A sample of the PTN SSEPs in chronic inflammatory polyneuropathy is seen in **Figure 31.**

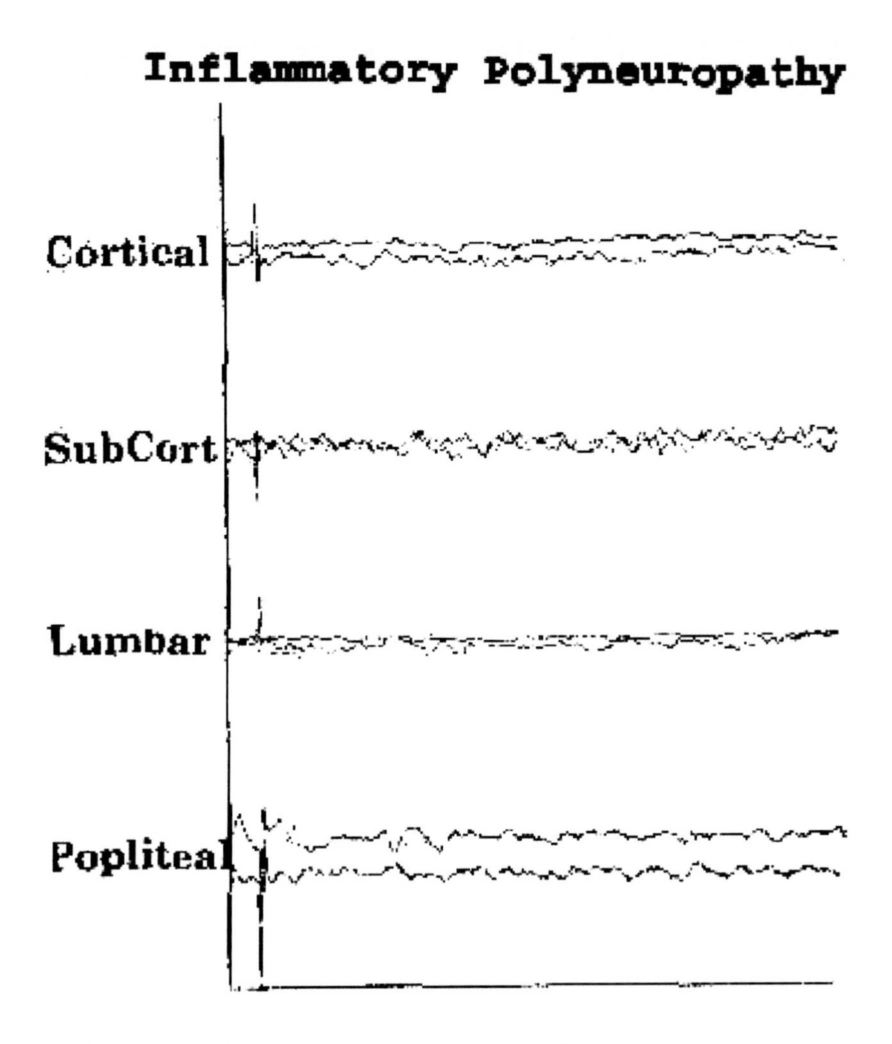

Figure 31. *This PTN SSEPs demonstrates a complete failure to obtain any waveform at any level tested. This is the pattern seen in either chronic inflammatory polyneuropathy or any other severe peripheral neuropathy.*

The mucopolysaccharidoses are a group of inherited disorders affecting the storage of mucopolysaccharides in connective tissues, such a ligaments and tendons. Most forms are seen in infancy and childhood, but certain forms occur during young adulthood. In Scheie's disease, which results from a mutation affecting the enzyme α-Iduronidase. A deficiency of α- Iduronidase in this age group leads to a complex disorder including normal intelligence, short stature, coarse facial features, limitation of joint movement and with time develop a spastic paraparesis. This spinal cord disorder is secondary to compression of the spinal cord by markedly enlarged anterior and posterior spinal ligaments. These enlarged ligaments are caused by the storage of mucopolysaccharide. Surgical decompression is frequently necessary to prevent paralysis. Because of the rigidity or the neck and non-extendible, these patients should be awake during intubation. **(Figure 32).**

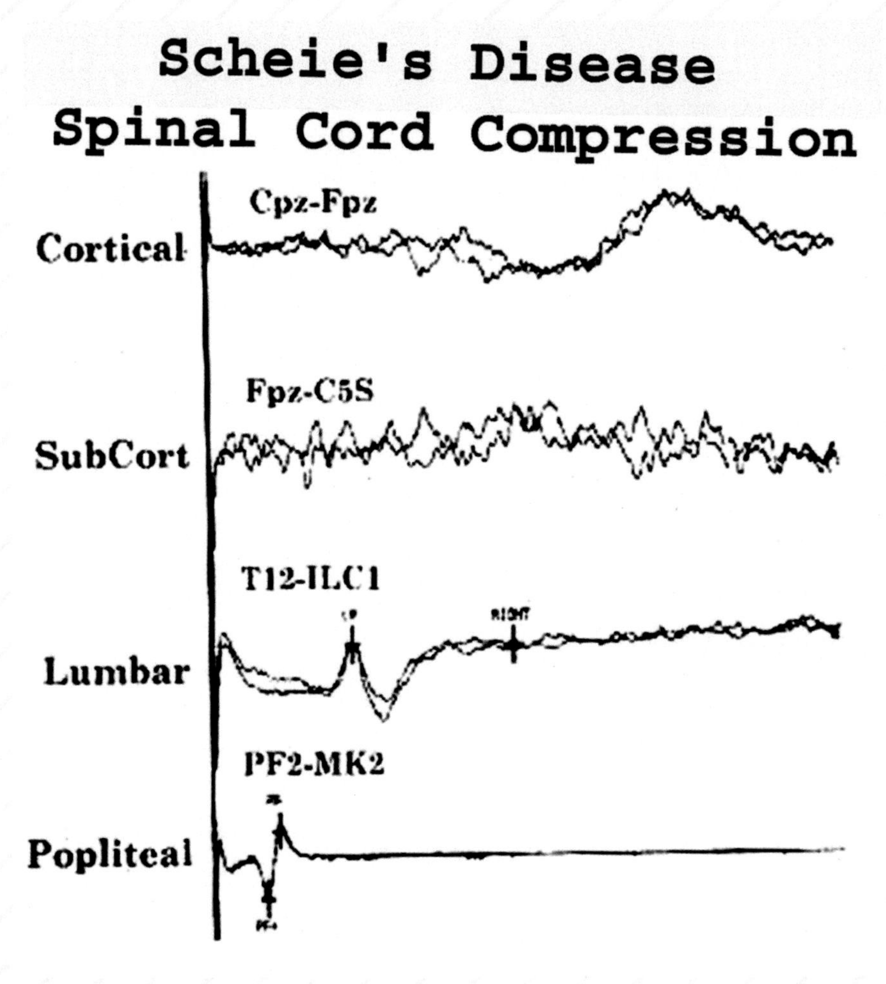

Figure 32. *PTN SSEPs in a 35 year old woman with Scheie' disease. Patient has very well developed peripheral and lumbar responses, but no responses above the lumbar response. Spinal compression leads to a failure to obtain either sub-cortical or cortical responses.*

Syringomyelia is a rare condition of the spinal cord causing a cystic structure to develop in the spinal cord which early in the course of the illness leads to a dissociation of sensation over the chest and shoulders. Crossing pain and temperature fibers are compromised and the patient develops a loss of those sensory modalities in the region. The syrinx may then expand into the dorsal columns to lead to loss of other sensory modalities or into the anterior horn to cause amyotrophy in the hands and arms. Cortical spinal signs and symptoms may also develop. A short stature and a curved spine are common. This complex disorder is frequently part of a Chiari malformation. **Figure 33 AND Figure 34).**

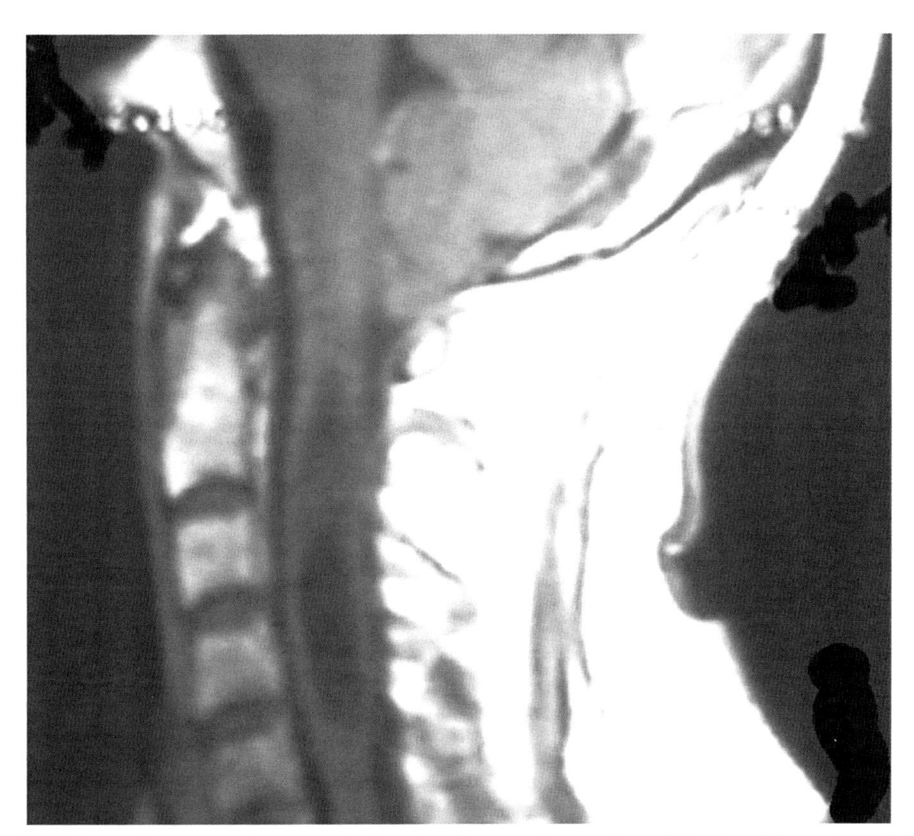

Figure 33. *MRI Scan demonstrating a cervical syrinx in a patient with a Chiari malformation of the cerebellum. Note, the cerebellar tonsils are impacted into the foramen magnum.*

PTN SSEPs-Syringomyelia

(Delayed Cortical Responses for height)

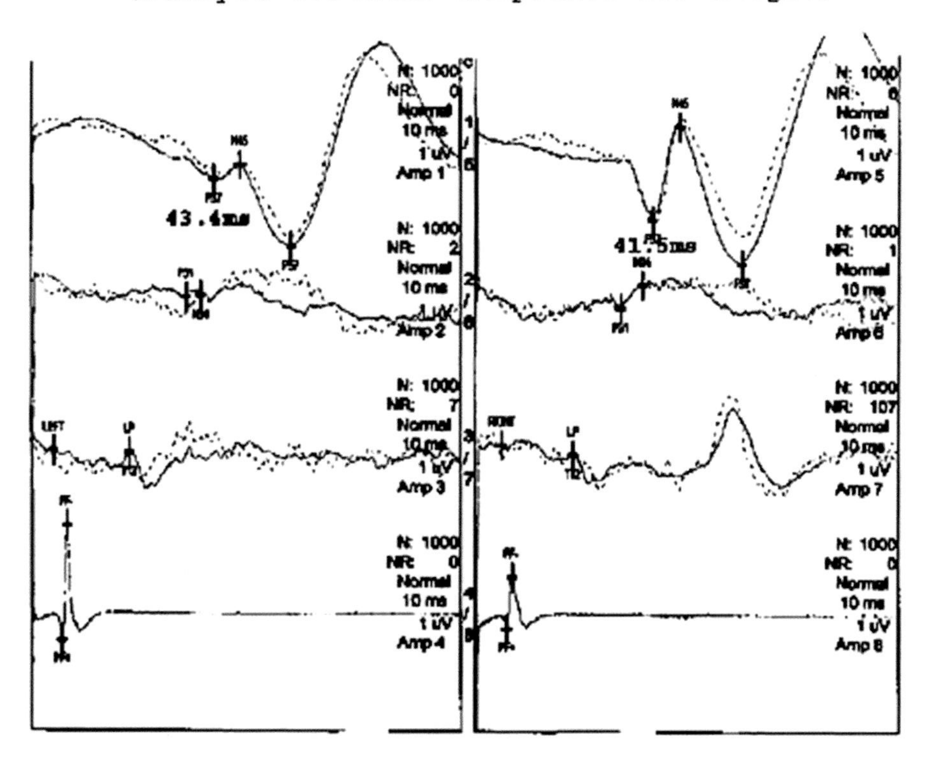

Figure 34. *PTN SSEPs in a patient with syringomyelia. The latencies for the peripheral and lumbar responses are normal. The latencies for the sub-cortical and cortical responses are delayed for height.*

Stretching of the lower spinal cord may result from a persistent short filum termanale. This clinical condition is known as a thethered cord syndrome. At times the persistent filum termanale is associated with a lipoma as well. Surgical section of the filum and removal of the lipoma may restore more normal function. **(Figure 35).**

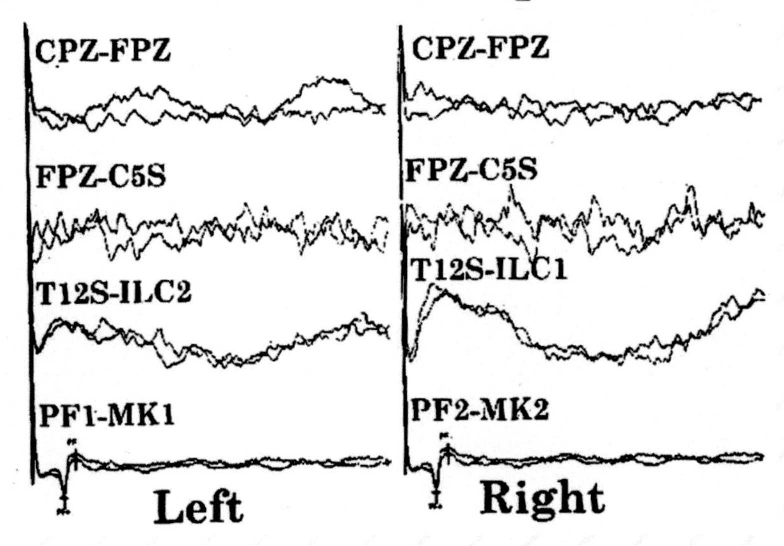

Figure 35. *PTN SSEPs in a young patient with a tethered cord syndrome. The peripheral response at the knee is normal, but no responses are obtained At the lumbar, sub-cortical, or cortical response.*

At times, it is not possible to use the standard region for stimulation to obtain SSEPs. If this is not possible one can substitute the common peroneal nerve at the knee for the PTN at the ankle. The common peroneal nerve SSEPs is nearly identical, except that the latency to the cortex is 10 milliseconds shorter. **See Figure 36.**

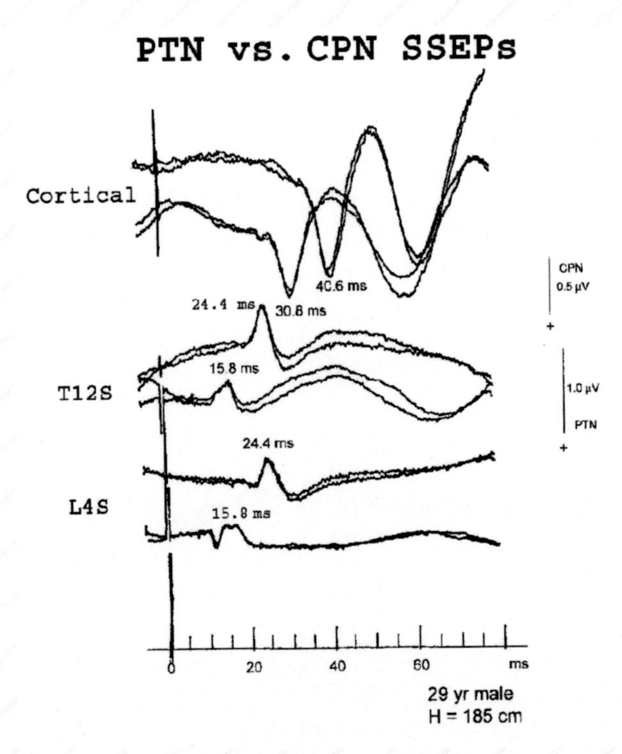

Figure 36. *Comparison of PTN SSEPs to Common peroneal SSEPs In all instances the latency for the waveforms is approximately 10 milliseconds shorter than the PTN SSEPs. The lumbar response is shown at two levels, L4S and T12S. The cortical response of overlaid and the shorter positivity is the common peroneal cortical response.*

Direct Cortical Responses to PTN Stimulation

Like median nerve SSEPs it is possible to record PTN SSEPs directly from the cortex. A 1 X 4 electrode strip is generally slipped into the interhemispheric fissure and the strip placed over the foot area of the Rolandic cortex on the side opposite the stimulus. The foot area of cortex is generally much smaller than the hand and usually not more than a couple of electrode contacts identify the cortical P37. **Figure 37** shows the typical response recorded directly from the cortex.

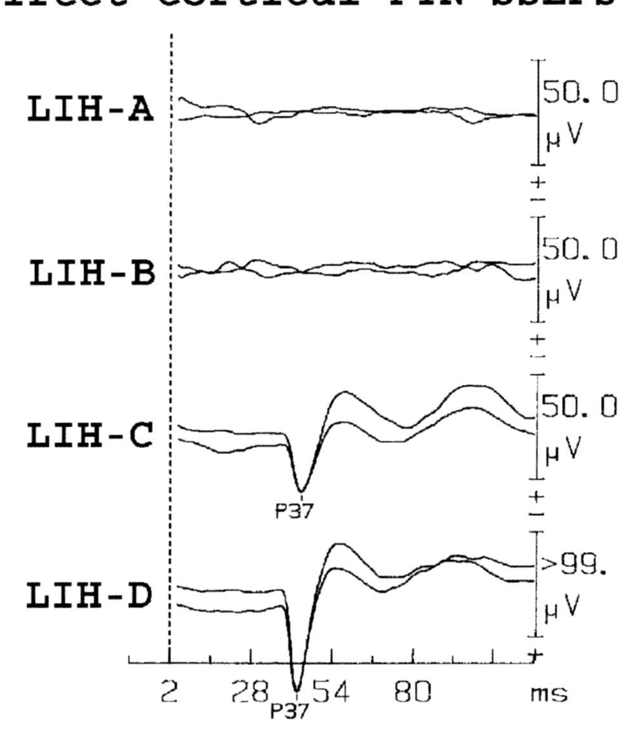

Figure 37. *Direct cortical recording of PTN SSEPs on a 4 contact strip placed over the foot area of the cerebral cortex. Largest P37 with shortest latency over LIH-D. The response seen over LIH-C is of lower amplitude and has a slightly longer latency.*

The cortical foot area may be identified in a area of cortex that extends from the area of brain just lateral to the interhemispheric fissure to an area 4 centimeters into the depth of the fissure.

In addition to its use as a diagnostic tool, the PTN SSEPs are also of great value in intra-operative monitoring. Serial PTN SSEPs are utilized to follow spinal cord function during a number of different types of surgery. This will be discussed fully in the section on Intra-operative monitoring.

SELECTED BIBLIOGRAPHY

Farrell, D.F., Burbank, N., Lettich,E., et. al. Individual Variations in the Motor-Sensory (Rolandic) Cortex of Man.. J. Clin. Neurophysiol 2007; 24: 286-293

Luders, H. Lesser, R.P., Hahn, J., et. al. Cortical somatosensory evoked potentials in response to hand stimulation. J Neurosurg 1983; 58: 885-894

Deiber, M.P., Giard, M.H., Mauguiere, F. Separate generators with distinct orientations for N20 and P22 somatosensory evoked potentials to finger stimulation. Electroencephalogr Clin Neurophysiol 1986; 65: 321-334

Mauguiere, F., Desmedt, J.E., Couron, J. Astereognosis and dissociated loss of frontal or parietal components of somatosensory evoked potentials in hemispheric lesions. Brain 1983; 106: 271-311

Slimp, J.C.,Tamas, L.B., Stolov, W.C., et. al. Somatosensory evoked potentials after removal of somatosensory cortex in man. Electroencephalogr Clin Neurophysiol 1986: 65: 111-117

Woolsey, C.N., Erickson, T.C., Gilson, W.E. Localization in somatic sensory and motor areas of human cerebral cortex as determined by direct recording of evoked potentials and electrical stimulation. J Neurosurg 1979; 51: 476-506

Guideline 9D: Guidelines on Short-Latency Somatosensory Evoked Potentials. J. Clin. Neurophysiol. 2006: 23: 168-179

CHAPTER 4
Dermatomal Sensory Evoked Potential

DERMATOMAL SEPS ARE at times a very valuable tool in the evaluation of of the spinal radiculopathies, brachial plexus injuries and both cervical and lumbar stenosis. The American Academy of Neurology in their Practice Guide rank Dermatomal SEPs as a Grade D. They have a negative recommendation based on inconclusive or conflicting Class II evidence. Even with this negative recommendation we have found dermatomal SEPs to be of clinical value in the above situations.

Dermatomal SEPs do have some short-comings that need to be kept in mind when interpreting this test. Unlike the Somatosensory Evoked Potentials, recording of the dermatomal SSEPs is confined to the cortical response. The response is not strong enough to be able to record at different levels of the neuraxis as in SSEPs. Before dermatomal SEPs are carried out, one must be sure that the spinal cord, brainstem and cortical pathways (central sensory pathways) are unaffected by any disease process. Any central sensory disorder would invalidate the results from dermatomal SEPs. Dermatomal SEPs are accomplished with methods that are similar to both the median nerve SSEPs and PTN SSEPs, except that skin dermatomal segments are stimulated and not peripheral nerves. The technologist carrying out this test should keep a figure of the human dermatomes so that stimulating electrodes can be appropriately placed.

Methods

Stimulation of the dermatomal segment uses a monophasic rectangular pulse with a pulse width of 100 to 300 milliseconds. The stimulus frequency is 3 to 5 Hz and the intensity 3 times sensory threshold. The scalp montage for cervical studies are CP3-CP4 for right sided stimulation and CP4-CP3 for left sided stimulation.. The montage for the lumbar and thoracic dermatomes is CPz-FPz; CP1-Ai; CPz-Ai; and CP2-Ai. The mid-line combination generally gives good reproducible responses. However, the lateral combinations may be necessary to allow recording of a response when it is not well seen over the midline. At least two replications are necessary to verify reproducibility, but more may be required in those individuals who have trouble relaxing (EMG artifact). The electrode impedence should be less than 5000 ohms and the band pass filters set at 30 for the low frequency filter and 3000 Hz for the high frequency filter at (-6dB/octave). The stimulating electrode (cathode) is placed over the appropriate dermatomal segment and the anode placed 2 centimeters distal to the cathode. The ground electrode is placed proximal on the upper leg, arm or high on the neck for upper cervical dermatomes. Analysis time is typically 40 milliseconds for the cervical dermatomes and 60 milliseconds for the thoracic and lumbar dermatomes. These times may need to be extended when latencies are greatly prolonged. Conservative interpretation should always be carried out. Changes in wave amplitude are of no clinical value as the amplitude varies with the clinical state of the individual. If the person becomes drowsy during the test there is likely to be a significant drop in amplitude of the waveforms or possibly even complete loss of waveforms. This is such a problem that dermatomal SEPs cannot be trusted in individuals who have undergone anesthesiology for surgery. The safest interpretation of an abnormality is to have a response completely absent or as a second best to have the latency shift greater than 3 standard deviations. If possible the technologist should also test levels above and below the lesion to make sure a normal response is obtained on either side of the lesion.

- Most levels of spinal cord function can be studied with this technique. The exceptions being C1, C2 and C3, thoracic levels are accomplished at every other segment level. A normal cervical dermatomal pattern, C5 to C8 is shown **in (Fig. 1).**

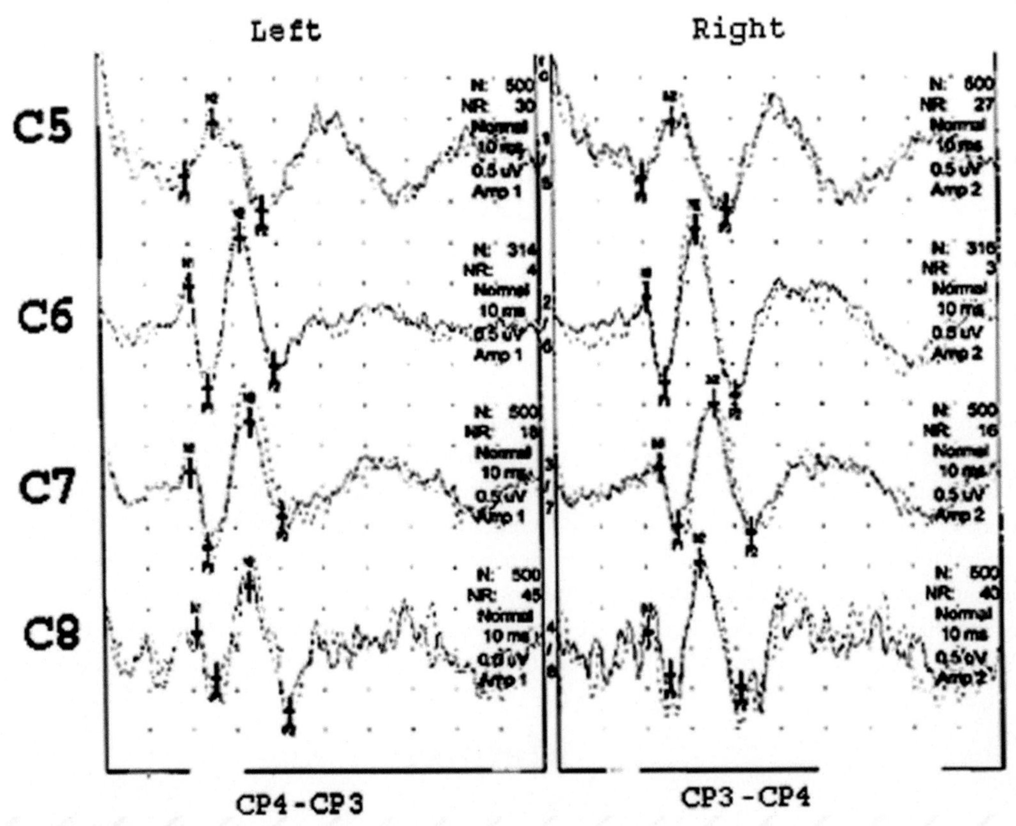

Figure 1. *The responses from the cervical dermatomal segments are shown. Note there is no initial N1 in any of the samples,.*

Another sample of normal cervical responses is shown in **(Fig. 2).**

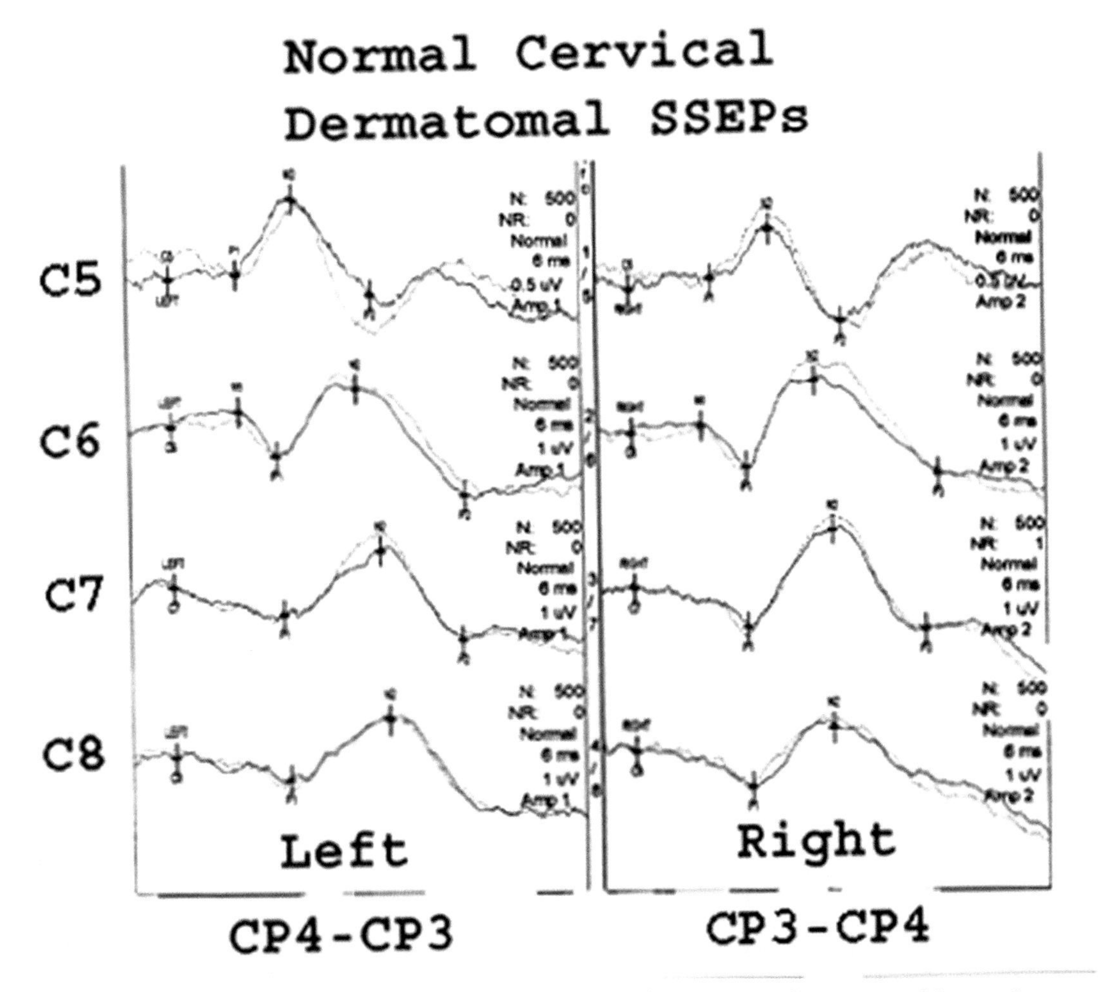

Figure 2. *Normal cervical dermatomal responses are shown. Note there is no initial N1 in any of the samples, especially the C5 and C6 levels, but there may be a trace at the C7 and C8 levels.*

Cervical Radicular Disease

A herniated Intervertebral disc into the lateral recess of the spine is by far the most common cause of this acute disorder. The part of the spine at most risk is at the C5-C6 interspace. This happens to be where the most flexibility of the spine occurs. The patient frequently hears a "pop" followed by intense pain radiating down the arm. This irradiated pain is frequently in the distribution of the affected dermatome. Intense muscle spasms of the neck develop which intensifies the pain. Conservative therapy is in order with use of a cervical collar, muscle relaxants and analgesics. Many cases of acute herniated disc will heal without the need for surgery. Usually a trial or 4-6 weeks of conservative therapy are in order before surgery is considered. One exception to this general rule is if there is early muscle weakness earlier surgery to decompress the nerve root is indicated. As an area, the acute herniated cervical disc has a lower spontaneous healing than found in the lumbar region. For single level disorders, an anterior spinal surgery approach is used and the recovery time and disability much less than posterior approaches. **(Fig. 3).** Demonstrates a radiculopathy at the right C5 level.

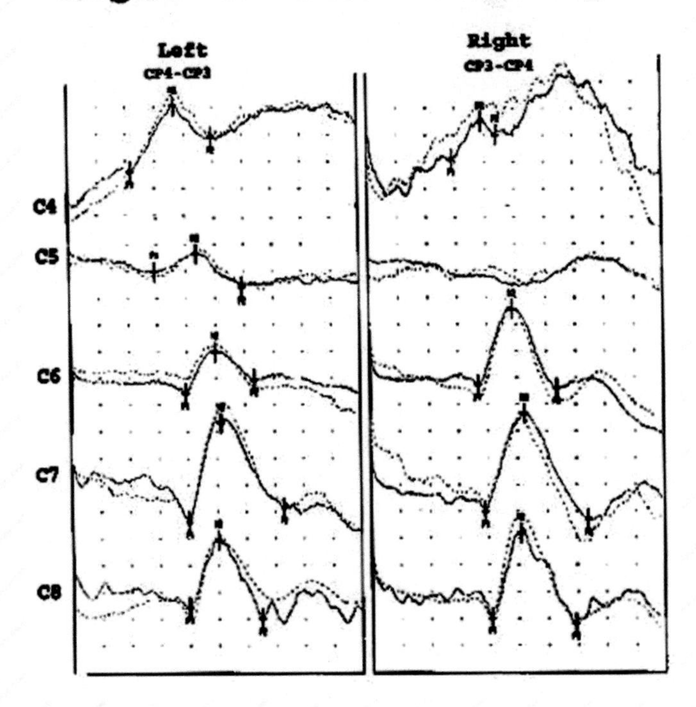

Figure 3. *This cervical dermatome study shows an absent response on the right at the C5 level. Note, the levels below and above have normal responses. The diagnosis is a right C5 radiculopathy.*

A more complicated case is shown in **(Fig. 4).**

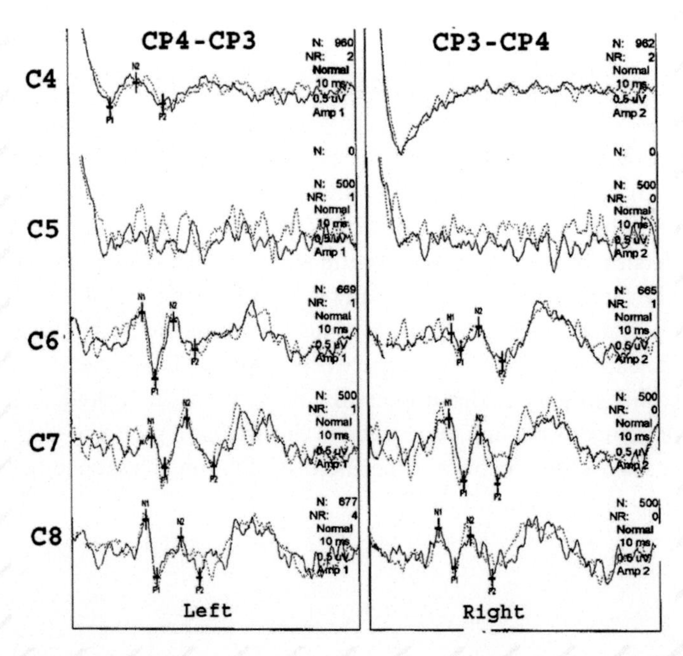

Figure 4. *This sample shows multiple levels of abnormality. The right C4 level and bilateral C5 levels are absent. The C6-C8 dermatomes are normal. This is one area in which data could not be collected above an abnormality. No responses above C4 can be obtained.*

Cervical Spondylosis

Cervical Spondylosis is a common disorder of the middle-aged to older-aged patient. It is by far the most common cause of a gait disturbance in the elderly. The pathophysiology of cervical spondylosis is complicated and at least two factors appear to play a role. One, the arthritic bone spur (bar) places pressure directly on the spinal cord, but this does not appear to be adequate to cause the disorder. The second component is a local vascular change that causes ischemia beyond the direct compressive area. Prior to CT scans and MRI the plane lateral X-ray of the cervical spine in flexion and extension was used to establish this diagnosis. The flexion and extension films frequently demonstrated the instability of the spine with the presence of subluxation. For many years arguments occurred as to whether there existed such a thing as cervical stenosis. Both asymptomatic and symptomatic cervical spondylosis occurred and the amount of arthritic spur did not always correlate with the clinical state. Careful anterior-posterior spinal measurements finally gave an indication as to which cases were likely symptomatic. It turned out if the A-P diameter of the spinal canal was less than 10 millimeters; the spondylosis was likely to be causative. If the measurements were greater than 13 millimeters it was unlikely to be the cause of the patient's spinal cord symptoms. It turns out that there is a predisposition to this condition. Individuals with a congenitally small spinal canal are at risk to develop this condition, whereas, those with a large spinal canal can have a large boney abnormality and still not compromise spinal cord function. Multiple level posterior decompressive surgeries (laminectomy) are generally indicated for symptomatic cases. This may require the use of hardware to stabilize the site. In general, one cannot expect improvement with this type of surgery, although it does occur. The surgery is done to prevent further damage to the spinal cord. **(Fig. 5).** is a sample of dermatomal SEPs in a case of cervical spondylosis.

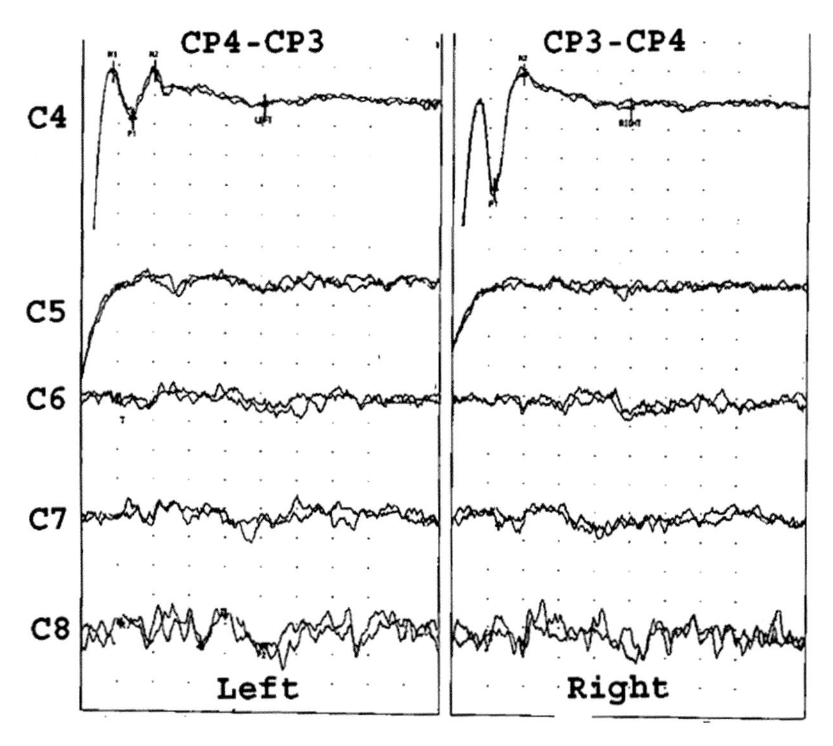

Figure 5. *Example of a dermatomal SEP study in a patient With cervical spondylosis. Note that no waveforms are generated Below the level of the spondylosis. There is however, a normal Response from the C4 dermatome bilaterally.*

Any time a symmetric loss of function occurs in the dermatomal SEP study one should suspect spondylosis as the cause.

Brachial Plexus Injuries

Dermatomal SEPs are a useful adjunct in the evaluation of brachial plexus injuries. Unlike the median and ulnar SSEPs, dermatomal SEPs will not identify whether the lesion includes root avulsions or the trauma is more distal in the plexus itself. The Dermatomal SEPs will however allow each root to be studied independently. **(Fig. 6)** is an example of such a brachial plexus injury.

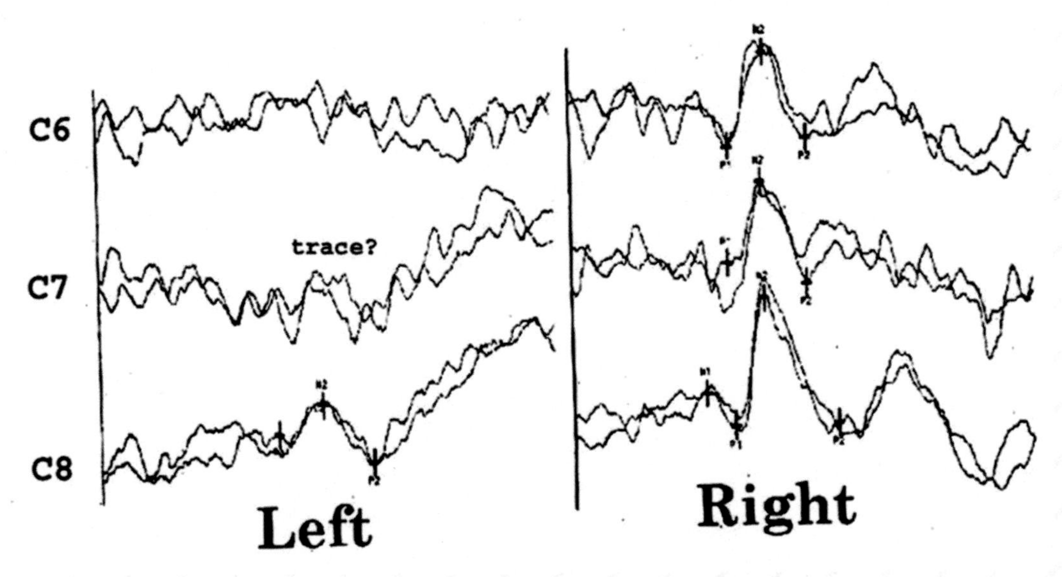

Figure 6. *Dermatomal SEPs of the cervical spine. Note that the right side is normal at all 3 levels, but on the left side the C6 and C7 dermatomes show no activity.*

THORACIC DERMATOMAL SEPS

Herniated Intervertebral discs involving the thoracic region of the spinal canal are rare in comparison to both cervical and lumbar regions. Special problems, primarily from the lack of space in this part of the spine lead to early spinal cord compression. The herniated discs can be either central where the spinal cord becomes compromised and surgical decompression should be carried out early. Lateral herniations involve roots and there is usually intense pain and discomfort in the distribution of the herniated disc. Thoracic dermatomal SEPs can be readily carried out. **(Fig. 7)** shows a normal study of the Thoracic dermatomes.

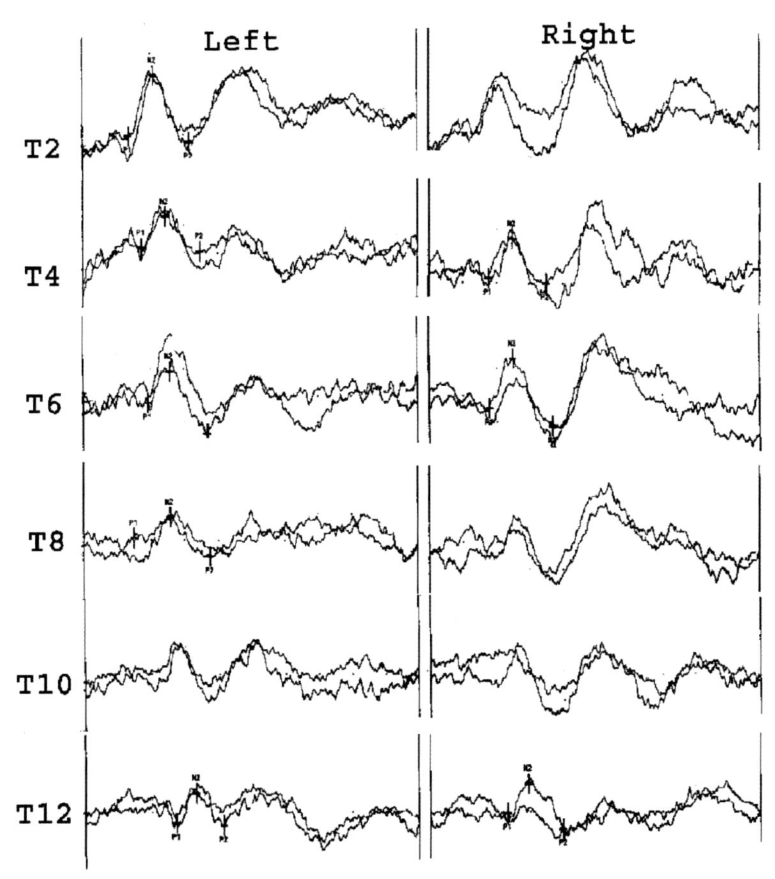

Figure 7. *Sample of Thoracic dermatomal SEPs from T2 to T12. Note the shortening of the latency as higher dermatomes are stimulated. Reproducibility is quite good.*

A thoracic dermatomal SEP study of a radiculopathy at T12 is shown in **(Fig. 8).**

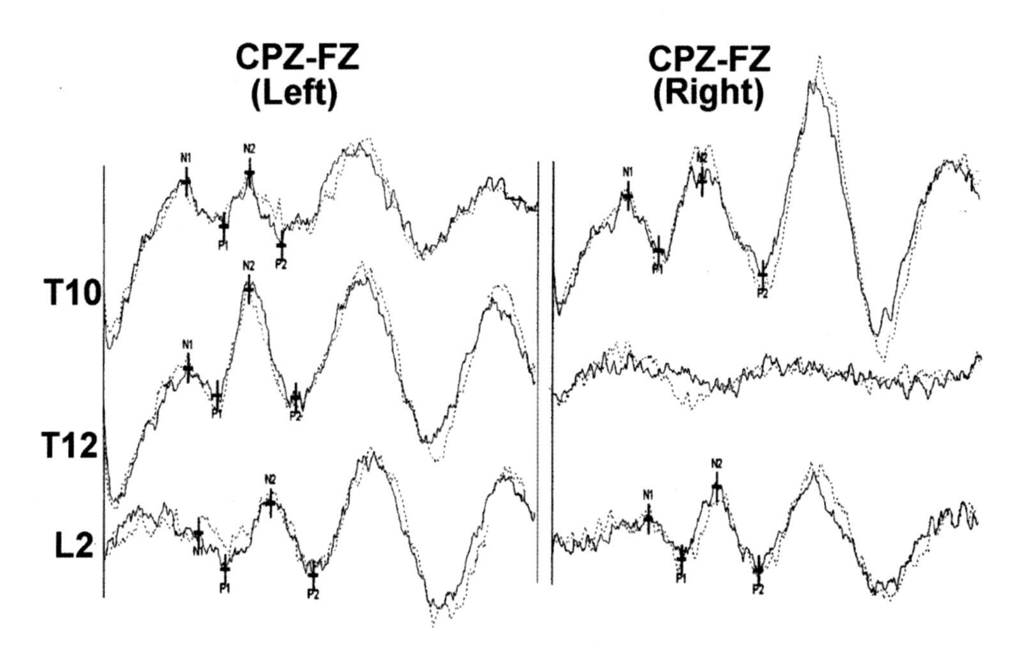

Figure 8. *Lower thoracic upper lumbar dermatomal SEPs showing a Completely absent response at the right T12 level. Dermatomal levels above and below the lesion are normal.*

DERMATOMAL SEPs OF THE LUMBOSACRAL REGION

Lumbar dermatomal SSEPs are particularly valuable in the identification of lateralized radiculopathies. **Figure 9** shows the normal responses obtained from stimulations of this region.

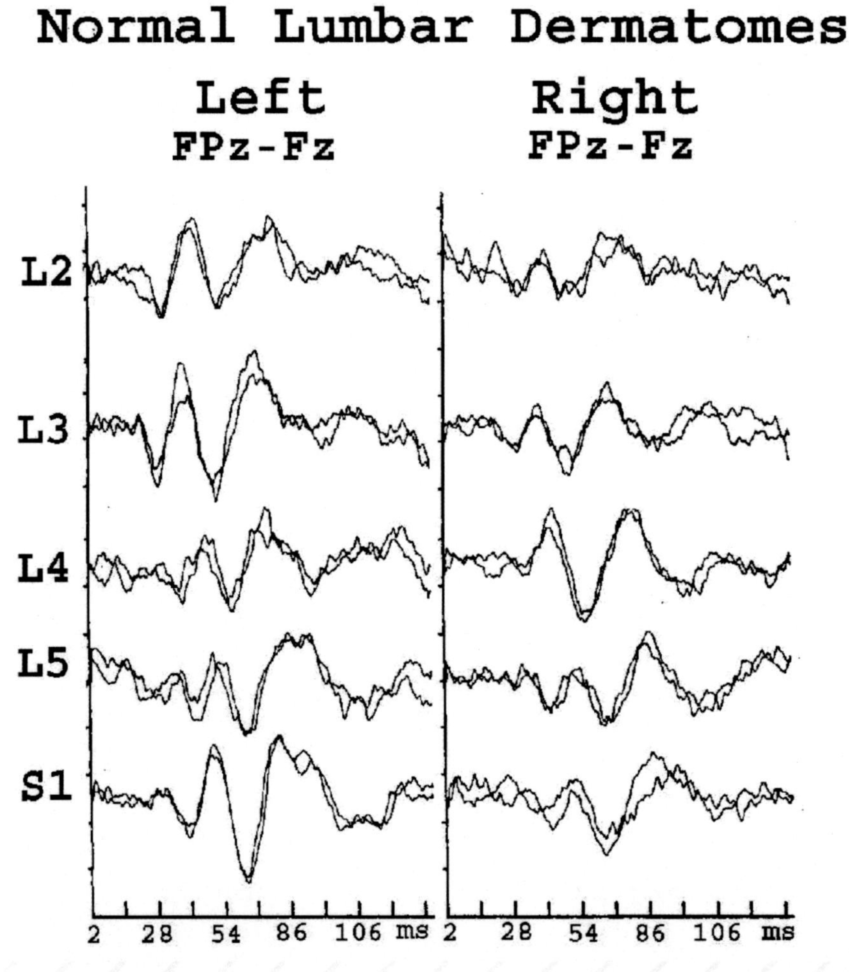

Figure 9. *The lowest level that can be tested is the S1 level and each level above that up to L2. As one ascends the latency of the response shortens. This sample has good reproducibility.*

Herniated Intervertebral discs involving the lumbar and sacral regions are even more common than those affecting the cervical region. At this level, there is little possibility of direct spinal cord compression as the spinal cord generally terminates between L1 and L2. The vast majority of acutely herniated discs affect L4 through S1. Again, at the onset the patient may hear a "pop" followed by pain irradiating down the leg. Different patterns of pain are relatively specific for the root being affected. This syndrome may induce severe muscle spasm of the low back which adds to the discomfort. This muscle spasm may be severe enough to either straighten the normal lumbar curvature or even in some instances reverse it. Conservative therapy is indicated with bed rest, muscle relaxants and analgesics being used. Approximately 70-80 percent of herniated discs will heal without the need for surgical intervention. An example of a single level, unilateral radiculopathy is shown in **Figure 10.**

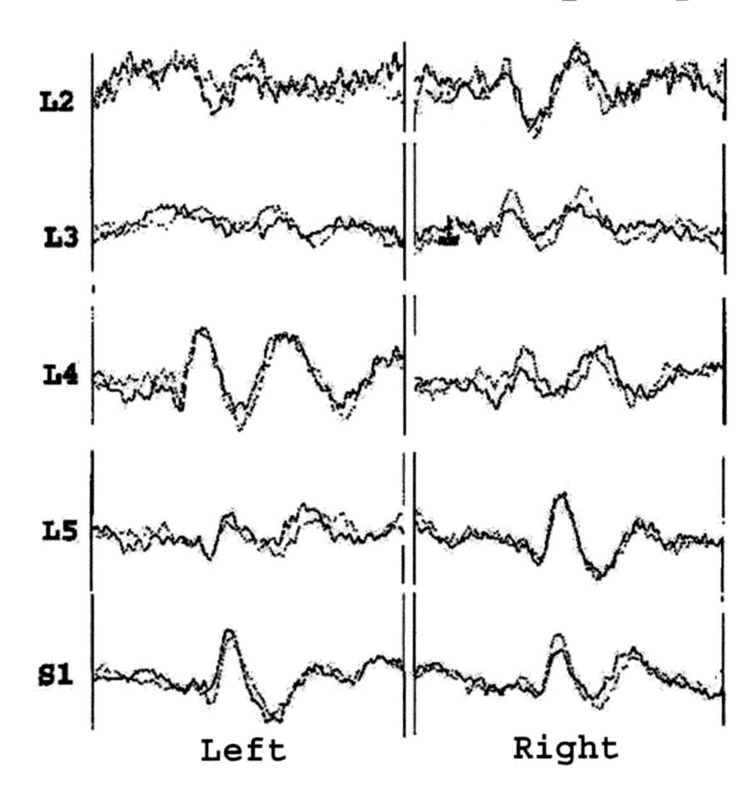

Figure 10. *This figure shows an isolated absent response at the left L3 level. All responses above and below this level are normal. Findings Consistent with a radiculopathy affecting left L3.*

An additional sample at a different level is shown in **(Fig. 11).**

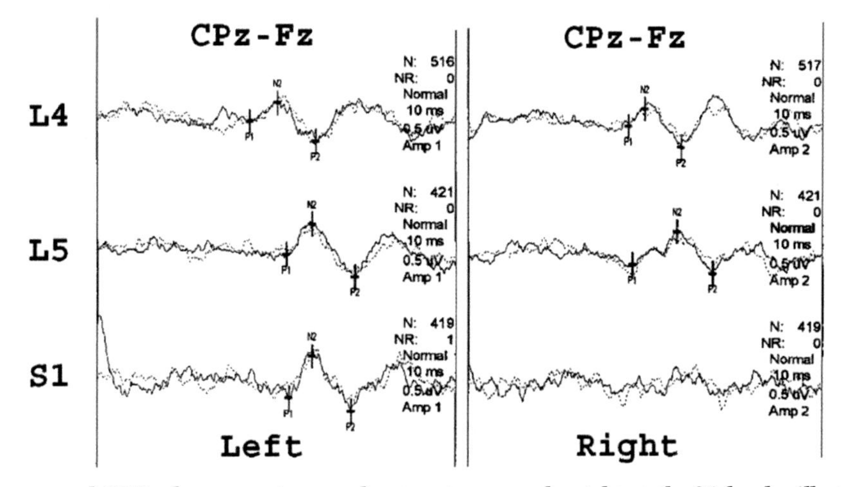

Figure 11. *Dermatomal SEPs demonstrating an absent response on the right at the S1 level. All other responses at higher levels and on the other side are normal.*

At times one will run into an individual who has bilateral disease. This may result from two independent events or from a single large disc that compresses both roots at the same level. **Figure 12** shows this type of change.

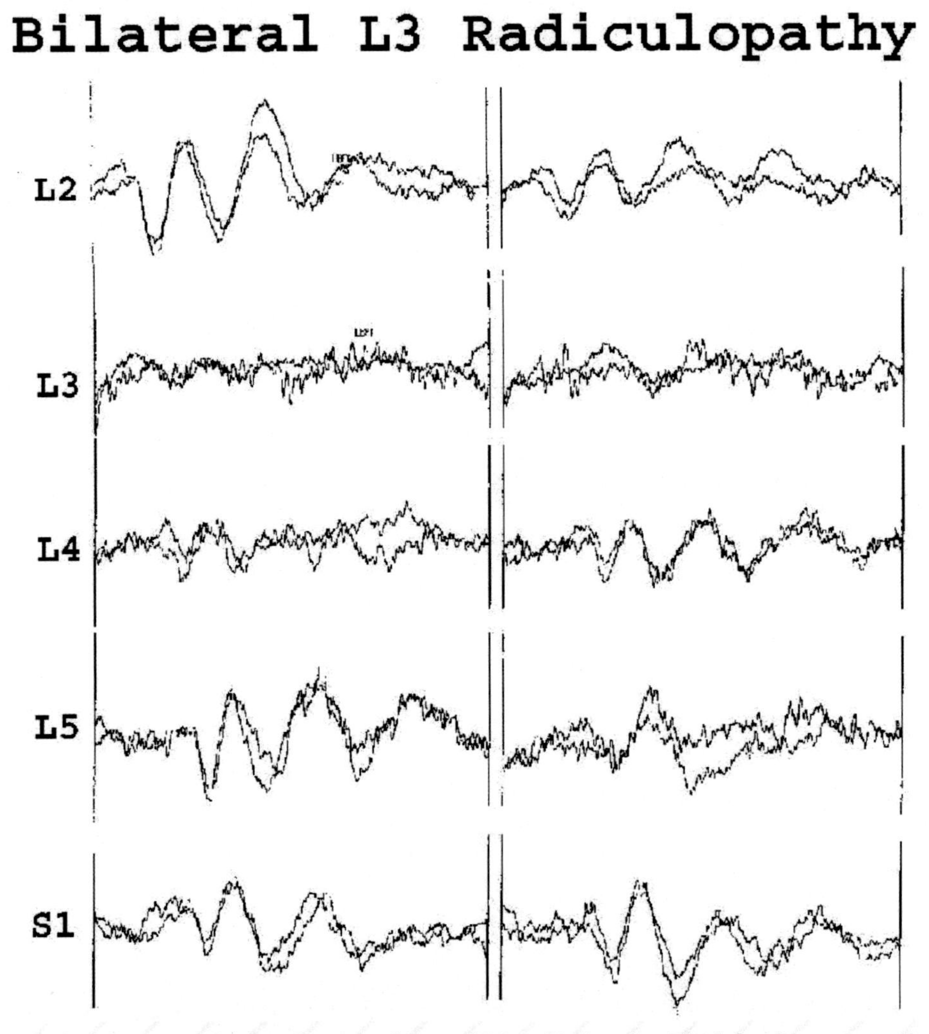

Figure 12. *Lumbar dermatomal SEPs showing a bilateral L3 radiculopathy. The left side appears to be worse than the right. However, the right side does not show a reproducible response.*

LUMBAR STENOSIS

Lumbar stenosis like its counterpart in the cervical spine is a common disorder. At this level, the spinal cord itself does not usually get compressed by the boney osteophytes, but nerve roots below L2 get compressed. The presentation is also different than in the cervical region. Lumbar stenosis is likely to cause claudication of the legs while walking for a short distance. The symptoms are nearly identical to the claudication that occurs from arterial insufficiency to the legs. Upon evaluation, the major arteries are open and pedal pulses are present. Posterior decompressive surgery usually solves the symptoms. At time metal appliances are needed to stabilize the surgical site. **(Fig. 13).**

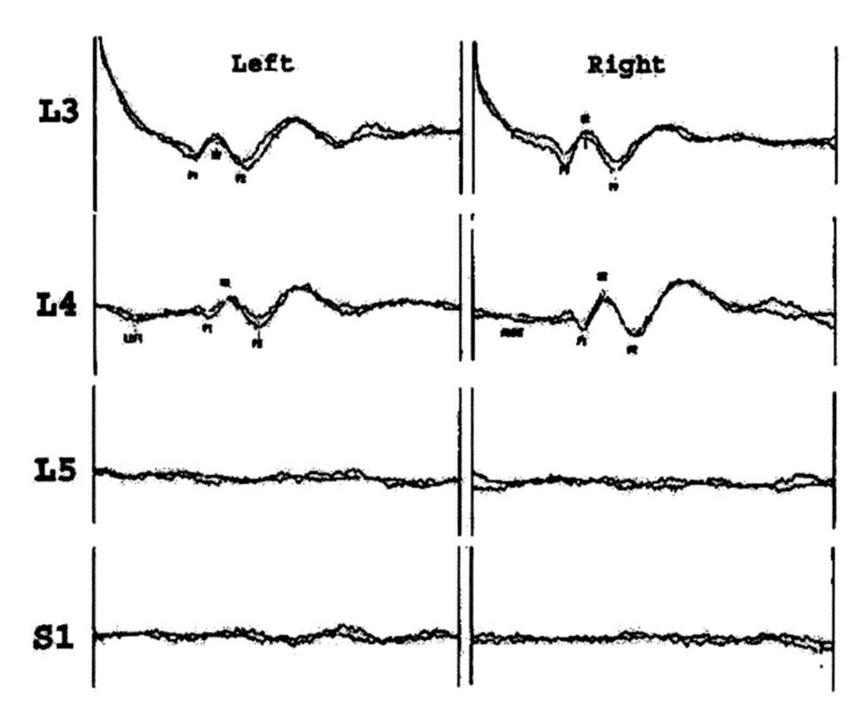

Figure 13. *Lumbar dermatomal SEPs showing no waveforms generated at L5 or S1. The responses from L4 and above are normal. This patient had a stenotic lesion at L5.*

Value of Dermatomal SEPs in the Evaluation of Back pain

Over a number of years a single neurosurgeon referred 110 cases of spine disease for evaluation by dermatomal SEPs. Each of these cases also underwent electromyographic examination and MRI scans. These cases had in common chronic back pain and various neurologic disabilities. Many of them had had earlier back surgery without relief of their symptoms. Over 50 percent of these cases had normal dermatomal SEPs and EMGs. Many of these individuals did have abnormalities on their MRI scans, but these changes did not correlate with their symptoms. Out of the 110 individuals all but 1 had an abnormal MRI scan. It appears that the MRI in these 110 patients was just too sensitive for abnormalities and not very specific in regards to the patient's symptoms. Those individuals with normal dermatomal SSEPs and EMGs were not considered candidates for additional surgery. The majority of the un-operated patients were referred to a pain clinic for management of their persistent low back pain.

Another group of patients were found to have persistent abnormalities in both the dermatomal SEPs and EMGs. Many of these individuals had also had previous back surgery, almost exclusively a simple laminectomy. The surgical failure in these cases appeared to result from the lateral recesses not being decompressed. Following re-operation with appropriate lateral recess decompression, this group all showed clinical improvement in not only in their pain, but neurologic dysfunction.

It has become clear that back surgery in which the sole symptom is pain does not improve with the surgery. However, those individuals who have pain plus signs of nerve root compression receive relief with appropriate surgery.

SELECTED BIBLIOGRAPHY

Eisner, A., Hirsch, M., and Moll, A. Evaluation of radiculopathies by segmental stimulation and somatosensory evoked potentials. Can J Neurol Sci (1983) 10: 178-182.

Aminoff, M.J., Goodin, D.S. Matter arising: Dermatomal somatosensory evoked potentials in Lumbosacral root compression. J. Neurol Neurosurg Psychiatry. (1988) 51: 740-741.

Slimp, J.C., Rubner, D.E., Snowden, M.D. and Stolov, W.C. Dermatomal somatosensory evoked potentials: cervical, thoracic and Lumbosacral. Electroencephalogr Clin Neurophysiol (1992) 84: 55-70.

Snowden, M.L., Haselkorn, J.K. Kraft, G.H., et al. Dermatomal somatosensory evoked potentials in the diagnosis of Lumbosacral spinal stenosis: comparison with imaging studies. Muscle Nerve (1992) 15: 1036-1044.

Monitoring Section II: Intra-operative Monitoring

Principles of Intra-operative Monitoring

This portion of the book is devoted to Intra-operative Monitoring. Many of the techniques involved with intra-operative monitoring are based on knowledge of short latency evoked potentials. Typically, BAEPs and Median nerve SSEPs are used in skull-based surgery. PTN SSEPs is used for most spinal surgery. Surgery in the upper cervical region and at the foramen magnum use MNSSEPs. Other types of monitoring include analysis of electromyography activity, both spontaneous and evoked through nerve stimulation by the neurologic surgeon. It is possible to monitor both cranial nerve innervated muscles and skeletal muscles depending upon the nature and site of the surgery. Increasingly, trans-cranial motor evoked potentials are being utilized to monitor the integrity of the cortical bulbar and spinal tract during surgery. This is most useful during spinal cord surgery, discussed in the next chapter. At this time, PREPs. and dermatomal SEPs are of little to no value during surgery. PREPs. require an awake and alert patient **who is** able to fixate on a target in the middle of the stimulating alternate checkerboard pattern and dermatomal SEPs are particularly labile to changes in the level of consciousness, especially those induced by anesthesia.

The development of skull-based surgery has depended upon certain technological advances. First, the operating microscope allows the surgeon to work using a stable platform that provides different levels of magnification of the surgical field. This plus the development of small specialized instruments allows the surgeon to remove tumors from both nerves and blood vessels. Secondly, advances in anesthesia allows the patient to remain anesthetized for prolonged periods of time, up to 24 hours in some instances.

A good working relationship with the anesthesiologist is imperative for quality intra-operative monitoring. A balanced anesthetic regimen is preferred. No single agent being used alone. High levels of a single anesthetic agent tend to suppress the cortical responses from the MN SSEPs and PTN SSEPs. High levels of nitrous oxide are notorious for inhibiting the cortical responses and no levels above 50 percent should be used during these surgeries. The BAEPs and sub-cortical responses of the somatosensory evoked potentials are generally resistant to anesthetic effect. In addition to balanced anesthesia the anesthesiologist should notify the monitoring team of any changes in anesthesia during the course of the surgery. A bolus of anesthesia is particularly harmful to evoked potentials. If muscle paralytic agents are used for induction, they must be of very short action so that the EMG activity can be monitored early in the course of the surgery, especially during opening. It must be remembered that the anesthesiologist has the final say as to the use of any anesthesetic agent in any given case that is necessary for the safety of the patient. Rarely, the selection of an anesthetic agent will be made for patient safety and that need over rides the needs of the monitoring team. The monitoring team also has a responsibility to the anesthiologist. And, at times, when the patient is lightening from not enough anesthesias, generalized spontaneous EMG activity will be recognized. The monitoring team needs to notify the anesthesiologist that the patient is lightening so that appropriate corrections can be made.

Unlike the laboratory, the operating room is an unfriendly environment for collecting evoked potentials. Electrical artifacts may be introduced from a wide variety of pieces of equipment in the OR and the technologist must be able to identify these, correct them when possible, or eliminate their influence by modifying how the data is collected.

Some of the worst offenders in the operating room include electrical cautery and the use of the Bovie electrosurgical generator, as well as electrical bone drills. These pieces of equipment generate very large electrical artifacts that will "saturate the amplifiers" causing them to block and not be able to process data. Much of the use of these types of equipment is intermittent and the technologist must pause the collection of data during these periods or have a cut-off switch so that each time the surgeon uses the equipment it will automatically pause the data collection. The use of the operating microscope with a slave TV for the monitoring team to watch the progress of the surgery allows the team to see and anticipate when the surgeon might use one of the offenders and pause the data collection without the surgeon having to remember to inform the team that he will be using a certain piece of equipment. Other artifact generators put out a constant artifact, which would include certain types of operating beds, which should be disconnected from an AC source during surgery and run on battery. Bottle warmers, electrical blankets, etc. use may have to be curtailed to collect data during surgery. Sometimes it is only necessary to modify the location or orientation of pieces of equipment to reduce the influence of that piece of equipment on the evoked potential. For example, the light generator for the surgeon's head lamp can be rotated so as to reduce the size of the interfering artifact (directional artifact). Electrical artifacts generated by the operating microscope may require a change in the stimulus frequency to reduce the influence of the artifact, for example changing a stimulus rate from 11.1 Hz to 11.3 Hz for the BAEPs may be enough to reduce the effect of such an artifact.

The first chapter (Chapter 5) in this section is devoted to skull-base surgery. By far the most common tumor at the skull base is the vestibular schwannoma, so that they will receive the most attention. However, there are a wide variety of other tumors that occur along the skull base and together make up a long list of histologic types and occur in a variety of locations. In addition to tumor removal, intra-operative monitoring is utilized in micro vascular decompression for trigeminal neuralgia and hemi-facial spasms and vestibular nerve section for medically intractable Meniere's disease.

The second chapter (Chapter 6) is devoted to spinal surgery. This will include monitoring for tumor removal, for decompressive and stabilization spine surgery, and scoliosis surgery.

All of these types of monitoring have one goal in mind and that is to make the patient's surgery safer by providing the surgeon up to date information about potential problems that are in the process of arising, to identify structures that may be misplaced because of a mass lesion or nerve fibers having been incorporated into a tumor capsule. For example, prior to the start of intra-operative monitoring 15-20 years ago virtually every patient undergoing surgery to remove a vestibular schwannoma developed a permanent facial palsy. With modern intra-operative monitoring the facial nerve is now spared in the vast majority of cases, in fact, it is unusual for such a patient to develop a permanent facial palsy. Monitoring also helps to preserve hearing in a large number of patients. About half of patients with tumors smaller than 2.5 centimeters will retain hearing, whereas the vast majority of patients with tumors greater than 2.5 centimeters will lose their hearing. Chapter 7 is devoted to the different types of monitoring used in Peripheral Nerve Surgery and Chapter 8 to differences in monitoring techniques used during Pediatric Surgery. The next chapter (Chapter 9) is devoted to Electrocorticography and its use in both Epilepsy surgery and neocortical tumor removal. Along with these types of surgery, language mapping and motor-sensory mapping are used to identify eloquent cerebral cortex. Eloquent cortex is that portion of the cerebral cortex that must not be damaged or removed during surgery or the patient will be left with a significant functional deficit.

The final chapter (Chapter 10) is devoted to monitoring during carotid endarterectomy. This type of monitoring has undergone radical changes over the years from straight EEG monitoring to monitoring an EEG derivative, power spectra derived from fast Fourier transformation. Spectral edge can be monitored continuously for changes in percent of the band-width. But, this to has been largely replaced with Transcranial Ultrasonography, a direct measure of blood flow velocity, rather than depending on ischemic changes in brain function which earlier the methods were dependent upon.

CHAPTER 5
Skull-Base Surgery

THE SETUP FOR intra-operative monitoring varies by the type of tumor and the location of the tumor. In addition to monitoring either BAEPs or median nerve SSEPs, muscles innervated by the different cranial nerves are used to help identify the cranial nerves in the area of the surgery. Fine wire electrodes are inserted into the muscle and the inserting needle removed leaving a fine wire electrode in place. This system is much safer for the patient than using a rigid needle electrode. The fine wire electrode is flexible and will not shift in location with exerted pressure by the surgeons or their appliances. In general, the fine wire electrodes are inserted into the appropriate muscles after the patient has undergone induction and is fully anesthetized. These fine wire electrodes are made in the laboratory as well as being commercially available. **(Fig. 1).**

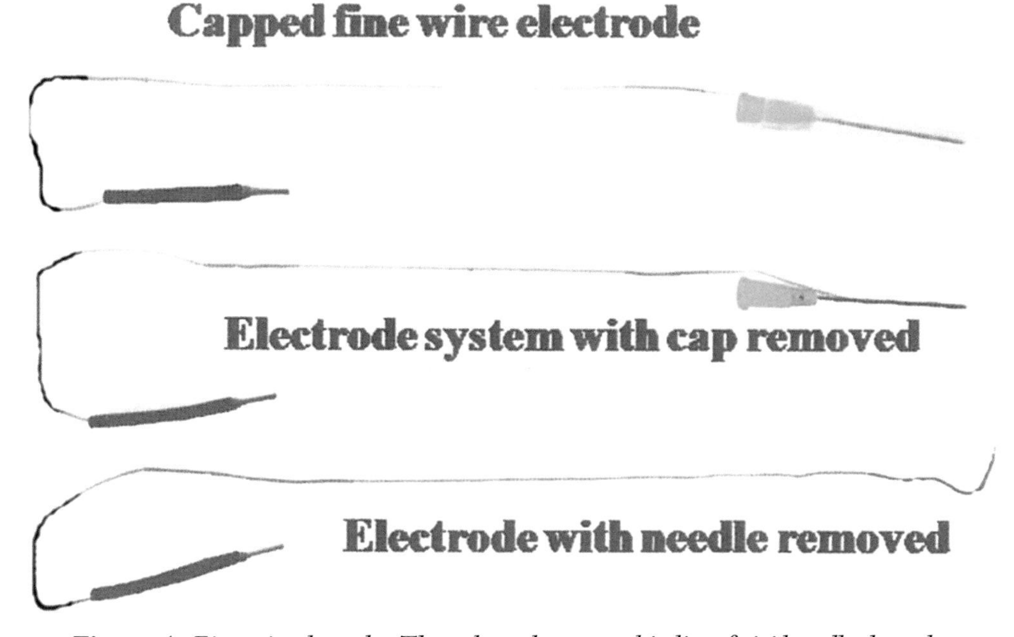

Figure 1. *Fine-wire electrodes. These electrodes are used in lieu of rigid needle electrodes.*

Fine wire electrodes are made using very fine insulated wire. Insulation is removed from each end of the wire. A pin plug is installed on one end and the other end stripped (about 3-5 millimeters) and is inserted into the barrel of a 21 gauge, 1 ½ inch disposable needle. The insulated wire is then wrapped around the needle 4 to 5 times and the cap replaced. Each electrode system is then packaged and gas sterilized before use.

The placement of these fine wire electrodes will be reviewed and simple schematics used to demonstrate their location. These fine wire electrodes have relatively high impedance, but are about the same for each electrode. Reference electrodes should be of the same type so as not to created an electrode imbalance and introduce electrical artifacts.

CRANIAL NERVES

For surgery involving the region of the cavernous sinus, Cranial Nerves III, IV and VI are at risk. Muscles innervated by these nerves are ocular muscles. For all of the muscles of interest, the muscle belly occupies an area that is in the posterior portion of the ocular globe and it is impossible to insert an electrode directly into the muscle, so the electrode is placed in the tissues near to a given muscle. It is possible to record muscle activity in these locations. The medial rectus muscle is innervated by the Oculomotor Nerve (III) and is used to monitor this nerve. The electrode is inserted through the skin at the medial canthus of the eye with the needle electrode being guided medially along the globe for a depth of 0.5 to 1.0 inch. The needle should be inserted through the skin just superior to the tear duct in a manner so as not to injure the duct. After insertion, the needle should be withdrawn and the wire electrode taped into place. The superior oblique muscle is innervated by the Trochlear Nerve (IV). The muscle is monitored by a fine wire electrode inserted superiorly at the middle of the eye. Again, the needle electrode is inserted through the skin with the needle pointed superiorly and posteriorly in the mid-line for a depth of 0.5 to 1.0 inches. The needle is then withdrawn and the electrode taped into place. Finally, the lateral rectus muscle is innervated by the Abducens Nerve (VI). The needle electrode is inserted through the skin at the lateral canthus of the eye. The needle is guided laterally and posteriorly for between 0.5 and 1.0 inches. The needle is then withdrawn and the fine wire electrode anchored in place with tape. A reference electrode is inserted in the opposite frontal region. The whole area is then covered with Tegaderm.

Because of the nature of tumors involving this region it may be necessary to use a bilateral setup. See the following photos for summary of electrode placements. **(Fig. 2, Fig. 3, and Fig. 4).**

X - Reference

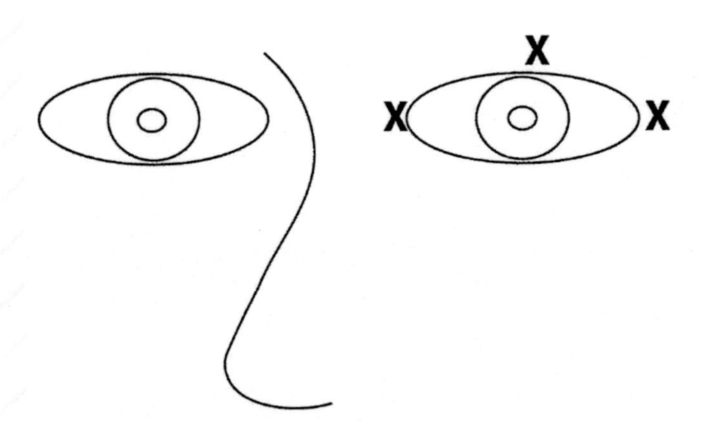

Placement of Electrodes to monitor CNs III, IV and VI.

Figure 2. *Points where electrodes are inserted to monitor the Medial rectus, Superior oblique and Lateral rectus muscles.*

Fine wire electrodes inserted in medial canthus to record from Medial rectus (CNIII), Superior orbit to record from superior oblique (CN IV), and Lateral canthus to record from Abducens nerve (CN VI).

Only the motor component of Cranial Nerve V (Trigeminal Nerve) can be monitored. Electrodes can be placed in either the Temporal Muscle or preferably the Masseter Muscle. The later is more useful as it is less likely to be in the field of surgery. Fine wire electrodes are inserted into the appropriate Masseter muscle by inserting the needle just anterior and superior to the angle of the jaw. In an awake patient the muscle can be identified by having the patient clench their jaw. To identify the Temporalis muscle in the awake patient this muscle can also be palpated while the patient makes a chewing movement. The electrodes to be inserted into the Temporalis muscle are inserted about 2 inches above the zygoma. One should recall that the motor component runs just below the sensory ganglion and accompanies the 3rd branch of the Trigeminal Nerve and may be compressed by large Cerebello-pontine angle tumors.

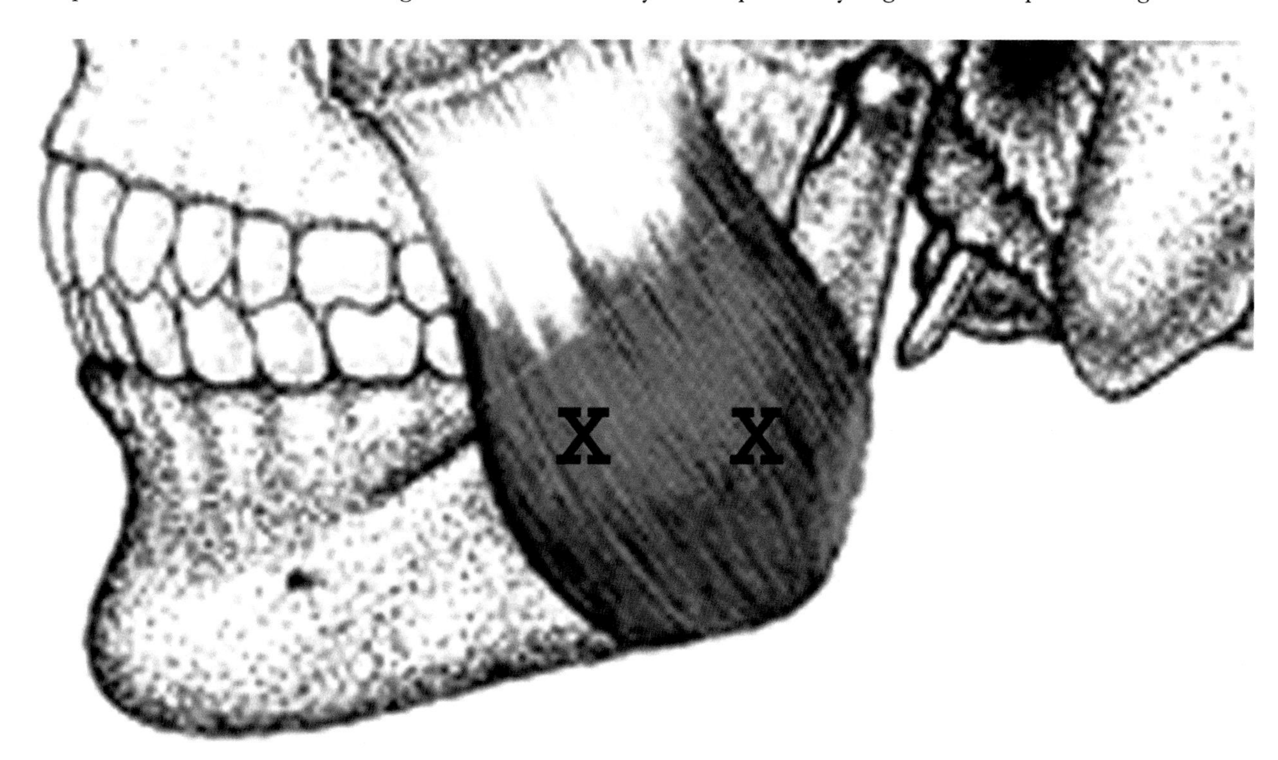

Figure 3. *Electrodes are inserted into the Masseter muscle to monitor Motor CN V.*

The Facial Nerve (VII) is by far the most frequent nerve to be monitored during skull based surgery. Three individual mimetic muscles are routinely monitored, they are: the Frontalis muscle, the Mentalis muscle and the Orbicularis oris muscle. Fine wire electrodes are placed in pairs to monitor both spontaneous activity (neurotonic discharges) and stimulated action potential (CMAPs) when the surgeon stimulates the Facial Nerve with a Prass probe. The schematic diagram below shows the location of the active electrodes.

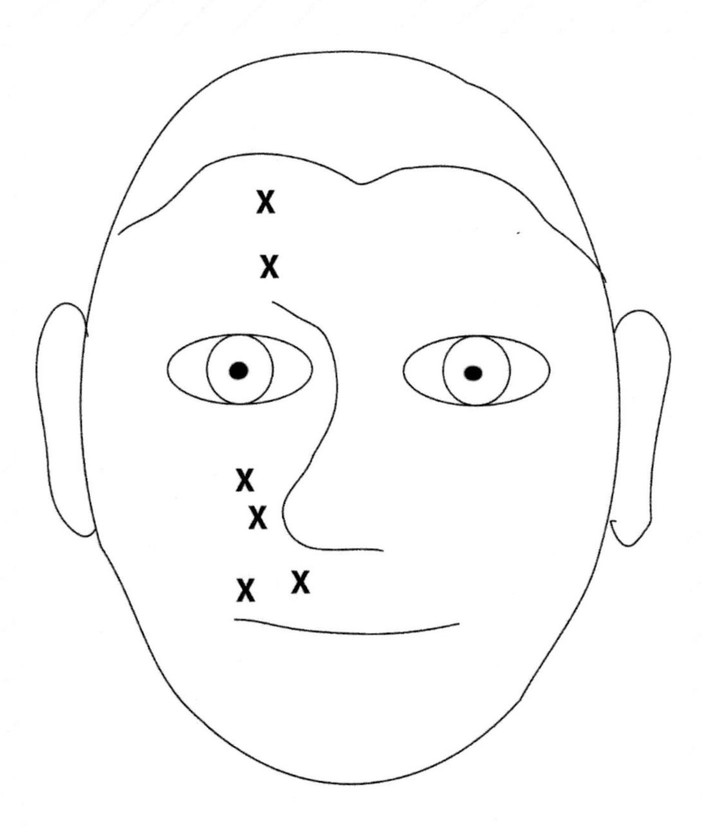

Electrode pairs in Frontalis, Mentalis, and Orbicularis oris to monitor CN VII.

Figure 4. *Electrode pairs inserted to monitor various muscles of CN VII. Electrodes in forehead to monitor Frontalis, electrodes in upper lip to monitor Orbicularis oris, and electrodes in chin to measure Mentalis. Reference inserted in opposite frontal region.*

An actual photograph of a patient with the facial electrodes in place is shown in **(Fig. 5)**..

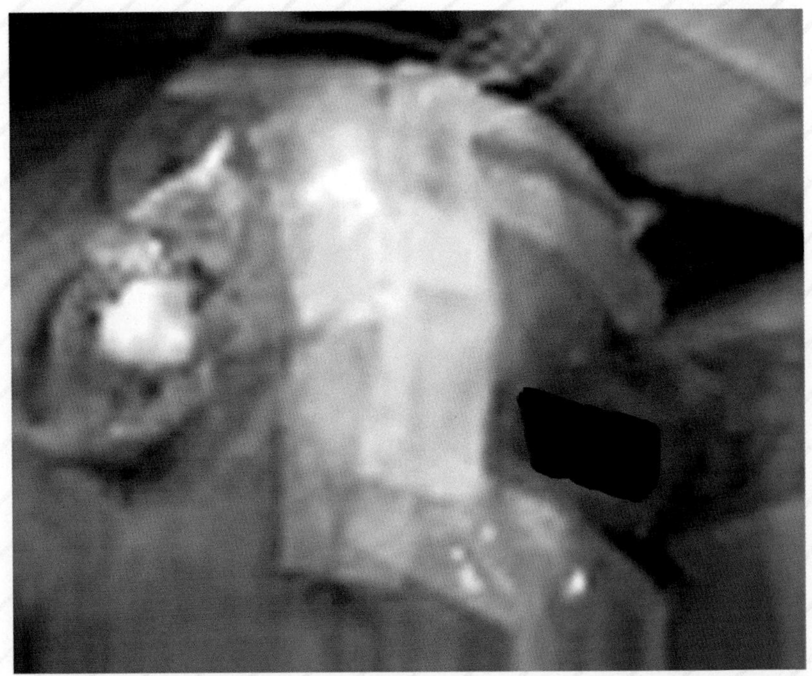

Figure 5. *Fine wire electrodes have been inserted and are now covered with tape and Tegaderm.*

Cranial Nerves IX and X are monitored by measuring action potentials generated by muscles of the false vocal cord. These fine wire electrodes are inserted through the trachea and are located on each side of the trachea. A schematic follows that shows the positioning of these electrodes. The thyroid cartilage is located by palpation of the neck.

The first ring cartilage of the trachea is then identified, and then the needle electrode is inserted in the mid-line between the first and second tracheal cartilage. After penetrating the membrane of the trachea the electrode is then angled laterally and advanced until resistance is felt. The procedure is then repeated for the other side. When carrying out the placement of these electrodes the anesthesiologist should deflate the endotracheal tube balloon so as not to puncture the balloon with a needle. Another alternative is for the anesthesiologist to use a laryngeal tube with surface electrodes available commercially. Surface skin electrodes cannot be utilized for monitoring the X nerve function as the platysma muscle (Facial Nerve) is just under the skin and much closer to the electrodes. **(Fig. 6).**

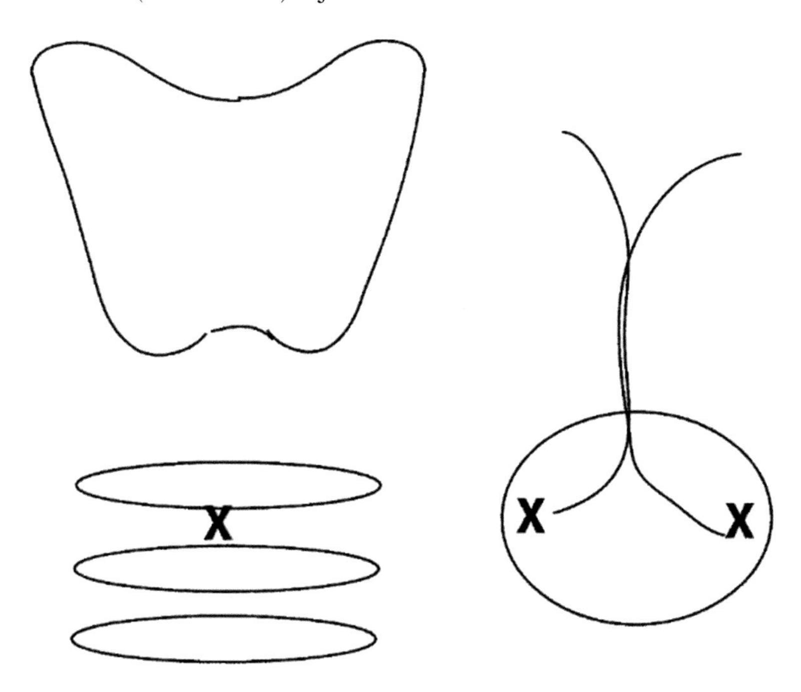

Electrodes in False Vocal Cord to Monitor CN IX/X.

Figgure 6. *Fine wire electrodes are inserted through the midline of the membrane between the first and second cartilagenous rings, the needle is then guided laterally to rest against each lateral wall. This allows for the monitoring of CN X.*

Cranial Nerve XI (Spinal Accessory Nerve) is monitored by inserting a pair of fine wire electrodes into the upper Trapeziums Muscle. This muscle is a very large muscle that runs from the base of the skull to the shoulder. A mid-muscle placement works well.

Cranial Nerve XII (Hypoglossal nerve) is monitored by inserting a pair of fine wire electrodes into the muscles of the tongue. There have been two approaches to these muscles, but I personally do not like to place a needle through the mucosa of the tongue, but prefer an external approach. The tongue is palpated externally behind the curvature of the jaw. A fine wire electrode is placed 1-2 centimeters lateral to the midline on each side. See the schematic for this insertion. **(Fig. 7).**

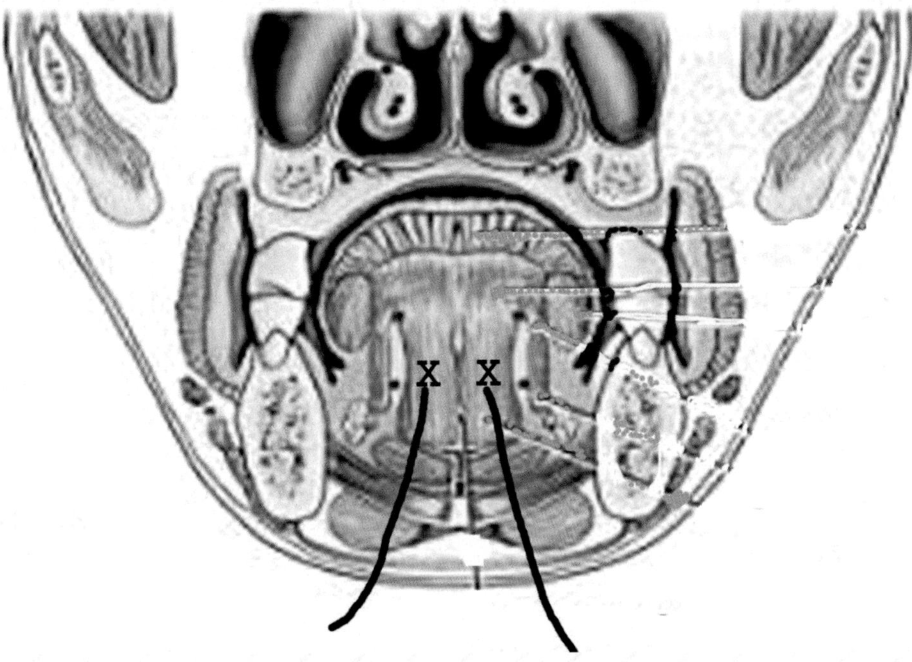

Figure 7. *Fine wire electrodes are inserted into the Tongue from under the jaw. The needles are inserted about one and one-half inches into the base of the Tongue to Monitor CN XII.*

BRAINSTEM AUDITORY EVOKED POTENTIALS IN INTRA-OPERATIVE MONITORING

Brainstem auditory evoked potentials is used very frequently to monitor hearing during the course of surgery of the cerebello-pontine angle. Intra-operative BAEPs are only slightly different than diagnostic BAEPs. Instead of using earphones to stimulate the cochlea, ear inserts are used. Two types of ear inserts can be used. A soft sponge insert with a tube running through it can be used. These inserts tend to dislodge during the course of a long surgery. An insert can be created for each individual patient (these are made of silicone that is injected into the external ear canal around one of the foam ear inserts – just used to secure the foam insert. These inserts are made prior to the surgery. If a silicone-foam insert has not been made prior to surgery, a foam insert may be held in place with bone-wax. These individually made ear inserts work quite well and do not dislodge easily. An ear insert is required for each ear. The tubes are then attached to a small clicker box which is taped to the chest. These clicker boxes provide a broad band click similar to an earphone. The active ear being tested receives a stimulus intensity of 120dB pe SPL while the non-active ear receives a white noise of 60 dB pe SPL. Some equipment is limited to 100dB pe SPL for the ipsilateral ear and 60dB pe SPL is provided in the contralateral ear as white noise. The source of the broad band click is further away from the ear than when earphones are used so that a latency of 1.0 millisecond is added to the latencies of the various waves obtained when compared to earphones. The stimulation frequency is usually about 10.1 Hz, but the frequency might have to modify depending upon interfering electrical artifacts. The ear is frequently folded over the insert and fixed in that position with surgical tape to keep the ear out of the surgical field. BAEPs are repeatedly recorded throughout the course of the surgery.

Early in the course of our intra-operative monitoring a small electrode ball was inserted into the middle ear, near the round window following a myringotomy by the neuro-otologic surgeon. This electrode is utilized to monitor the electro-cochleogram along with the BAEPs; however this electrode is very prone to complications and frequently causes bleeding into the middle ear which caused a loss of BAEPs, especially in long surgeries where the patient might partially deplete their clotting factors. We have substituted a direct recording from the acoustic nerve to monitor nerve function during the course of the surgery. During the course of a long surgery it may be necessary to increase the frequency of the stimulus up to 60-80 Hz in order to identify Wave V. If this maneuver is used, Wave I will typically disappear if present at the start and the wave V will be enhanced.

The surgeon is to be notified if the following changes occur, if the latency of Wave V becomes prolonged by 0.5 milliseconds, or if the amplitude of Wave V drops by 50 percent. The neurologic surgeon should also be notified if "neurotonic" discharges occur in the spontaneous EMG channels being monitored. This response tells the surgeon that the facial nerve is being irritated. This occurs with either traction of the nerve or when tumor capsule is being removed from the nerve. At times, during opening, before the dura is opened, the patient will show spontaneous neurotonic discharges. This provides evidence that the tumor itself has irritated the facial nerve. There is no good data regarding the potential harmfulness of neurotonic discharges, but it is probably best to let them settle down and become quiet when they are induced.

Intra-operative Somatosensory Evoked Potentials

Both the median nerve and posterior nerve SSEPs are accomplished just like the diagnostic SSEPs. The frequency of the stimulus is generally 5.4 Hz, but may have to be adjusted for certain types of electrical artifacts. Motor threshold intensity is used to obtain a maximum evoked potential response in an anesthetized patient. Again, these tests are run continuously during the course of the surgery. During critical parts of the surgery, it may be necessary to stop a given test and restart the data collection immediately to provide the latest and most accurate information.

The surgeon should be notified in there is an increased latency of the cortical responses (N20 and P37) by 10 percent or a drop in the Wave amplitude by 50 percent.

Stimulators are used by the neurosurgeon or neuro-otologic surgeon during the course of the tumor removal. The Prass probe is a fine tipped semi-flexible stimulator that the surgeon can use to identify motor nerves in the region of the surgery. We have added a "clicker box" developed locally to provide an auditory click with each stimulation. This clicker box has a threshold of 0.047 milliamps, which means a current greater than this must be present for the clicker function to work. The second feature is that there are no clicks when the Prass probe is not in contact with bodily tissues. It is the contact with tissues that completes the circuit and allows the click with each stimulus to occur. Just the click is heard when the probe is in contact with bodily fluids and tissues. When a motor nerve is stimulated, not only is the click heard by the surgeon, but the typical sound of a compound motor action potential is heard. The Kartush stimulating dissector is also used with the clicker box. This allows the surgeon to know when current is being delivered by the dissector. Only the click is heard when not in contact with a motor nerve, but when a motor nerve is approached the click is followed by the sound of a compound motor action potential (CMAP). With both types of stimulators, not only is the CMAP heard, but is observed on the monitoring screen. **(Fig. 8).**

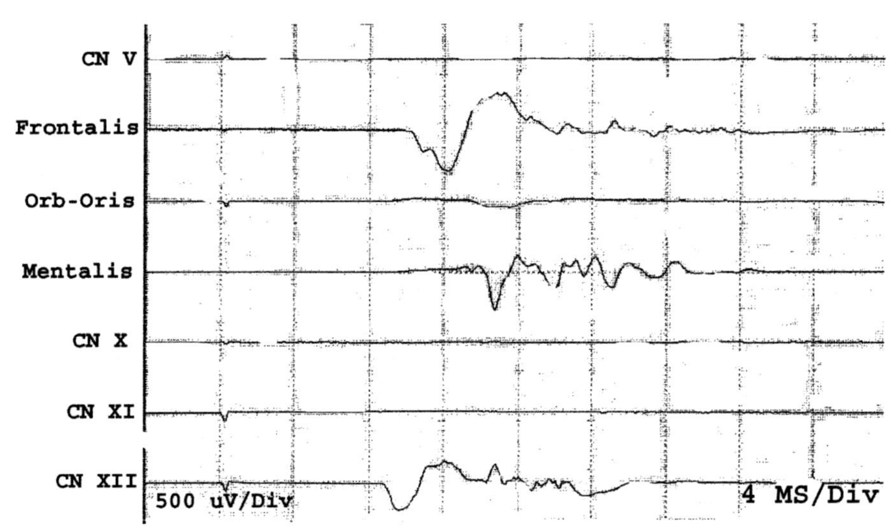

Figure 8. *CMAP following stimulation of the Facial Nerve with a Prass stimulator. Note only the Frontalis and Mentalis muscles show the CMAP. Very frequentlya CMAP is also seen in the Tongue when the Facial nerve is stimulated. This is thought to result from coordinating central connections between the facial nerve and the hypoglossal nerve.*

Transcranial Electrocortical Stimulation to Monitor Facial Nerve Motor Function during Skull Base Surgery.

The newest technique to monitor Facial Nerve function is the Transcranial Electrocortical Stimulation of the Facial Nerve. This type of testing cannot be used continuously, but is of great value in predicting Facial Nerve outcome following cerebello-pontine angle surgery. This type of testing can be used intermittently during surgery and is likely to be of greatest value to measure Facial Nerve integrity following prolonged runs of neurotonic discharge which result from Facial Nerve irritation and at the end of the surgery. Stimulating needle electrodes are inserted at C3 or C4 (Electrode 10-20 system) and at CZ depending on whether the right or left side of the brain is to be stimulated. Recording electrodes are the same facial electrodes (Frontalis, Orbicularis oris and Mentalis) used in measuring spontaneous facial muscle activity. Stimulation always occurs over the contralateral hemisphere from the operated side. The stimulus is provided as a series of rectangular pulses from 3 to 5 with a voltage from 200 to 500V, at a 50 µs pulse duration and an interstimulus interval of 2 ms. Band pass filters are set from 150 to 3000 Hz. Stable responses during the course and at the end or surgery provide optimistic outcomes for facial nerve function. A drop in amplitude between 35 and 80 percent from baseline should be used as a warning sign for altering surgical strategies to avoid definitive facial nerve damage.

Vestibular Schwannoma

Vestibular Schwannoma are a rare benign tumor that originates on the vestibular nerve while it is in the auditory canal. The vast majority of these tumors are proximal in the canal and eventually erodes the temporal bone, enlarges the porous acoustica, and grows into the subarachnoid space. A small percentage of the tumors are more distal in the auditory canal and do not decompress themselves and the increased pressure leads to sudden catastrophic hearing loss in the affected ear. This loss is presumed to result from compression of the auditory artery resulting in infarction of the cochlea and permanent hearing loss. The more proximal tumors do not cause this sudden loss, but the patient may have gradual loss of hearing, especially in the hearing range of the spoken word. Speech discrimination problems are the hallmark of these tumors. This is confirmed in the audiogram where "rollover" is a common finding. In this test, the word recognition gets worse as the intensity of the word stimulus increases. The BAEPs may show a number of different patterns, the most specific is the increase in the Wave I to III interval. Other changes that may occur include a small percentage of normal responses in very small tumors, a prolonged I to V interval with an absence of Wave II and a prolonged I to III interval and a prolonged III to V interval in large tumors compressing the brainstem. Rarely all BAEP waveforms may be absent. The MRI scan shows the location and size of the tumor so that planning of surgery can be accomplished. **(Fig. 9, Fig. 10, and Fig. 11).**

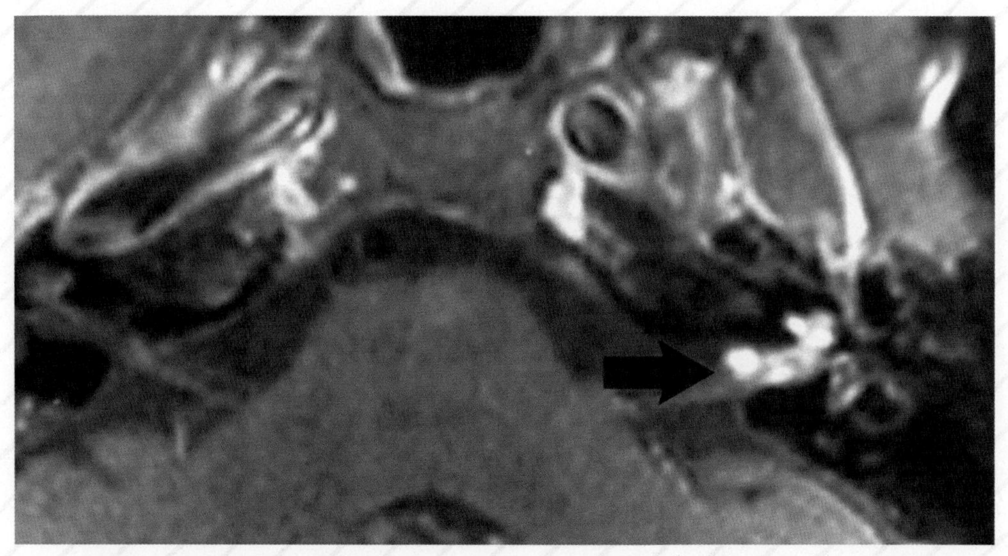

Figure 9. *MRI scan showing a very small intracanilicular vestibular schwannoma.*

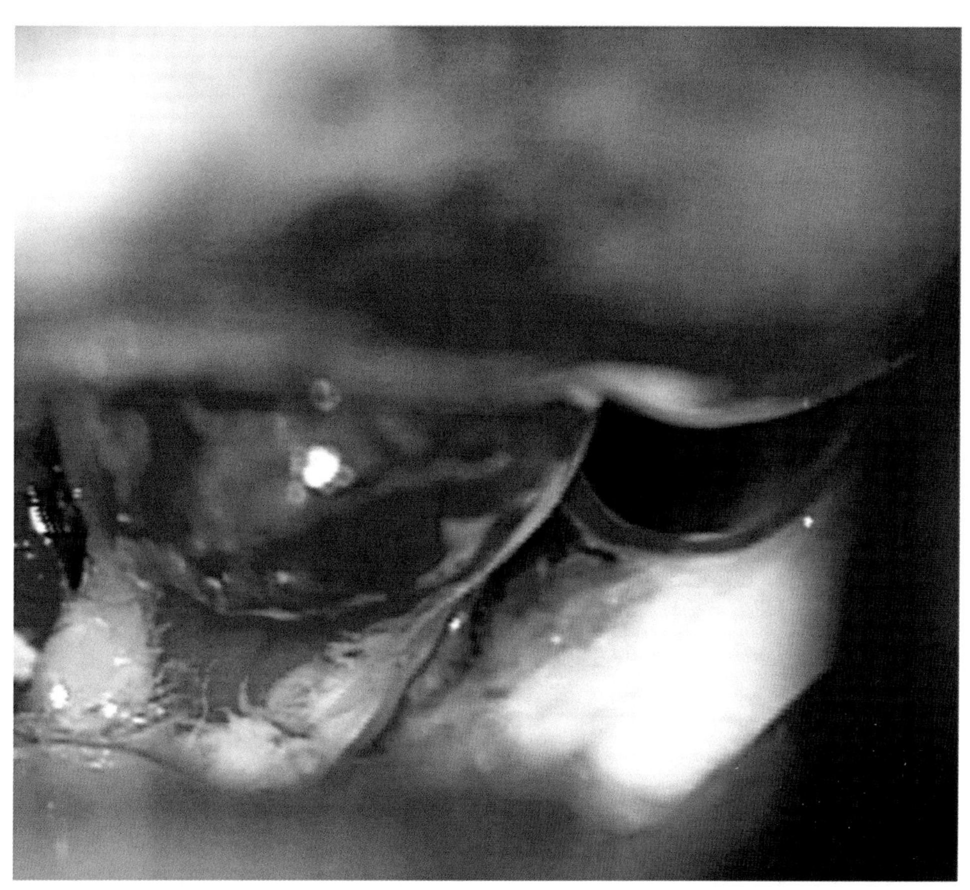

Figure 10. *This specimen demonstrates a small vestibular schwannoma just exiting from the internal Acoustic canal. The Vestibular and Auditory nerves are seen just under the tumor. The Facial nerve is not seen here, as it is deep to the VIII nerve in this surgical approach.*

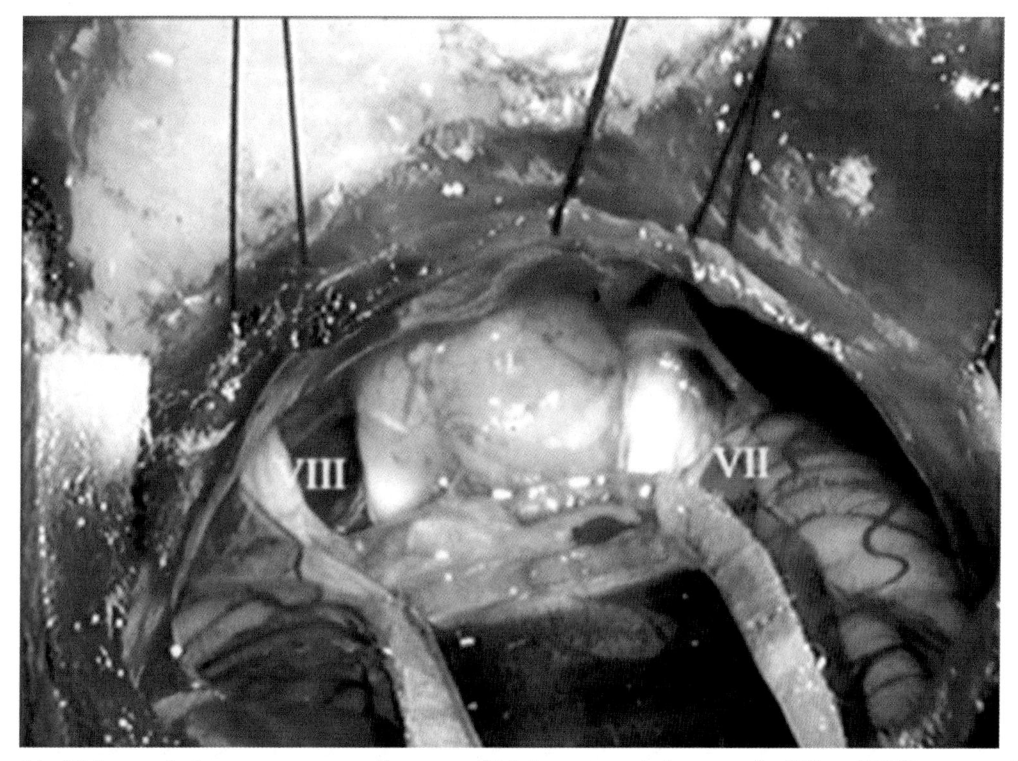

Figure 11. *This sample demonstrates a small tumor which has grown in between the VII and VIII nerve and rotated the nerves into this position. This is an uncommon finding.*

Larger tumors are even more common and many of them compress the brainstem and when the surgery begins the VIII and VII are deep to the tumor. **(Fig. 12).**

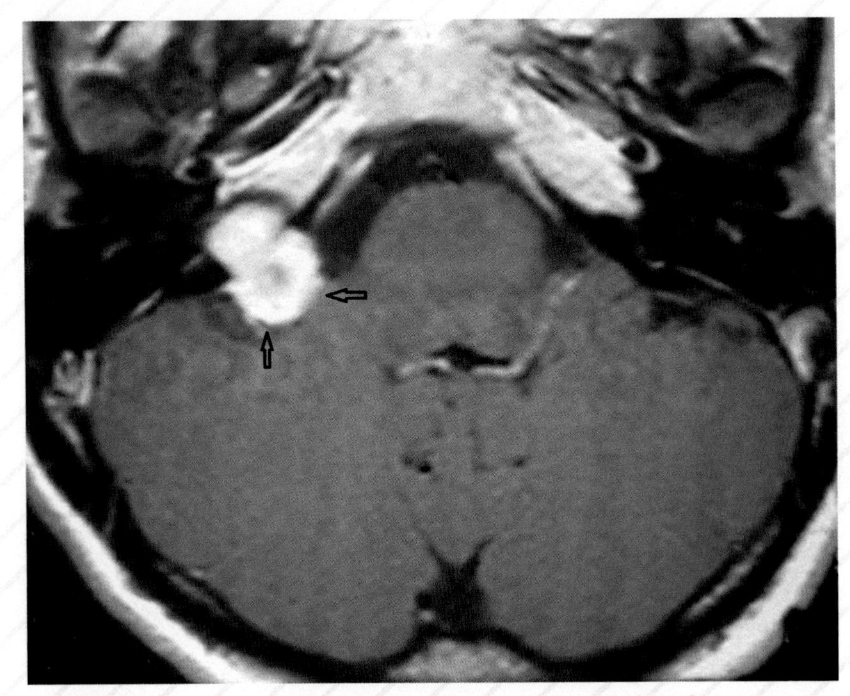

Figure 12. *MRI scan showing a moderate sized vestibular schwannoma.*

Giant tumors not only cause loss of hearing, but frequently cause an unsteady gait is an example. **(Fig. 13 and Fig. 14).**

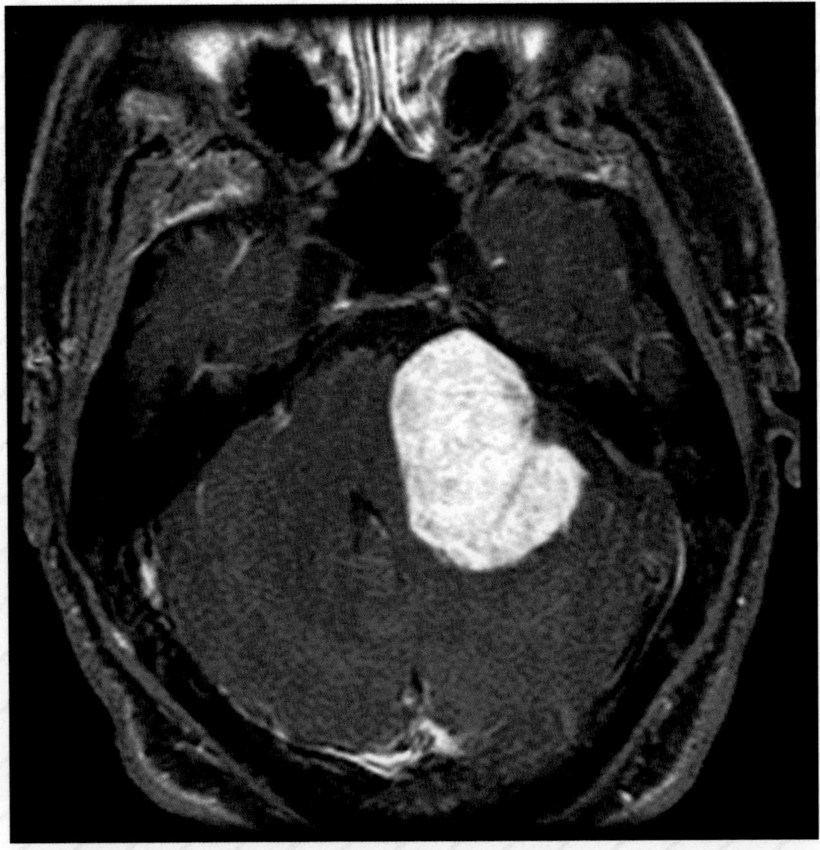

Figure 13. *This MRI scan reveals a giant vestibular schwannoma.*

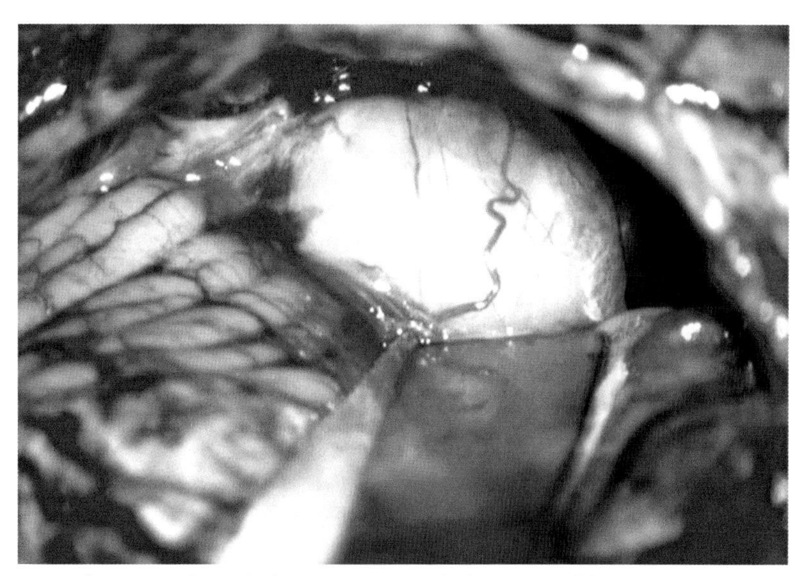

Figure 14. *This is a moderate sized vestibular schwannoma before removal has started. The VIII and VII nerves are deep to the tumor and cannot be seen at this stage in the surgery.*

When the tumor is small and not compressing the brainstem the only evoked potential studies necessary to monitor the surgery is that of the BAEPs, the EMG of the facial muscles is also monitored. However, if the brainstem is compressed by the tumor or if there is no BAEP response or if it is anticipated that the BAEPs will be lost during surgery then MN SSEPs are added to the monitoring regimen.

Remember when monitoring during removal of a vestibular schwannoma, the BAEPs is invariably abnormal. See **(Figure 15, Fig. 16, and Fig. 17).**

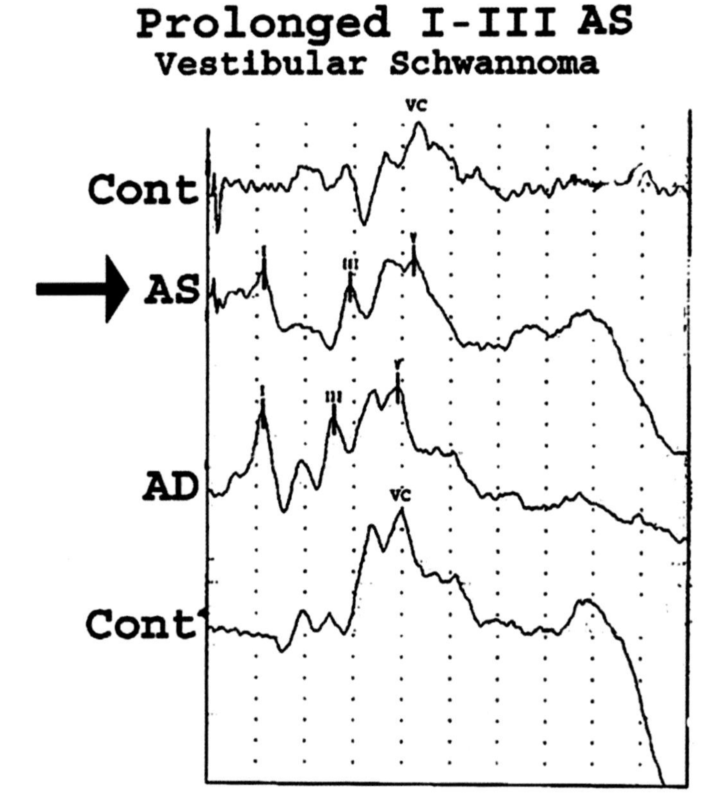

Figure 15. *The left sided BAEP shows a prolonged I to III interval compared to the right. These waves are well developed and should be able to monitor during surgery.*

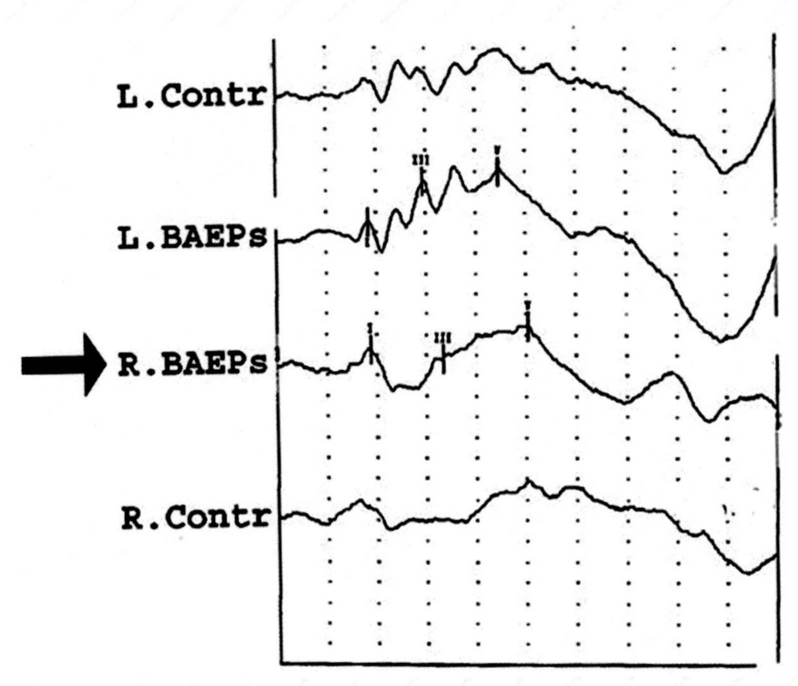

Figure 16. *BAEPs showing prolonged interval between Waves I and III, Waves III and V, and Waves I to V.*

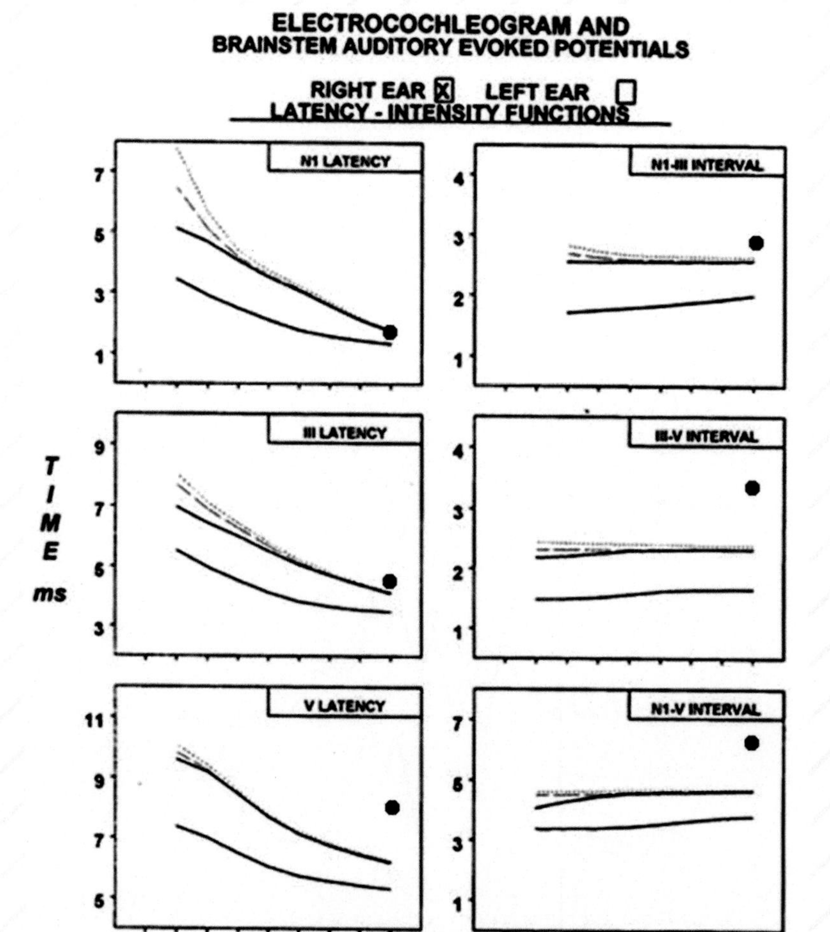

Figure 17. *This report shows the prolonged latencies for Waves I and V. All the intervals I to III, III to V, and I to V are prolonged showing that the tumor slows conduction in the nerve and centrally by compressing the brainstem.*

The above BAEP sample shows a greatly prolonged Wave V latency. In general a latency of Wave V over 7.5 milliseconds makes it unlikely that hearing will survive the surgery.

Once the Auditory nerve is identified, continuous monitoring of the direct nerve action potential can be accomplished. Figure 12 shows the placement of the recording electrode on the auditory component of the VIII nerve. **(Fig. 18).**

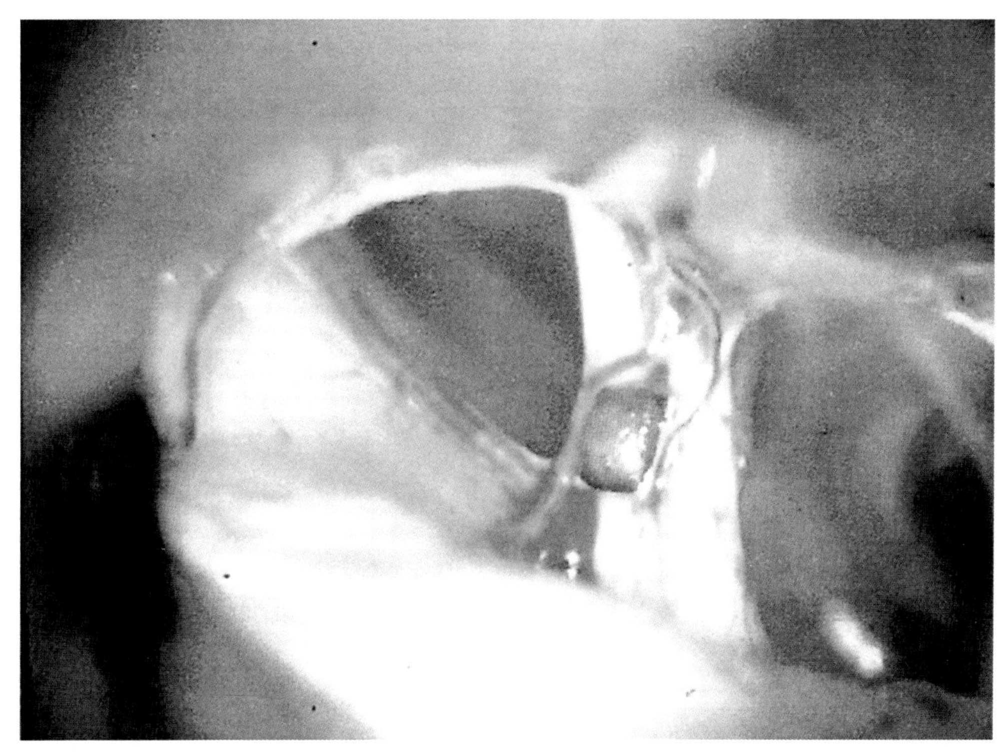

Figure 18. *Intra-operative photography showing electrode placed on Auditory nerve for long term monitoring of the Action Potential.*

(Figure 19and Figure 20) demonstrates the typical waveform and latency of the direct auditory compound action potential.

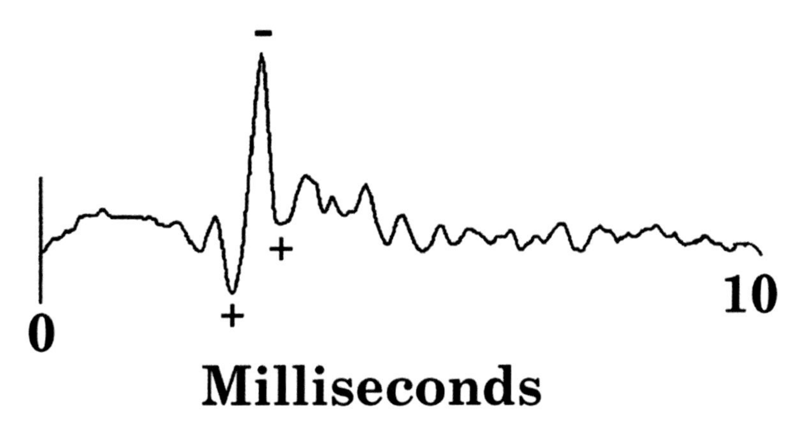

Figure 19. *Action potential of the acoustic nerve upon stimulation of the cochlea with broad band clicks. This is a typical waveform for any nerve with the electrode on the external surface of the nerve.*

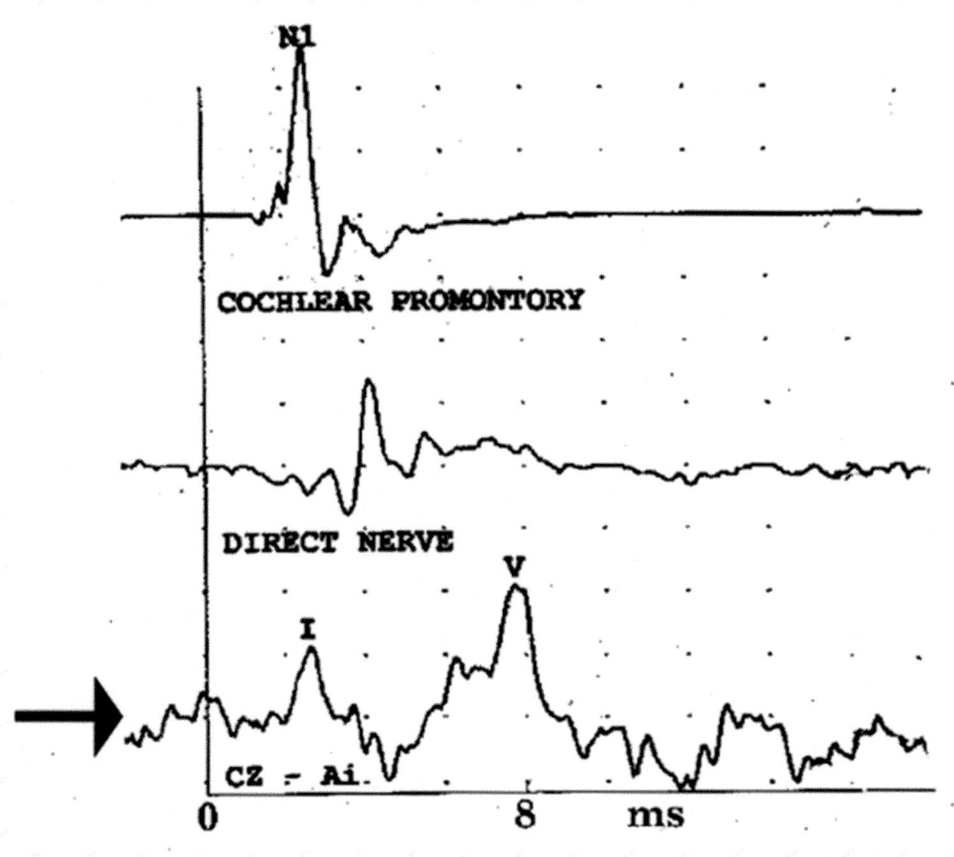

Figure 20. *The electro-cochleogram and direct nerve action potential are normal. The latency of Wave V is almost 8.0 milliseconds. Latencies of this length generally indicate that Wave will not withstand the minor traumas associated with surgery.*

The type of monitoring depends upon the surgical approach to tumor removal. With a small tumor isolated to the internal auditory canal and a complete loss of hearing, the approach is generally "translabyrinthine". With this approach, BAEPs are of no use as the cochlea will be destroyed during surgery, and only facial nerve monitoring is of value. During drilling for this approach the facial nerve is identified in the facial canal by stimulation with a Prass probe.

The "transmastoid approach" however is for individuals with at least some hearing remaining and the typical monitoring setup includes BAEPs, Facial nerve and motor V. In those cases where the tumor compresses the brainstem, then MN SSEPs are added.

A series of baseline studies are run during opening. This data takes time to collect as there are frequent pauses necessary during drilling and cautery so as not to have the amplifiers saturated by artifact and block the collection of data. In general, 4-6 complete evoked potential studies can be done during this time and the data provides a range of baselines to compare with during the remainder of the surgery. The first critical measurements occur following retraction. At this stage the VIII nerve frequently cannot be seen, but may be stretched. Any prolongation in the latency of Wave V should be reported to the surgeon so that the tension on the nerve from retraction can be corrected.

Once the retractors are in place the surgeon stimulates the exposed tumor capsule with a Prass probe to make sure there are no facial nerve components trapped in the exposed capsule. After this is verified, an entry site into the tumor is prepared and, using a CUSA Ultrasonic aspirator the inside of the tumor is debulked. **(Fig. 21).**

Figure 21. *After removing the capsule at the operative site, the fatty looking tumor can be removed from inside the remainder of the capsule. A CUSA ultrasonic aspirator is used for this part of the surgery.*

After debulking the inside of the tumor, it is time to start to remove the capsule of the tumor. The tumor capsule to be removed is first stimulated with a Prass probe to make sure no facial nerve components are present. That portion of the capsule is then removed. The next segment of capsule is then stimulated and the process repeated until the capsule has been removed from the brainstem. The facial nerve may remain as a single nerve bundle or may be split up into a number of smaller bundles trapped in the tumor capsule. The state of the facial nerve can only be ascertained after partial removal of the capsule. As the tumor is removed, irritation of the Facial nerve is indicated by the presence of neurotonic discharges. **(Fig. 22).**

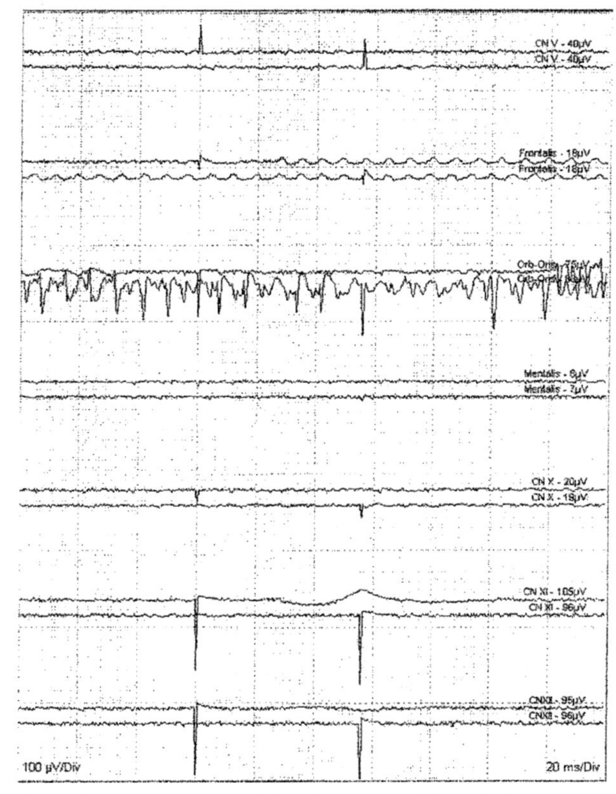

Figure 22. *Run of neurotonic discharges involving Orbicularis oris. These discharges result from irritation of the Facial nerve during tumor removal. They usually resolve when the nerve is allowed to rest.*

This generally occurs when the surgeon is putting traction of the Facial nerve by removing the capsule from the nerve. The surgeon should be notified immediately when neurotonic discharges begin. A short period of rest will usually allow the Facial nerve to calm down and become electrically quiet. After as much of the tumor can be removed in the subarachnoid space, the next stage of the surgery is to remove the tumor from the internal auditory canal. The temporal bone is then drilled to provide access to the tumor remaining in the canal. See **(Fig. 23)** for a photo of that stage.

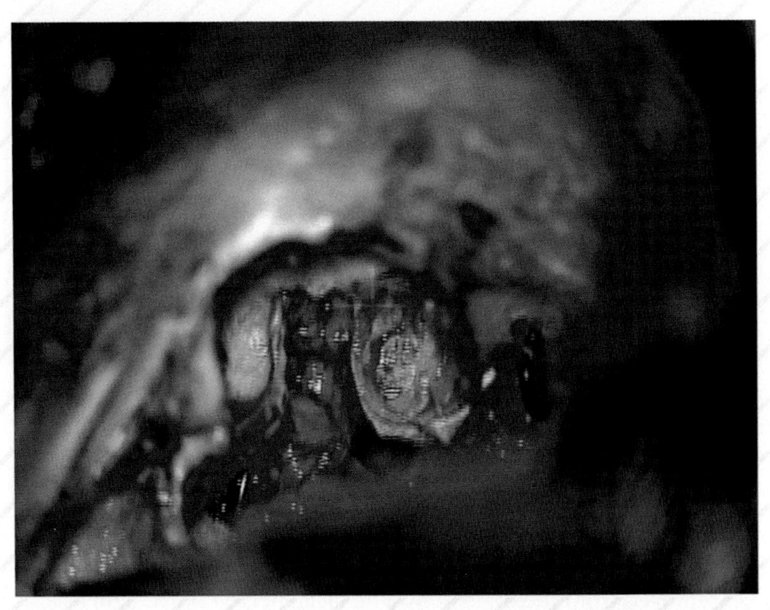

Figure 23. *After removing the tumor in the subarachnoid space, the temporal bone is drilled to expose the intracanicular tumor. The tumor is then taken off the underlying nerves with a stimulating dissector, including a sickle knife and small ring.*

The neuro-otologic surgeon almost always utilizes stimulating dissectors (Kartush) to remove the intracanalicular portion of the tumor. The most frequently used instruments include a small sickle knife and a small ring. Each time these instruments get close to the facial nerve, a compound motor action potential will be heard and recognized in the facial muscle channels. **(Fig. 24).**

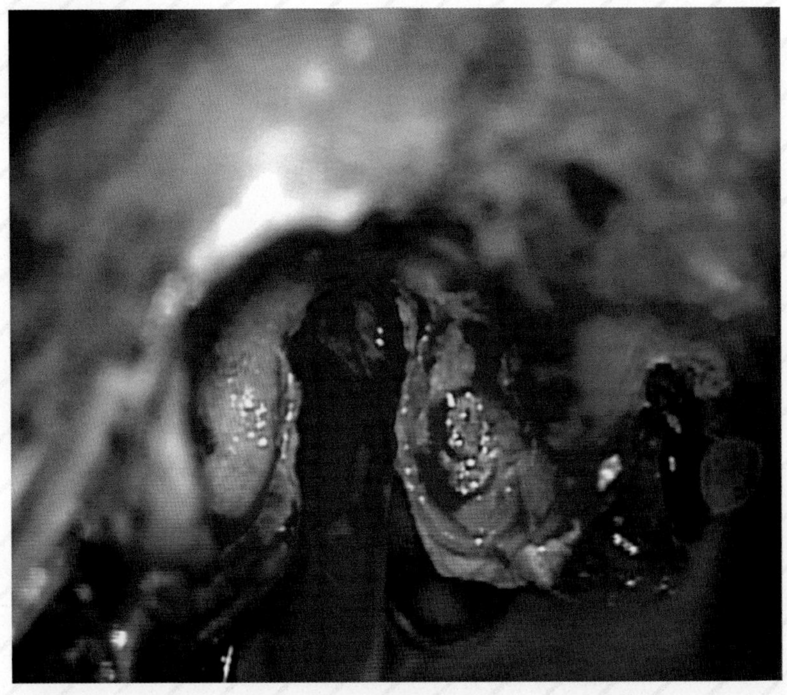

Figure 24. *Tumor removed. The VII and VIII nerves are intact following tumor removal.*

A muscle plug is then used to close the hole in the temporal bone. The area is then sealed with "glue" to make sure there are no spinal fluid leaks. **(Fig. 25).**

Figure 25 shows the final closure of the canal.

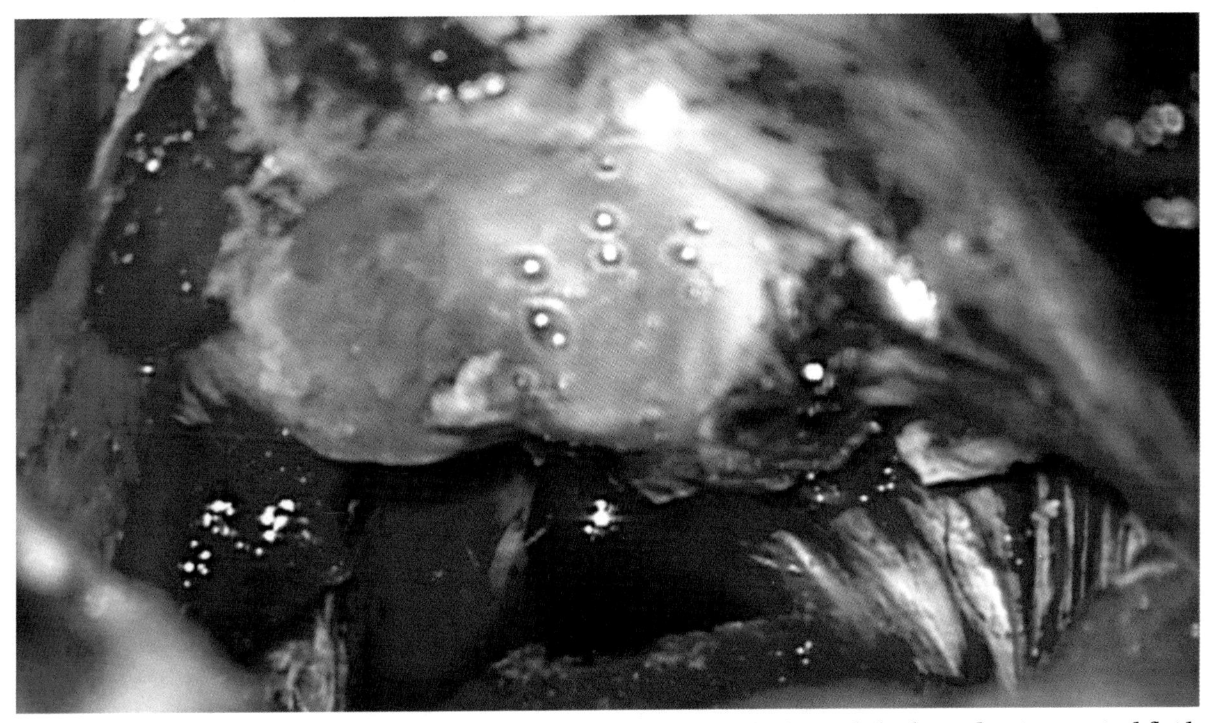

Figure 25. *Drilled hole in temporal bone has been plugged with a muscle plug and glued to seal region to spinal fluid leaks.*

Following this procedure, closure begins and monitoring continues at least until the dura is closed.

Results: If the patient has any residual hearing, every attempt should be made to salvage that hearing. Hearing will be salvaged in approximately 50 percent of the cases when the tumor is 2 centimeters or less. For tumors over 2.5 centimeters useful hearing is usually not preserved. If the Wave V latency is 7.5 milliseconds or longer, hearing is generally not preserved. On the other hand, facial nerve function is almost always salvaged. In a very small percentage of cases a delayed facial palsy develops, but the prognosis for recovery over the next few weeks is excellent.

Other tumors occur at the Cerebello-pontine angle. The most common is the meningioma. The surgical approach and monitoring is virtually identical to that with a vestibular schwannoma. In general, meningiomas are embolized the day before surgery to reduce their blood supply and make removal easier and with less blood loss.

Multiple Neurofibromatosis, Type II

The surgical management of Multiple Neurofibromatosis, Type II constitutes a very special problem. These individuals have bilateral vestibular schwannomas and maintaining hearing for as long as possible is very important. In addition, these individuals frequently have other tumors including schwannomas along the spinal neuraxis and meningiomas. The growth rate of the tumors varies greatly, and some individuals will need debulking of their tumor every couple of years. Complete removal of the larger tumors is unlikely so the goal is to debulk the tumor allowing the individual maximum function for the longest period of time. **(Figure 26)** is an example of an individual with NF II.

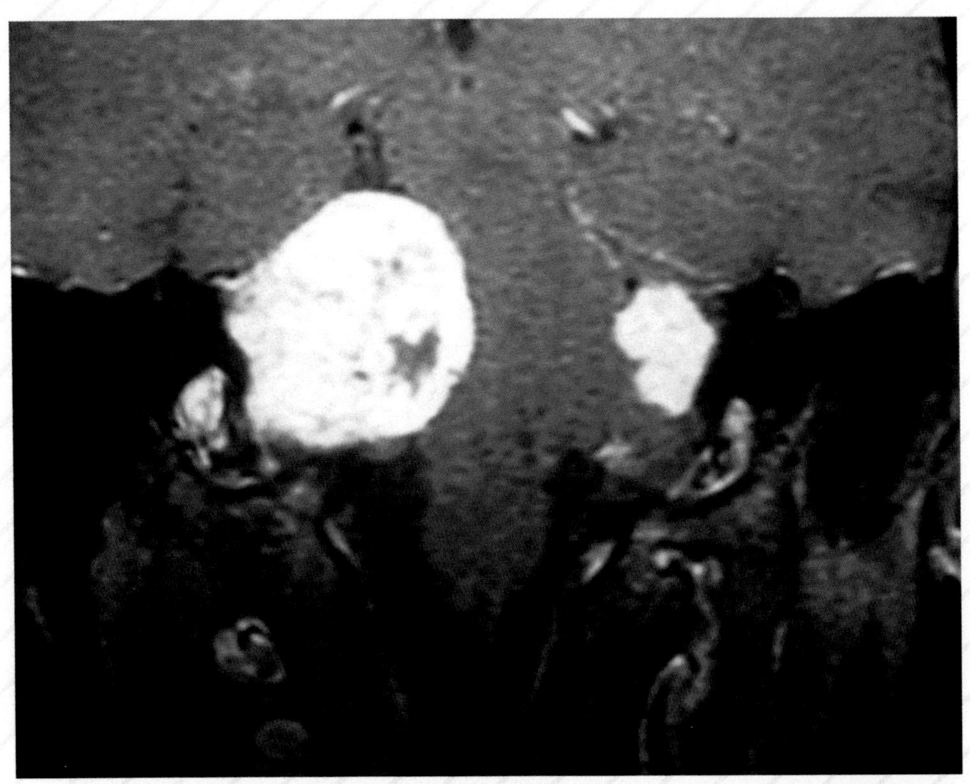

Figure 26. *MRI scan demonstrating bilateral Vestibular Schwannomas in a young patient with Neurofibromatosis, Type II. In this individual there is a large size difference between the two sides.*

The next Figure shows an individual with bilateral Vestibular Schwannomas and a meningioma compressing the cerebellum. **(Fig. 27).**

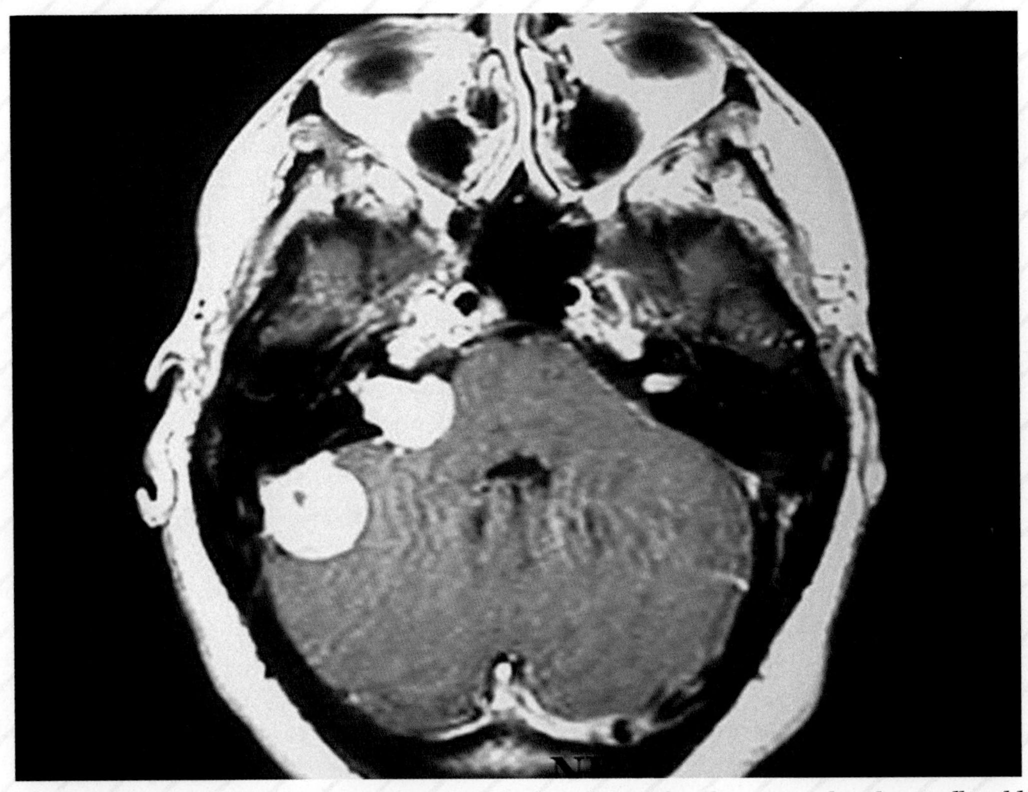

Figure 27. *MRI scan showing bilateral Vestibular Schwannomas, one of moderate size, the other small and localized to the auditory canal. A meningioma is also present and compresses the cerebellum.*

The brainstem auditory evoked potentials generally show bilateral abnormalities, as seen in **(Fig. 28).**

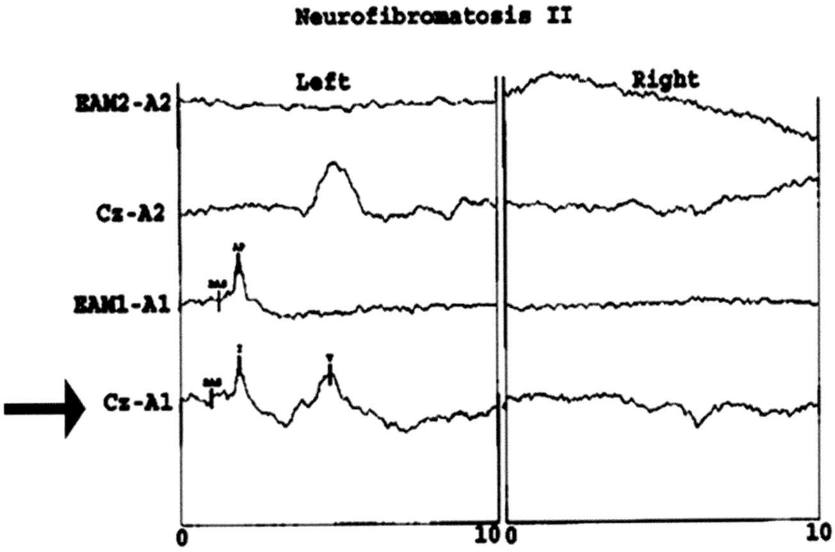

Figure 28. *BAEPs demonstrates no response in the right ear and a prolonged Wave I to V interval. With large tumor the Wave III to V interval may also be prolonged.*

Compression of the brainstem is a common finding in Neurofibromatosis, Type II. The following median nerve SSEPs demonstrates the results of this compression. **(Fig. 29).**

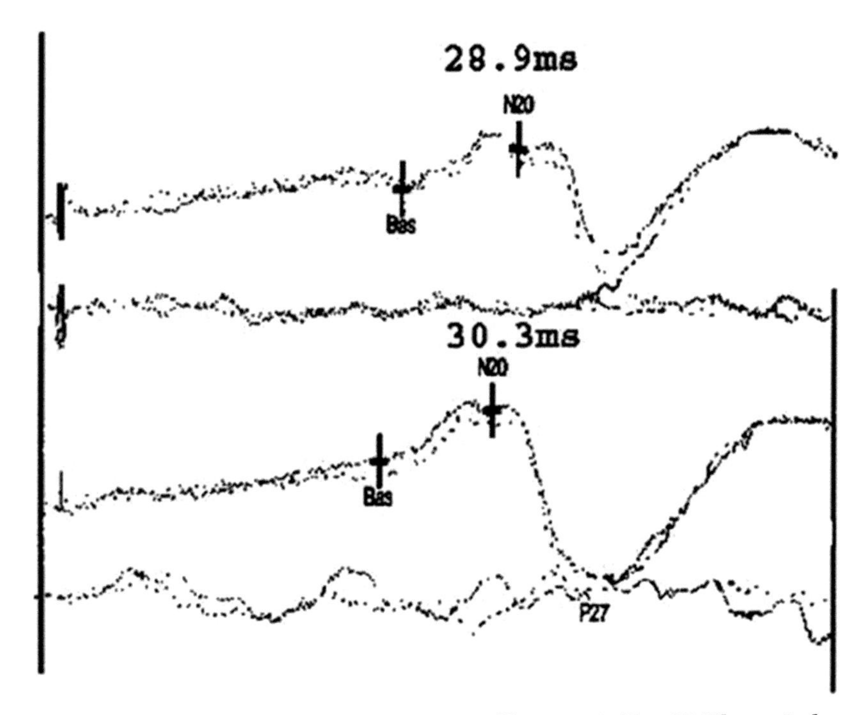

Figure 29. *Median Nerve SSEPs in an individual with Neurofibromatosis, Type II. The cortical responses as severely prolonged. This finding demonstrates that thereis a moderately severe compression of the brainstem by the tumors. The normal latencyfor the N20 is about 20 milliseconds.*

Because of the presence of bilateral tumors, it is sometimes necessary to alter the surgical approach by entering from the posterior midline so as to make sure that no significant brainstem shifts occur with unilateral decompression. Eventually, total hearing loss occurs in these patients and more aggressive decompression may be possible. Non-tumor conditions affecting the C-P angle

Vascular Compression

Two vascular compression syndromes occur at the Cerebello-pontine angle. The first is Trigeminal Neuralgia . Trigeminal Neuralgia is primarily a disorder of the older patient with one major exception, it can occur in a younger patient with Multiple Sclerosis, but with a different pathophysiology. Trigeminal Neuralgia occurring in the older population results from aberrant blood vessels compressing the sensory components of the trigeminal nerve close to where it enters the brainstem. This syndrome consists of paroxysms of lancinating pain in the distribution of one of the branches of the Trigeminal Nerve. A trigger point is universally present and may be on the lip, gum or tongue. The fear of setting of a paroxysm of pain makes the patient avoid certain activities such as brushing of the teeth, shaving, or even eating. Large weight losses can occur. The initial treatment consists of medical management with one of the following agents, carbamazepine, lomotrogene, phenytoin and others. Most cases can be managed successfully with these agents. A small percentage of patients break through their medical management and depending upon their age and physical health may undergo microvascular decompression. In the infirm elderly patient percutaneous radiofrequency destruction of the Gasserian ganglion may be the treatment of choice. Monitoring is only utilized for the microvascular decompression procedure. The Trigeminal Nerve lies just superior to Cranial nerves VII and VIII so these nerves are within the field of surgery. Monitoring includes BAEPs and Motor Branch of V along with the Facial Nerve. The surgical approach is retrosigmoid to enter the Cerebello-pontine angle of the skull. The aberrant vessel is identified and the arachnoid membrane is removed from the vessel or vessels and the vessel is then separated from the proximal nerve. See **(Fig. 30).**

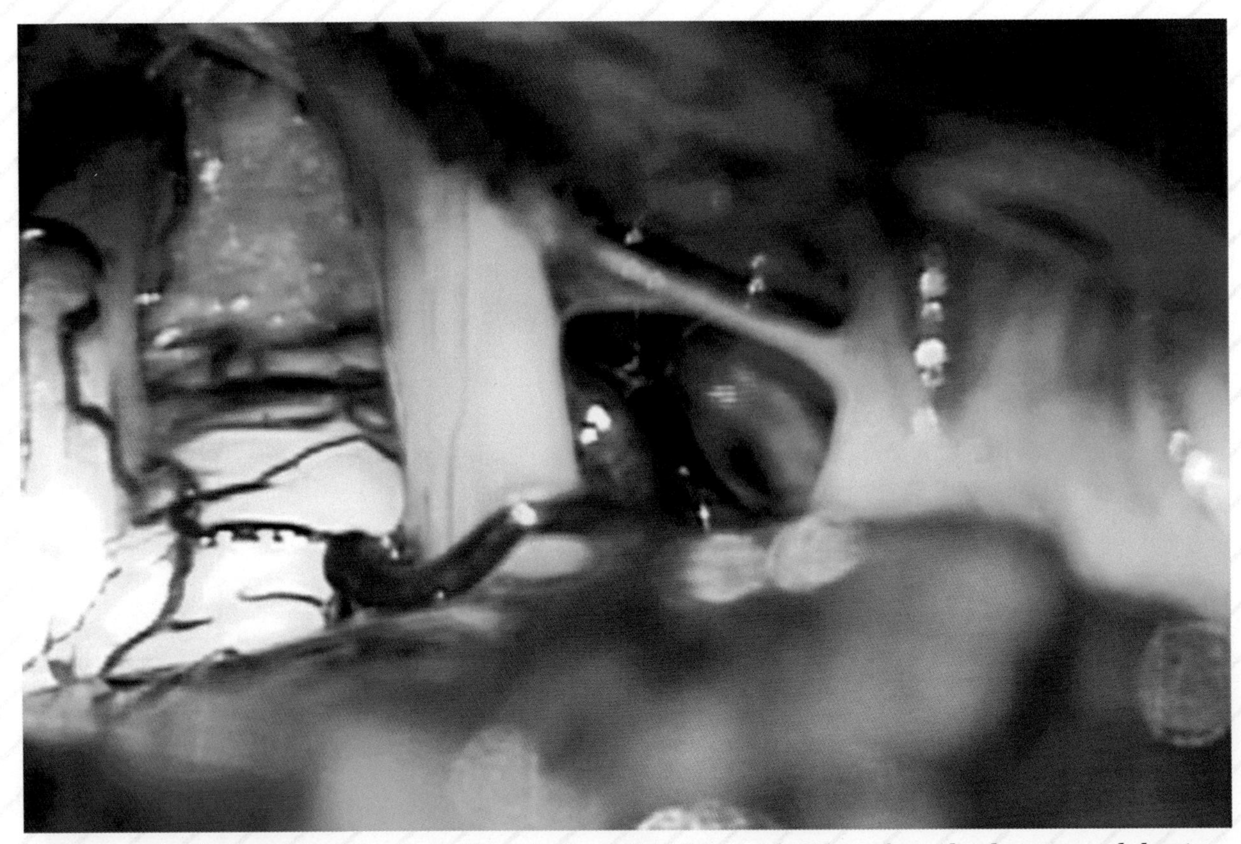

Figure 30. *Photo at time of surgery for Trigeminal neuralgia. The arachnoid membrane has been removed showing vessel surrounding Cranial nerve V. One sponge has already been placed in upper left corner of photo.*

The offending vessel or vessels are separated from the nerve and pledgets are placed so as to keep the vessel separated from the nerve. **(Fig. 31).**

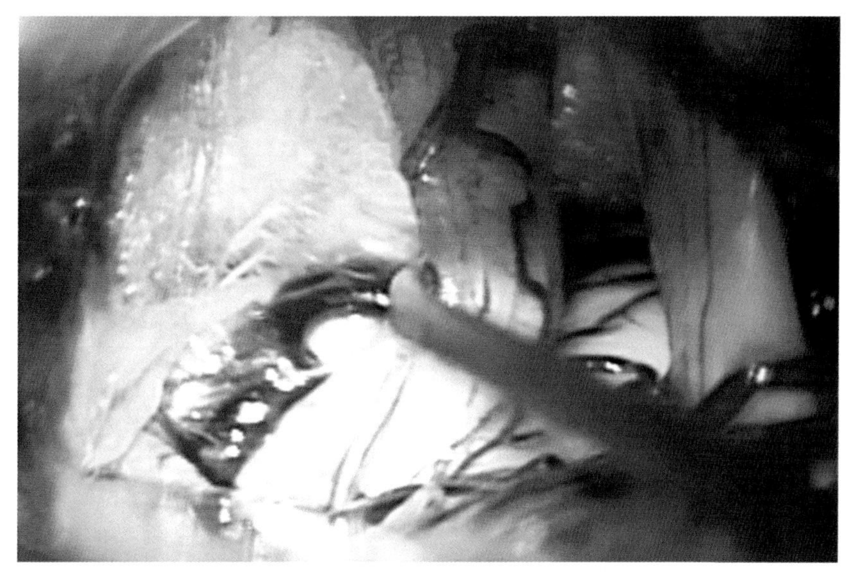

Figure 31. *Photo shows pledgets being placed between Trigeminal nerve and the compressing artery.*

Once the pledgets are in place separating the vessel from the nerve, the paroxysms of lancinating pain almost always cease. **(Fig. 32).**

Hemi-facial Spasm

Hemi-facial spasm is a rare disorder that results from the compression of the Facial Nerve by vessels close to the area where the nerve leaves the brainstem. Patients with this disorder develop continuous muscle spasms or contractions of the face on one side. The contractions are not only continuous, but regular in frequency and do not disappear with sleep. Unlike Trigeminal neuralgia, pharmacological therapy does not help this condition. Temporary relief may be obtained by partial paralysis of the affected facial muscles with injections of botulinum toxin. This generally reduces the intensity of the contractions, but does not obliterate them. To obliterate them completely with botulinum toxin would require doses that cause complete paralysis of the muscles. This is generally not an acceptable outcome. This form of therapy should be reserved for the elderly or medically fragile patients. In those in good health, surgical decompression offers the possibility of complete relief from these symptoms. In addition to the clinical features, MRI scans may show the presence of the aberrant blood vessels. Upon opening the dura, the offending vessels can usually be seen being bound to the nerve by thickened arachnoid membranes. See **(Fig. 32).**

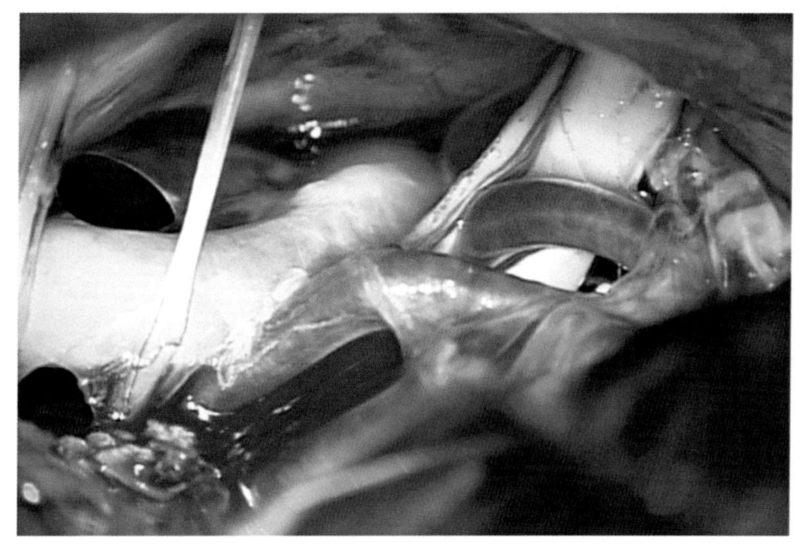

Figure 32. Intra-operative photo showing blood vessels compressing Cranial nerves VII and VIII. The thickened arachnoid membrane has been partially removed from the vessels.

After removing the thickened arachnoid membrane and freeing the offending vessels from the nerves, pledgets are inserted between the vessel and the nerve. See **(Fig. 33).**

Figure 33. *Intra-operative photo demonstrating the pledgets in place and separating the offending vessel from the nerve.*

A number of different variations in the relationship of the vessels to the nerve exist. **(Fig. 34)** shows a different variation and **(Fig. 35) shows** the placement of the pledgets.

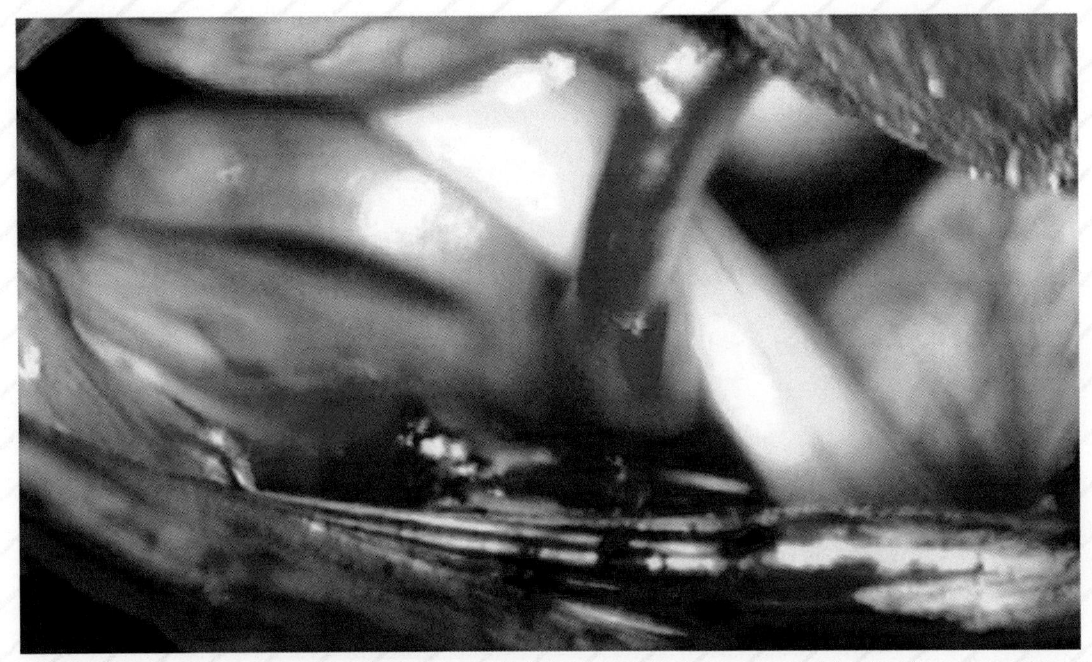

Figure .34. *Intra-operative photograph shows vessel indenting Cranial nerve VIII. The facial nerve lies anterior to VIII and is trapped by this same vessel.*

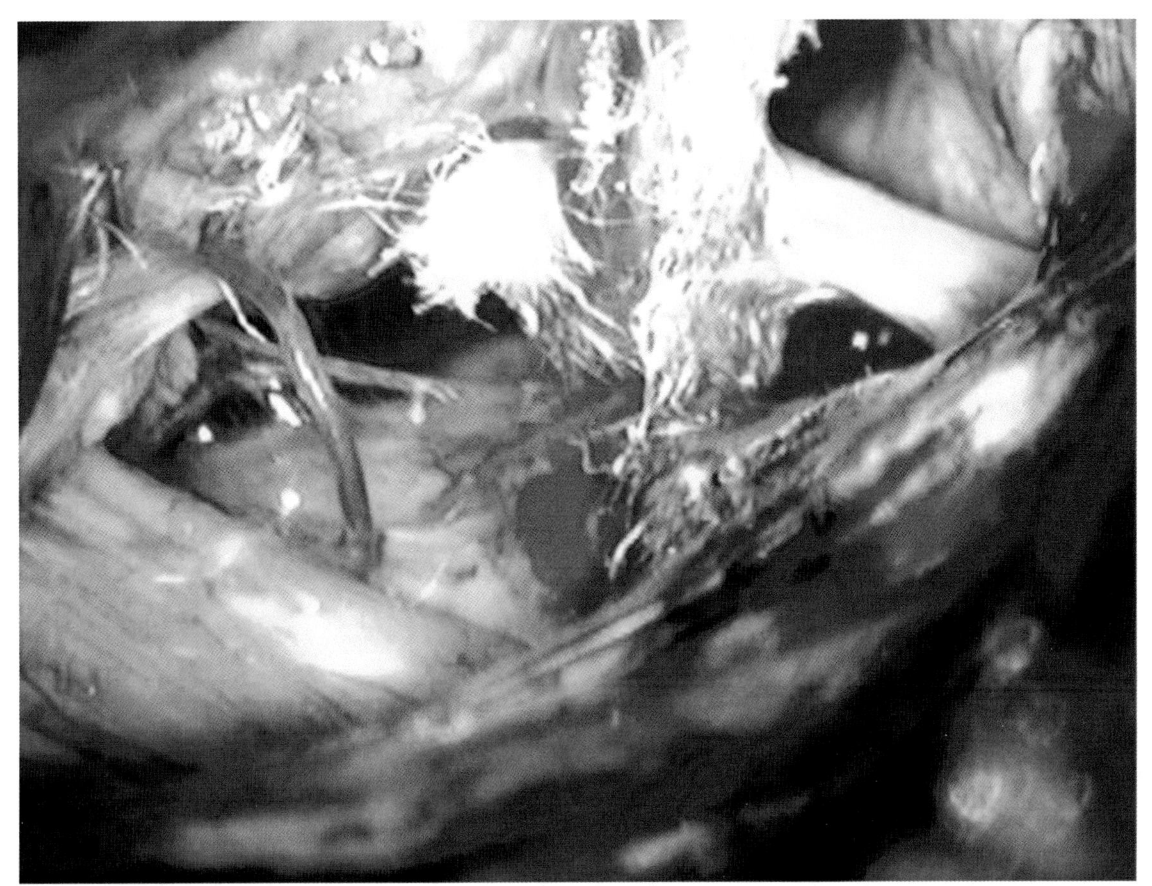

Figure 35. *Intra-operative photograph showing vessel separated from Cranial nerves VII and VIII with a pledget in place.*

The hemi-facial spasms cease almost immediately with the proper placement of the pledgets.

Vestibular Neurectomy for Medically Intractable Meniere's Disease

Meniere's disease or hydrops of the labyrinth is a sporadic disorder of the middle-age or elderly patient. The symptoms include intense paroxysms of vertigo. These symptoms are so severe that a given patient may fall to the floor. The paroxysms occur at irregular intervals and last for various amounts of time. Pathophysiologically, the labyrinth accumulates an excessive amount of fluid (hydrops). Medical therapy includes moderate to high doses of the diuretic, Diamox, a carbonic anhydrase inhibitor. Minor surgical procedures include the formation of a surgical window to drain the excessive fluid.

Once Meniere's disease has been determined to be intractable to both medical and minor surgical procedures Vestibular Neurectomy can provide relief from the attacks of vertigo. A retrosigmoid approach to the Cerebello-pontine angle is used. It is important to guard against any damage to the auditory component of the VIII and the Facial Nerve. Monitoring consists of continuous BAEPs, Facial nerve monitoring and once the dura has been opened, direct Auditory nerve action nerve potentials. In the typical exposure, the two components of Cranial nerve VIII are seen in the field and the facial nerve lies deep to the VIII th nerve. Figure 36 shows the typical anatomic relationship. **(Fig. 36) and (Fig. 37),**

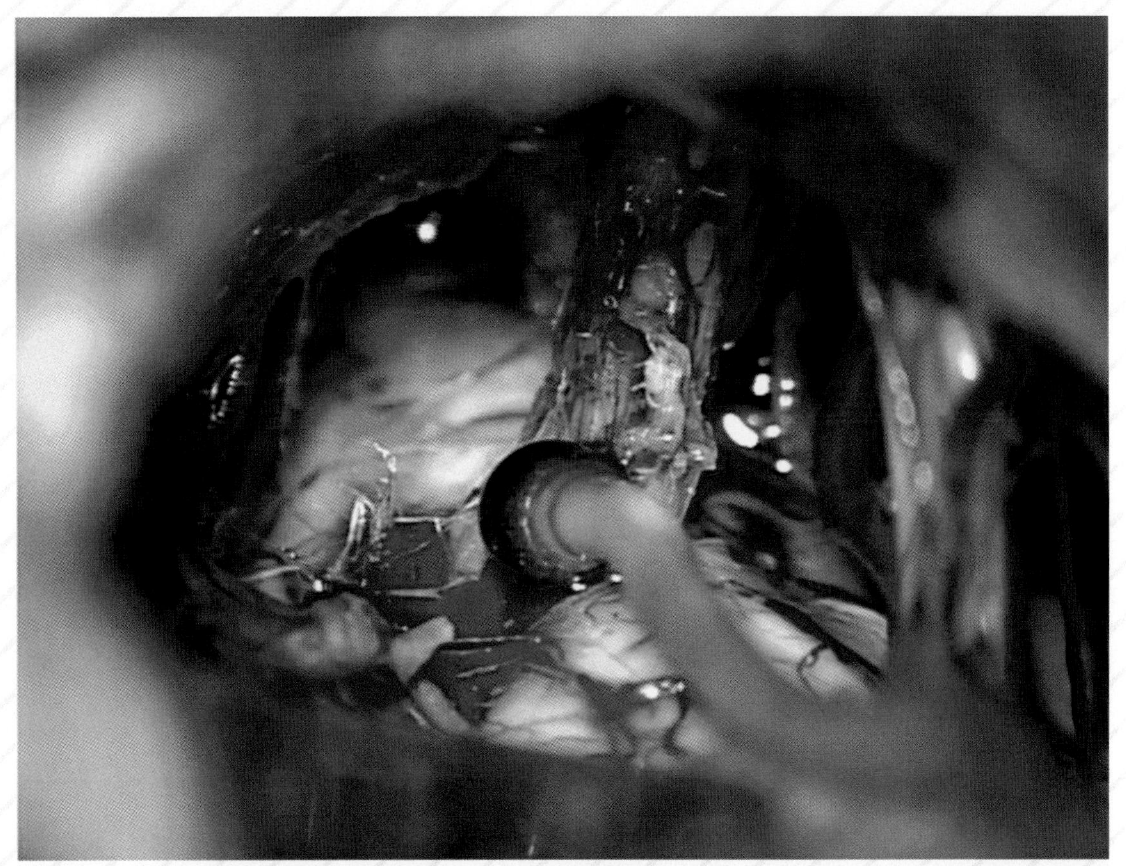

Figure 36. *Cranial nerve VIII is shown with a hand held electrode over the auditory component of the nerve. There is commonly a small artery that separates the two components of the nerve. The vestibular component is on the right side of the electrode.*

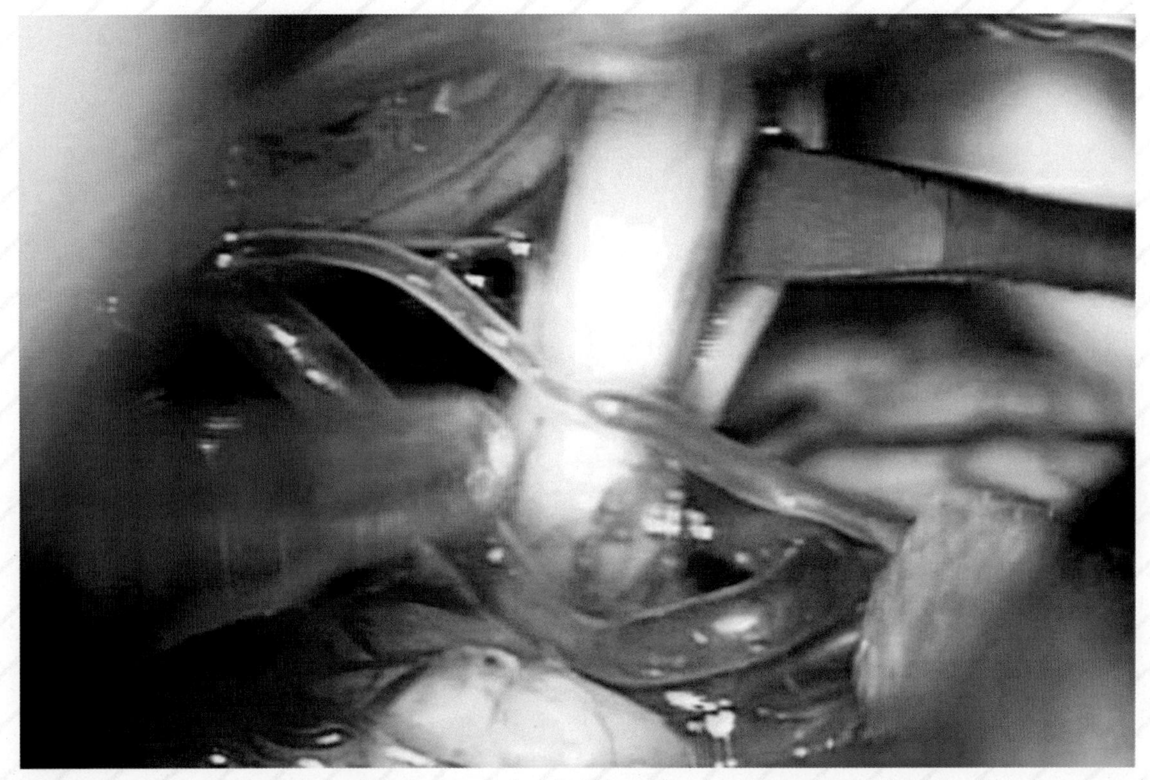

Figure 37. *Intra-operative photograph demonstrating the blunt dissection (separation) of the Vestibular nerve on the right from the larger Auditory nerve on the left.*

After the vestibular nerve is separated from the adjacent auditory nerve and the underlying Facial nerve, a section of the vestibular nerve is removed.

Typically the action potential (AP) seen in the auditory nerve is the same as seen in any peripheral nerve. It consists of a leading positive wave followed by a negative wave, then an additional positive wave as seen in Figure 38. The negative wave approximates the latency for wave II of the BAEPs.

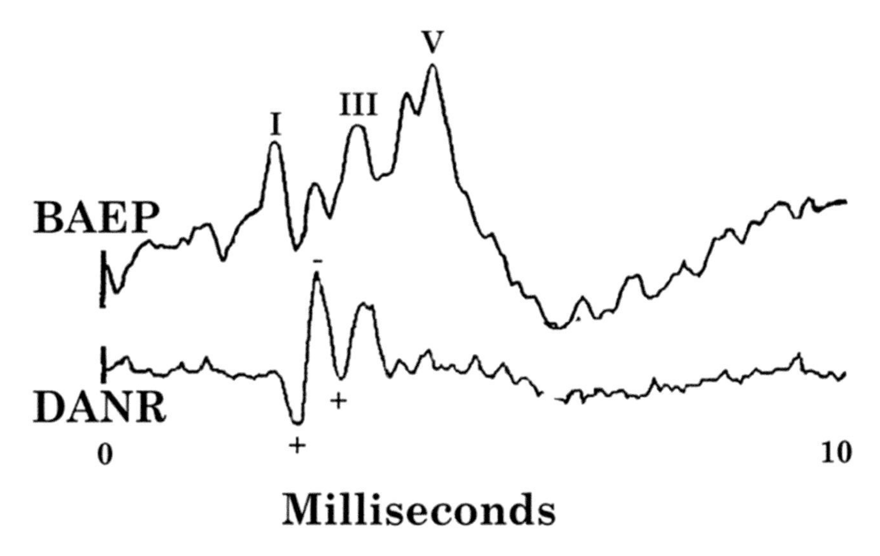

Figure 38. *The upper channel shows the normal BAEPs in this patient with Meniere's disease. The lower channel shows the typical action potential obtained with a direct Auditory nerve recording.*

After identifying the components of the VIII nerve and confirming the location of the auditory nerve with physiologic techniques, the two nerves are carefully separated from one another as shown in **(Fig. 39)**..

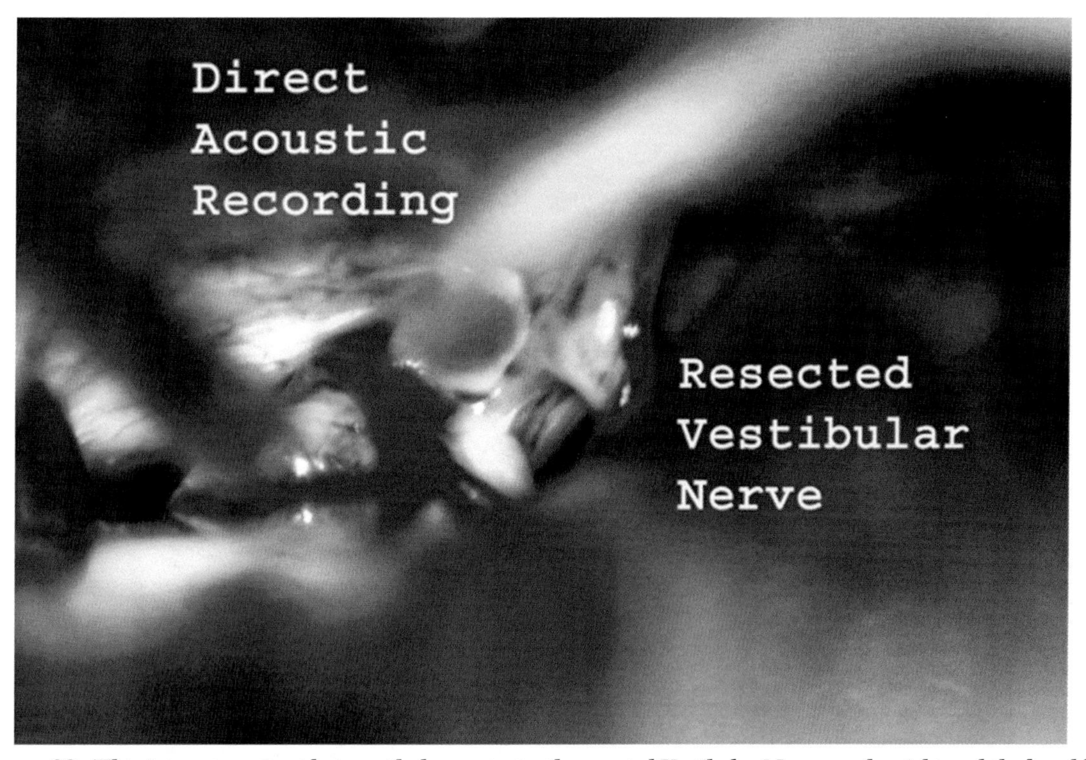

Figure 39. *This intra-operative photograph demonstrates the resected Vestibular Nerve on the right and the hand held electrode on the Auditory Nerve to confirm that no damage occurred during the time of resection. The facial nerve can be seen deep to the resected vestibular nerve.*

(Fig. 40) shows the BAEPs and action potential of a direct Auditory Nerve recording following the removal of a section of Vestibular Nerve.

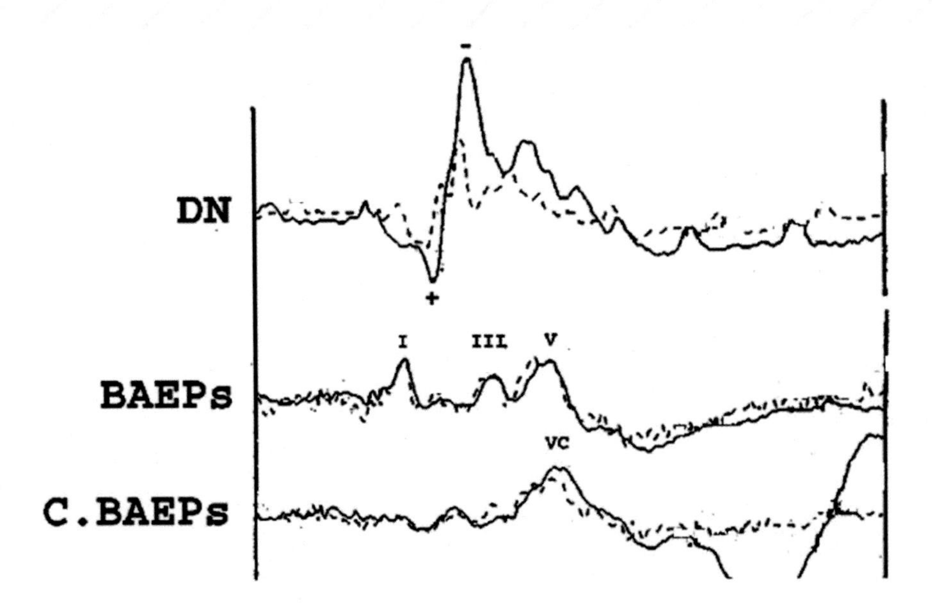

Figure 40. *This figure demonstrates the intact action potential obtained from a direct Auditory Nerve following Vestibular Nerve resection. The BAEPS remain normal following the procedure.*

Immediately, following neurectomy the patient will suffer from moderately severe continuous vertigo that is treated with sedatives such as diazepam. After a few days, an accommodation to the loss of one vestibular nerve occurs and this compensation allows the patient to become free from their vertiginous symptoms.

Tumors in the Middle Fossa and Clivus

A myriad of different types of tumors and cysts occur in the middle fossa and along the clivus. The mass lesions include both tumors, primary and metastatic, vascular lesions such as aneurisms, and cysts such as dermoid or epidermoid. The symptoms vary from being asymptomatic to the development of headaches and visual disturbances, generally double vision secondary to involvement of one of the nerves that innervate the muscles of the eye. There are three nerves that innervate the various muscles of the eye. Cranial nerve III (Oculomotor nerve) innervates the following muscles: medial, superior, inferior rectus muscles and the inferior oblique muscle. Cranial nerve IV (Trochlear nerve) innervates the superior oblique muscle. Cranial nerve VI (Abducens nerve) innervates the lateral rectus muscle. The double vision that results from one of these nerves being involved is different for each of the nerves. All of these cranial nerves pass through the wall of the cavernous sinus on their way to the orbit to innervate their appropriate muscles. The cranial nerves that innervate the Oculomotor muscles can be monitored measuring Compound Motor Action Potentials (CMAPs) generated when the surgeon stimulates one of these cranial nerves. An example of the responses are shown in **(Fig. 41)**.

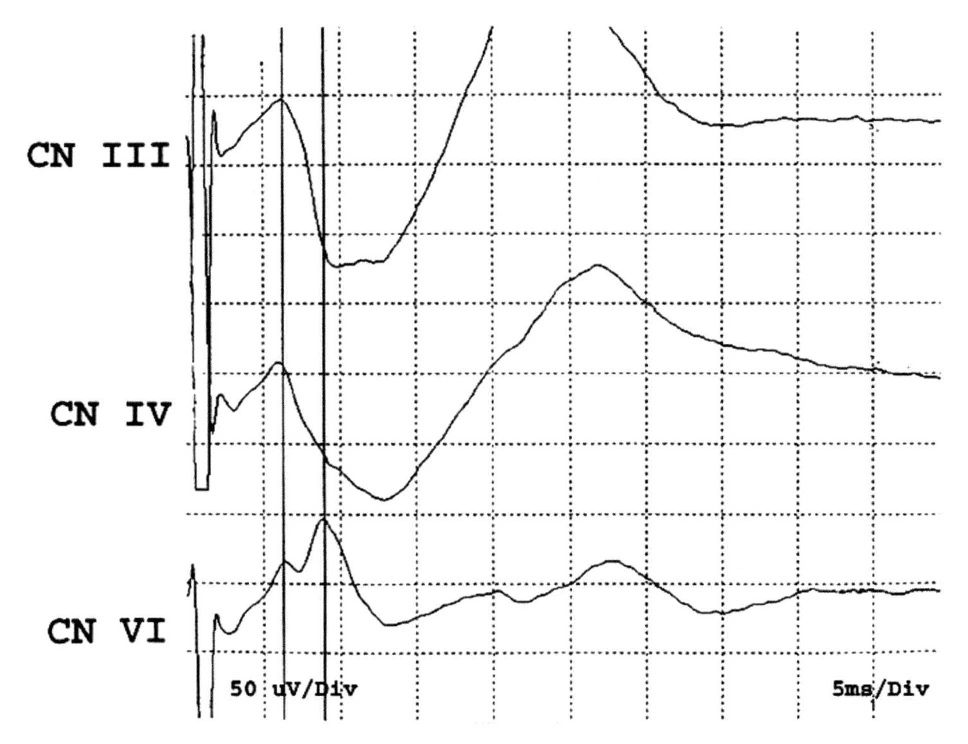

Figure 41. *CMAPs from muscles innervated by Cranial nerves III, IV and VI. The responses are complex and begin about 2 milliseconds after the stimulus. Each muscle shows an individual pattern. The first major negative peak occurs at about 6 milliseconds for CN III and CN IV, whereas the largest early peak occurs at about 8 milliseconds for CN VI.*

Additionally, the ophthalmic branch of the Trigeminal nerve also passes through the wall of the cavernous sinus just inferior to the Oculomotor and Trochlear nerves. In addition to these cranial nerves, the carotid artery passes through the cavernous sinus before its trifurcation.

A variety of surgical approaches are used to remove lesions in this region of skull. Which of the approaches is utilized depends upon the exact location of the lesion. The most common is probably a lateral sub-temporal approach. This would be best for tumors or cysts that are lateralized to one side. The trans-sphenoid approach has been used primarily for surgery on pituitary gland tumors. Finally, the trans-oral approach requires entry through the mouth. This approach is for midline or lesions that are bilateral and are close to the midline. **(Fig. 42)** shows an example of this later approach.

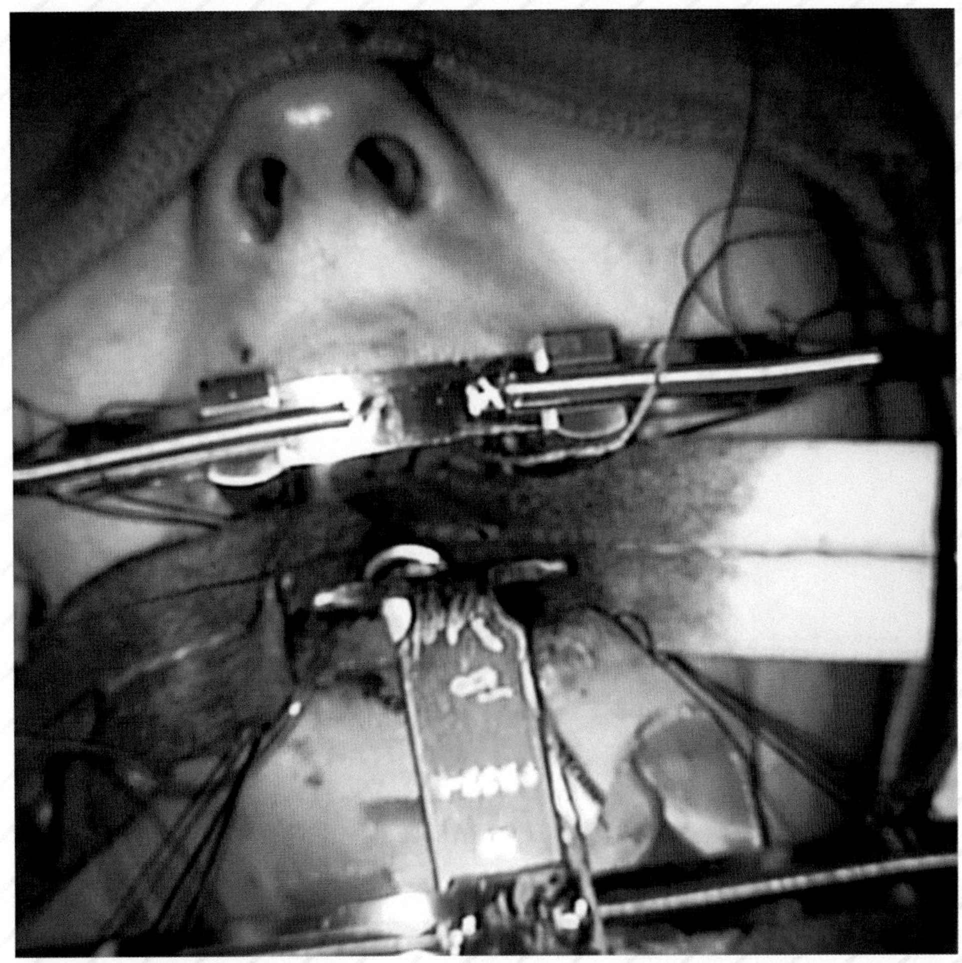

Figure 42. *Transoral approach to the clivus. The mandible may have to be separated to gain access to the pharynx. The posterior pharynx is opened and the bone drilled to gain access to the tumor.*

After drilling the bone, access to the midline tumor is demonstrated in **(Fig. 43).**

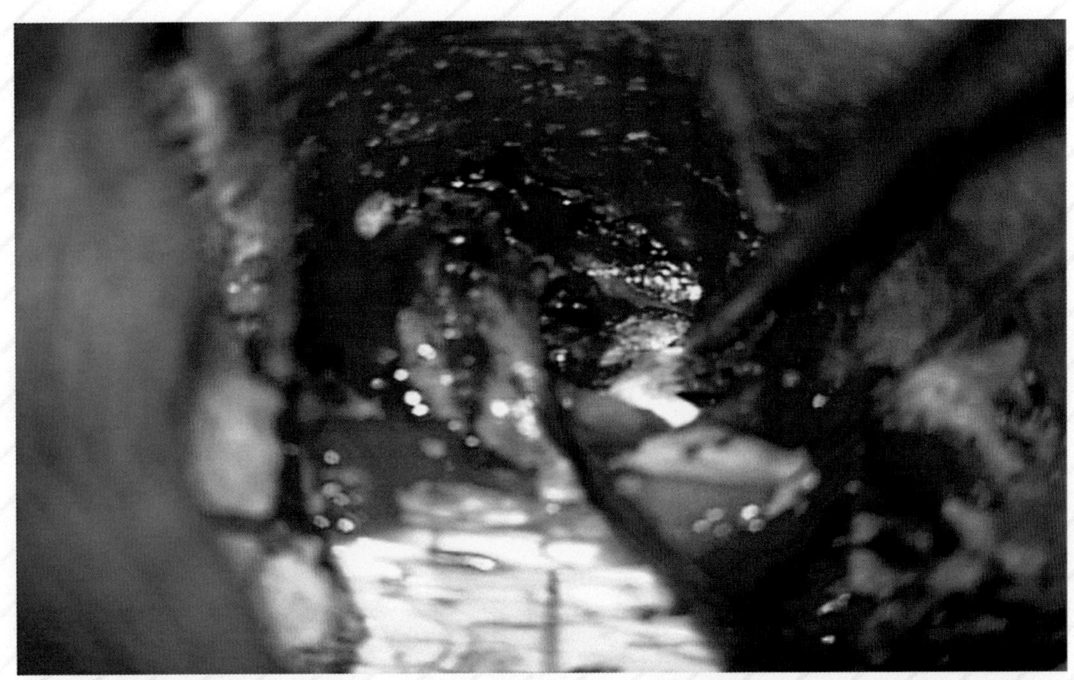

Figure 43. *Intra-operative photograph showing tumor removal using a Transoral approach.*

Probably the most common approach to this region is via the sub temporal. This approach is typically used to operate on lateralized tumors or even small midline lesions.

See this approach in **(Fig. 44).**

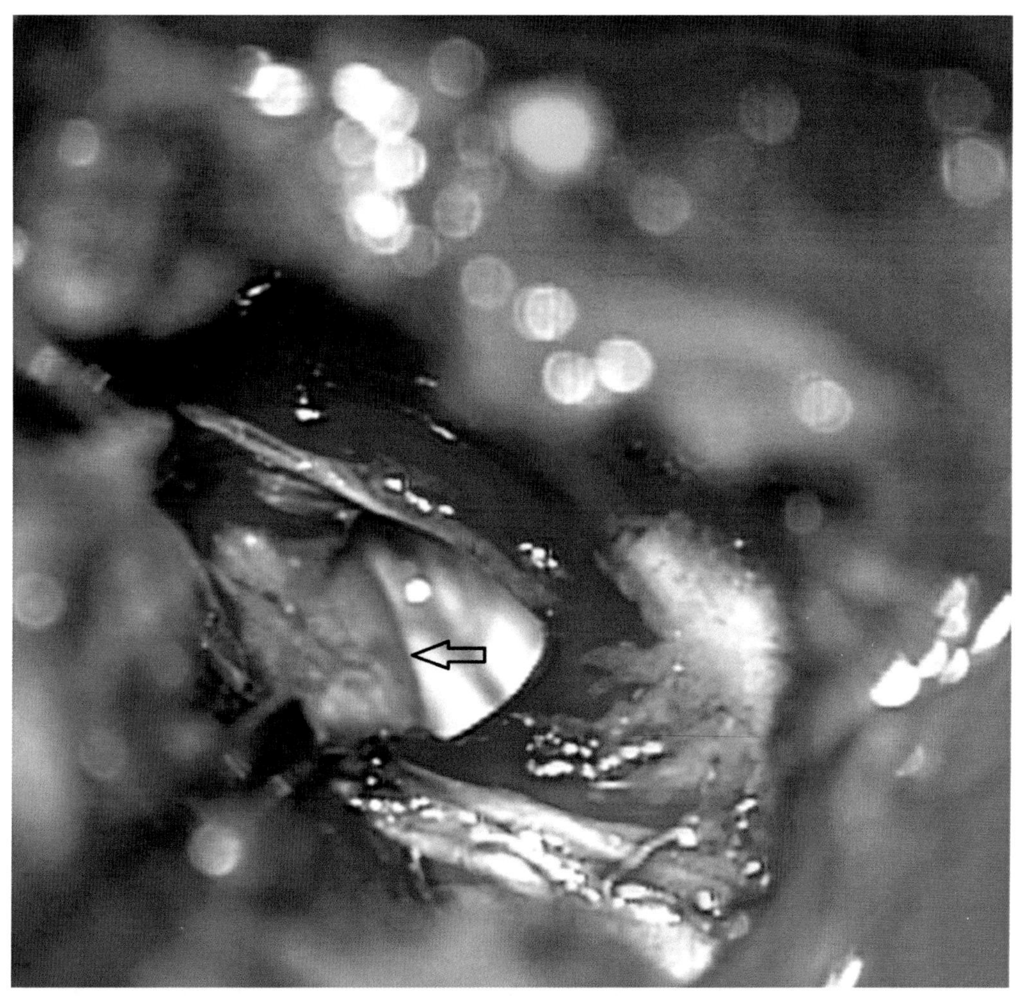

Figure 44. *Intra-operative photograph showing opening of the cavernous sinus from a sub-temporal approach. Note the grayish tumor in the anterior inferior portion of the opening.*

Meningiomas are one of the common tumors found in this region of the skull. Meningiomas are derived from the meninges and are usually attached to the dura. These tumors vary greatly in their aggressiveness, some being well encapsulated others being "en plaque". Meningiomas tend to be benign histologically, but a small percentage of them tend to spread outside of the nervous system and may metastasize to lung, bone or even liver. At times, the location of this tumor makes it impossible to completely remove; so that recurrence of the tumor mass is a problem. These tumors are all quite vascular and derive their blood supply from numerous local sources.To make the removal of meningiomas safer, they are typically embolized the day prior to theirsurgical removal. Meningiomas tend to induce local bone growth known as hyperostosis. Thevast majority of the damage resulting from meningiomas is by compression. These tumors may present with an onset of new headache, and unilateral cranial nerve abnormalities.

Visual losssecondary to compression of the optic nerve is common in sphenoid wing tumors.

(Fig. 45) shows a sphenoid wing meningioma.

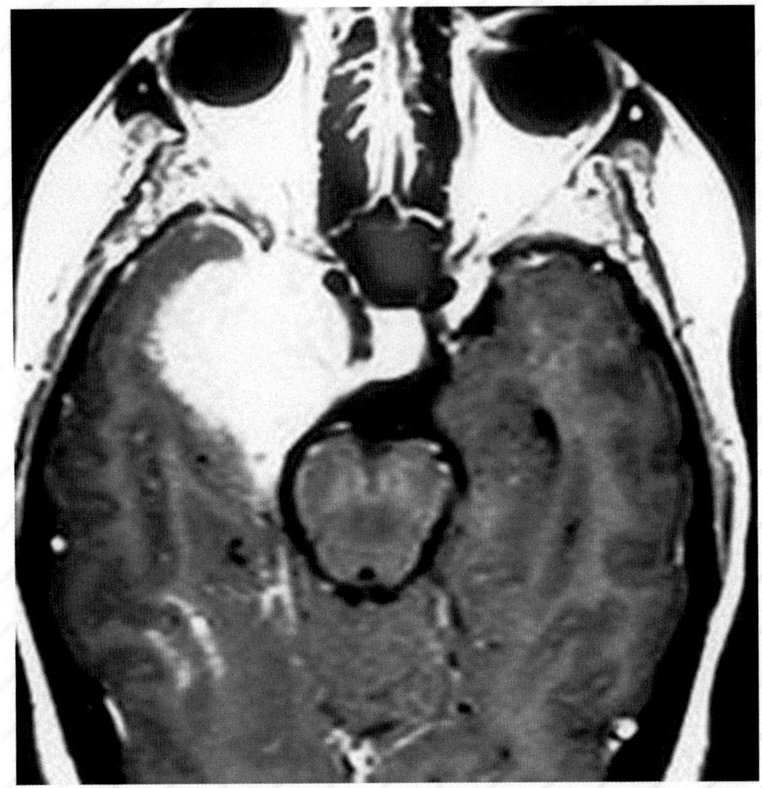

Figure 45. *MRI of a sphenoid wing meningioma. This tumor compresses the mesial temporal lobe and surrounds the carotid artery in the cavernous sinus. The optic nerve is compressed by the tumor.*

Meningiomas may also be found in other locations in the middle fossa of the skull as demonstrated in **(Fig. 46).**

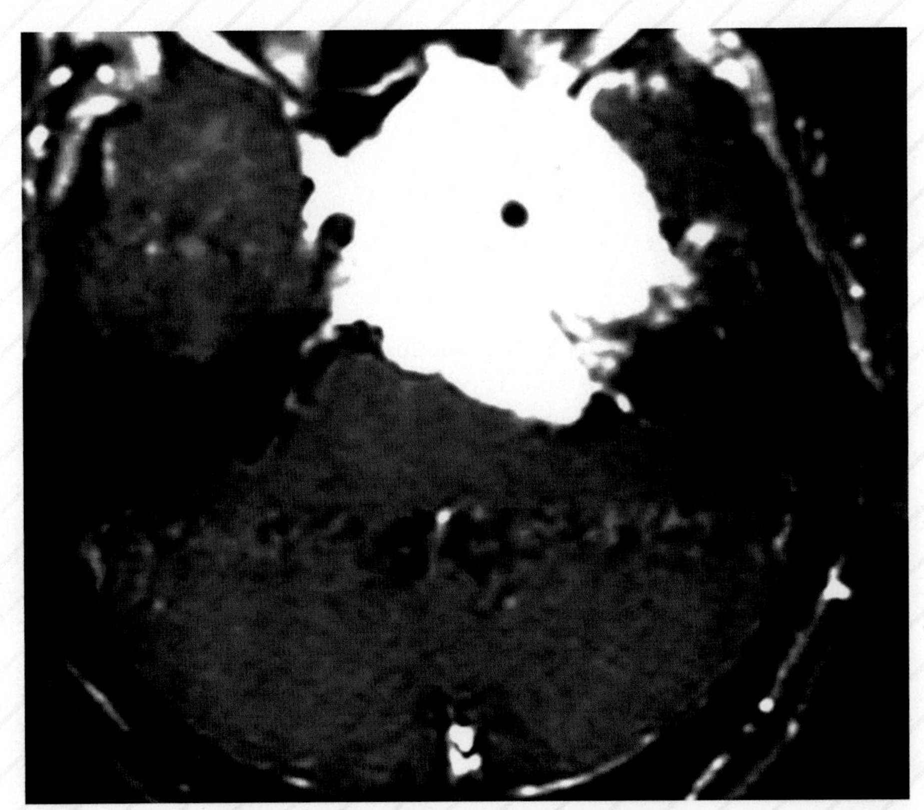

Figure 46. *MRI scan reveals a large meningioma encompassing both carotid arteries and extending down the clivus. Tumor compresses both the temporal lobe and Pons.*

The surgical approach to both of these examples is by entering via a sub temporal location. Monitoring includes Median nerve SSEPs to monitor brainstem function and Cranial Nerves III, IV, and VI. At times, more than one surgical approach is necessary and staging of the surgery may have to be accomplished. In this instance, a posterior fossa approach may be necessary to remove that part of the tumor below the tentorium. In this instance, additional monitoring will require BAEPs, motor V and facial nerve monitoring.

Schwannoma

Vestibular schwannoma have already been discussed, but this tumor type may also affect other cranial and spinal nerves as well. Both cranial nerve V and VII can be involved. Schwannoma of the Facial nerve are indistinguishable for the Vestibular schwannoma prior to surgery. It is only at the time of surgery that this diagnosis is made. When the facial nerve is involved, simple debulking of the tumor is indicated with every attempt being made to salvage facial nerve function for as long as possible.

Schwannomas of the Vth cranial nerve are rare. They generally present with numbness of one side of the face. **(Fig. 47)** is an example of a large schwannoma of the Vth cranial nerve.

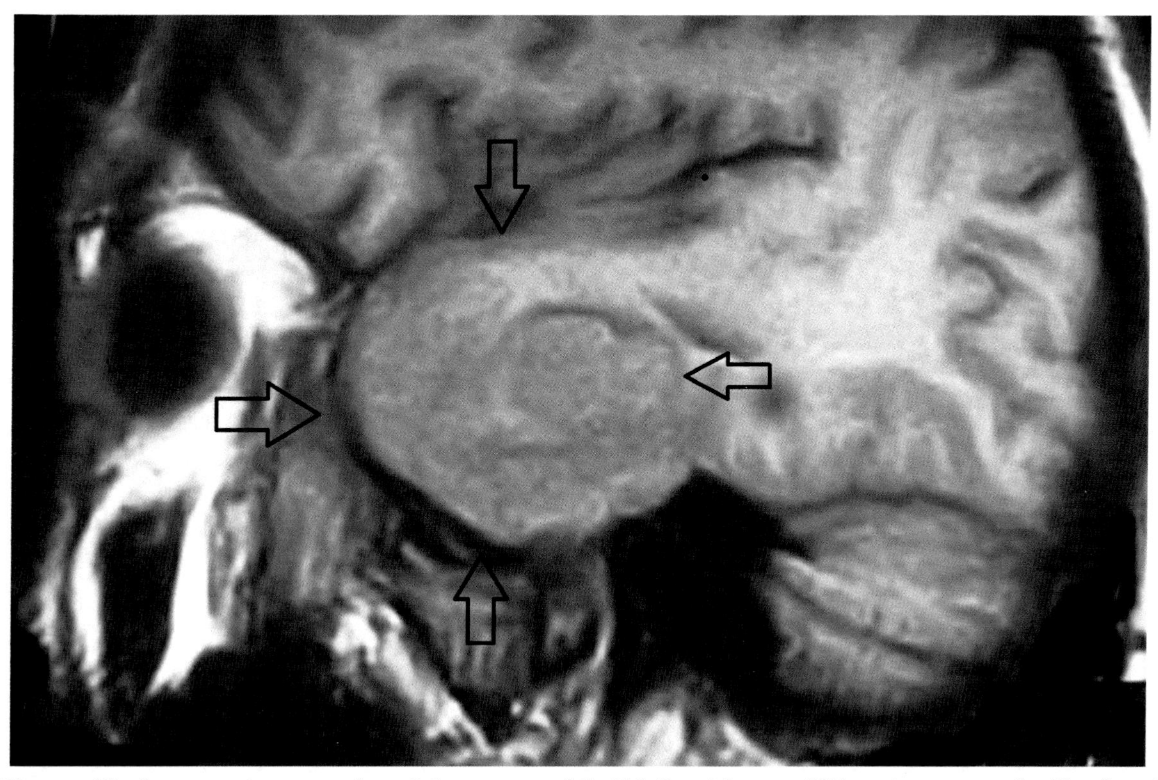

Figure 48. *demonstrating a very large Schwannoma of the Vth Cranial nerve. This patient presented with a loss of sensation on half of the face. All three divisions were affected.*

Cordomas

Cordomas are a rare malignant tumors derived from the primitive notochord. They are always midline and about 35 percent involve the clivus, 15 percent the spine and 50 percent the sacrum and coccyx region. Cordomas occur in excess in males with an M: F ratio of 2:1. They generally cause symptoms by compression and the most common cranial nerve involved is the Abducens nerve (VI). Headache is another common symptom. The VIth nerve is followed by involvement of the Facial nerve and Auditory nerve. Monitoring is tailored to the patient. For those tumors arising in the upper clivus and involving the 6th cranial nerve, the EMG of Cranial nerves III, IV and VI and median nerve SSEPS are needed for monitoring. When the Facial nerve and Auditory nerve are affected or at risk, the BAEPs and Facial nerve and median nerve SSEPs are needed for monitoring. Recurrence of this tumor is common so that in most instances high dosage irradiation is used as an ancillary form of therapy. Average survival with this tumor is 4.1 years. **(Fig. 48)** is an example of a Cordoma of the clivus.

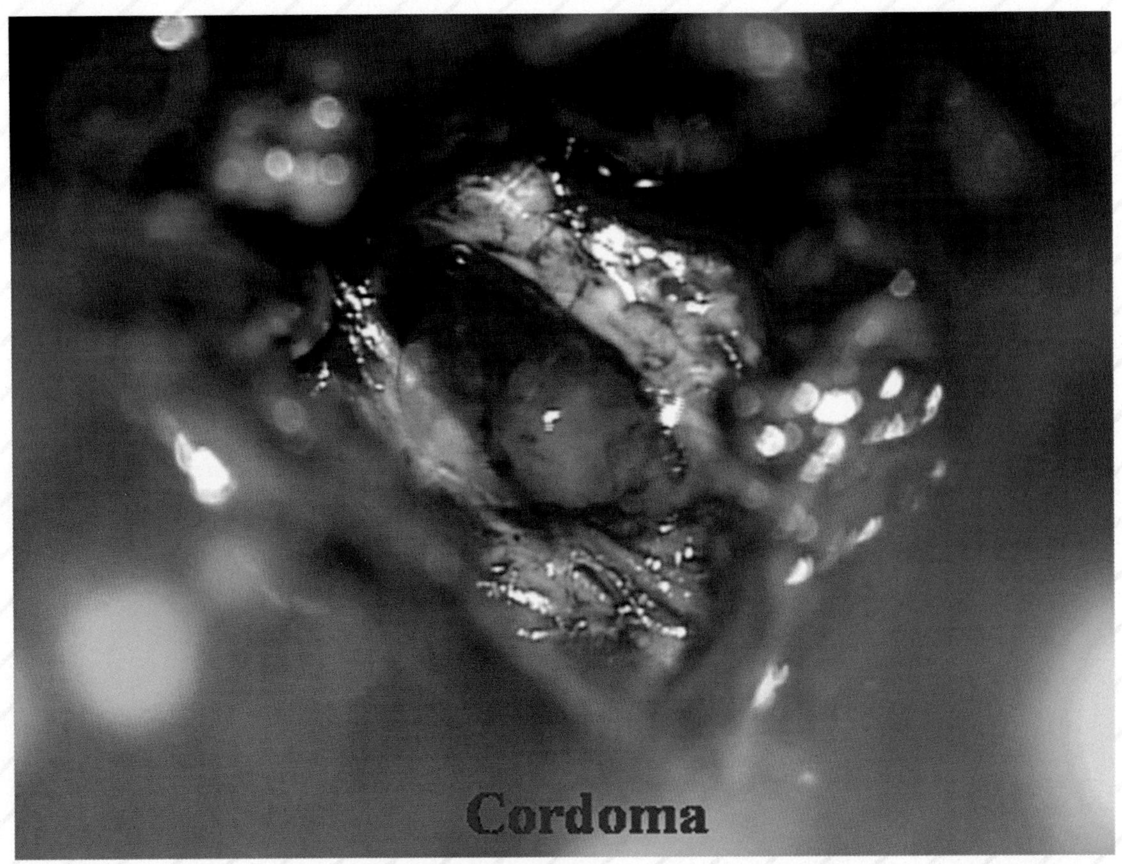

Figure 48. *Intra-operative photograph of a Cordoma of the clivus. These tumors are typically partially calcified. This is a moderately small tumor, but at times they can become quite large and aggressive.*

Chondrosarcomas

Chondrosarcomas, like Cordomas are very rare tumors that tend to arise from synchondrosis near the temporo-occipital junction. The distribution of these tumors include: 66 percent temporo-occipital, 28 percent clivus and 6 percent spheno-ethmoid. Chondrosarcomas are mostly low grade malignant tumors derived from cartilaginous structures at the base of the skull. These tumors frequently present much like the Cordomas with headache and lateralized cranial nerve findings, the most common being the Abducens nerve (Cranial nerve VI). About a third of patients develop decreased hearing, vertigo and tinnitus. Complete resection of this tumor is important for long term survival. At times, resection followed by high dose irradiation is recommended. However, the total number of cases is small and best therapy has not been determined. Survival rate with a 5 year recurrence-free period is 90 percent. Adjunctive Adriamycin and Cis-platinum have been used for this tumor with reported good results. Again, the total numbers are small and it is not clear whether the efficacy of this form of therapy is best or whether other agents might be more useful. **(Fig. 49 and Fig. 50)** are examples of Chondrosarcomas of the middle fossa.

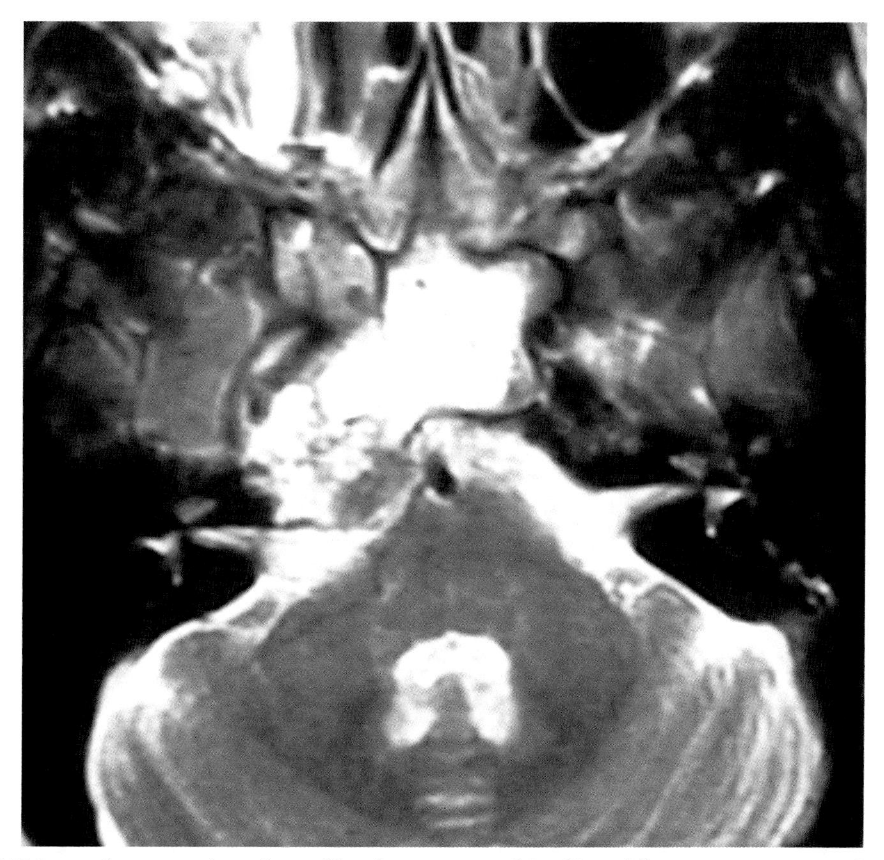

Figure 49. *MRI scan demonstrating a large Chondrosarcomas of the clivus. This tumor compresses the peduncle at the junction of the pons and midbrain. The tumor also encases the basilar artery.*

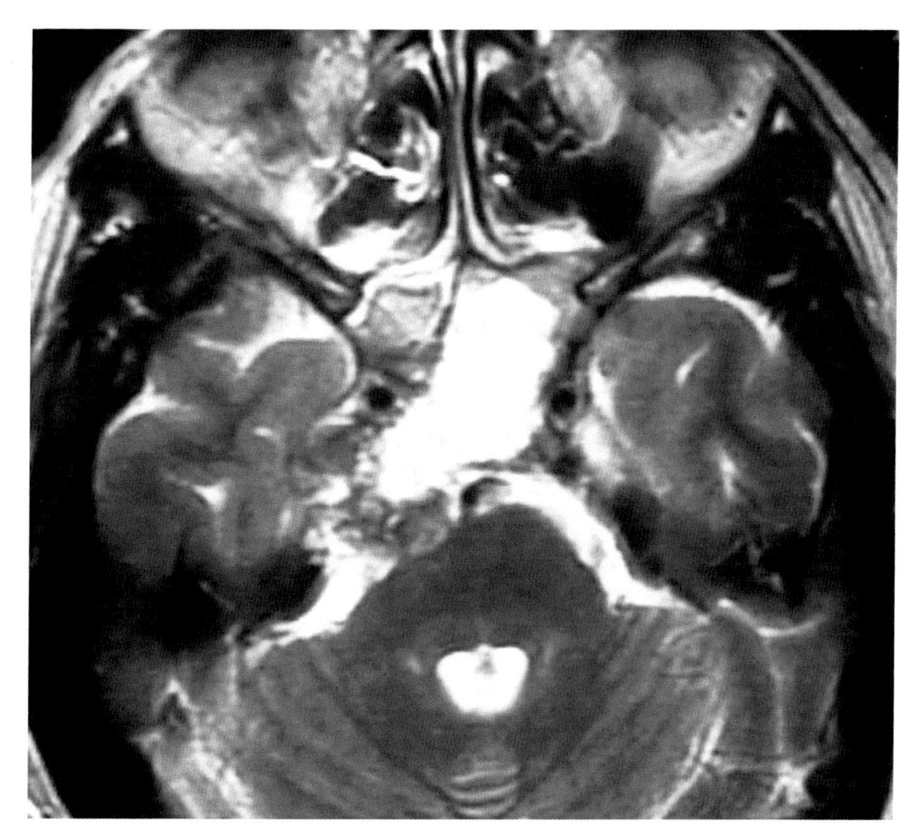

Figure 50. *MRI scan demonstrating a Chondrosarcomas at a higher level. Note compression of the pons.*

Because of the location of these tumors, monitoring can be extensive and must be tailored to each case. It is not uncommon to monitor BAEPs and Median nerve SSEPs plus numerous cranial nerves from III to IX/X. **(Fig. 51).**

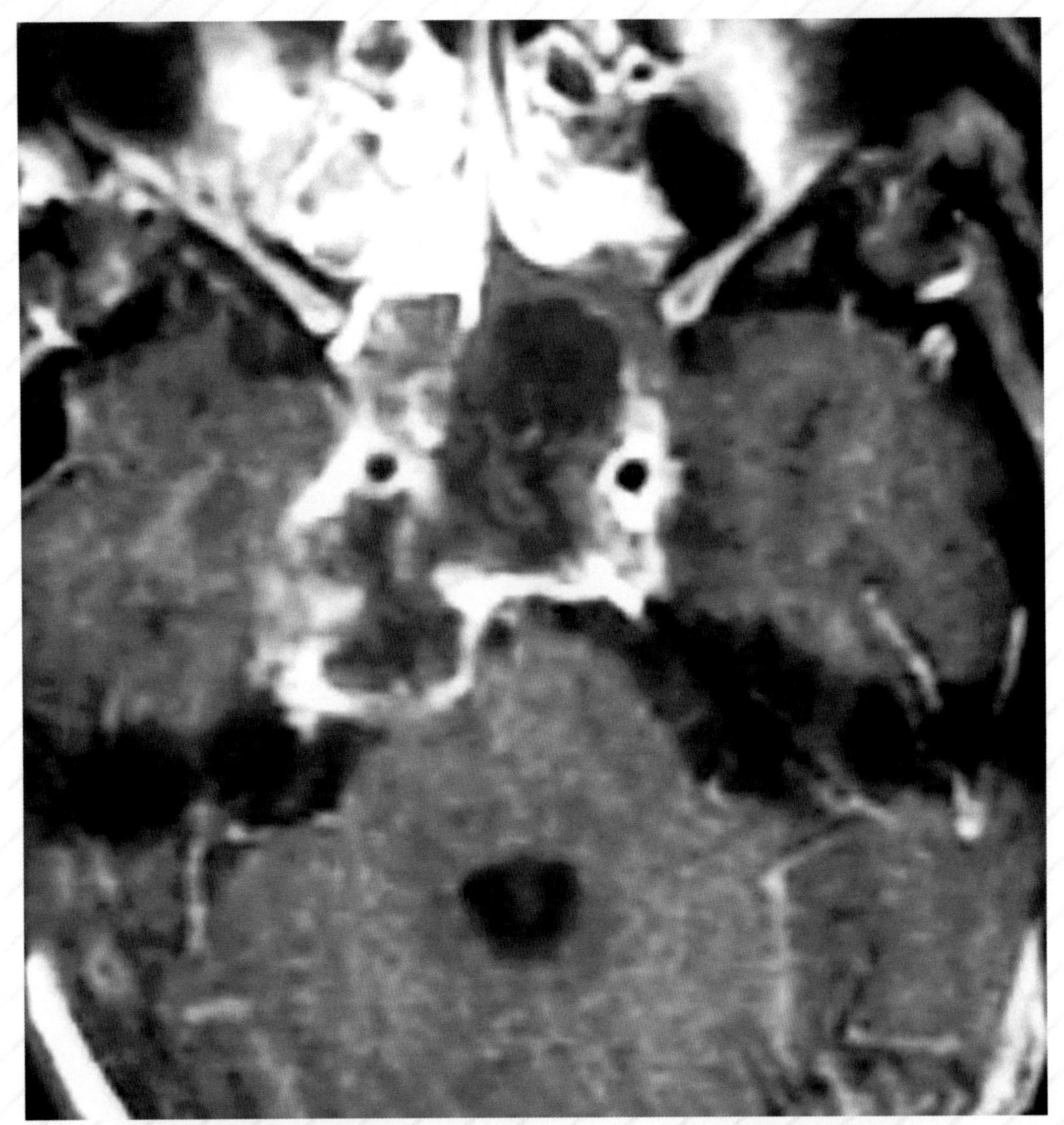

Figure 51. *demonstrating the post-operative surgical site with the vast majority of the tumor removed. Post-operative irradiation is used to treat any residual tumor remaining after resection.*

Dermoid Cysts

Dermoid cysts are derived from a developmental abnormality. They consist of a white capsule and are filled with numerous skin components including hair, teeth, and skin glands. These cysts continue to enlarge over time as more elements are secreted into the cyst. They are almost always found in or near the midline of the central nervous system. Their clinical presentation depends of the specific location of the cyst. Common presenting findings include unilateral cranial nerve abnormalities, seizures, long tract sensory or motor abnormalities. Occasionally, these cysts rupture and are the source for recurrent bouts of chemical meningitis. In this instance severe headache, fever and neck stiffness are prominent. An intense cellular response occurs in the cerebrospinal fluid as does an increase in protein. CSF cultures remain sterile. During removal care must be taken not to spill the contents of these cysts into the subarachnoid spaces. Both the contents of the cyst and the capsule must be totally removed. These cysts do not respond to either irradiation or chemotherapy.

(Fig. 52) demonstrates an example of a midline dermoid cyst.

Figure 52. *Intra-operative photo showing a midline dermoid cyst. Note the whiteness of the Cyst wall. Dermoid cysts act as a mass lesion, but are not true tumors. They cause symptoms by either compression or by leakage.*

Intra-operative monitoring must be individually tailored for Dermoid cysts. Those at the upper clivus will require median nerve SSEPs and EMG of the extra ocular muscles while lower down MN SSEPs, BAEPs, facial nerve monitoring or even lower cranial nerve monitoring (IX/X, XI and XII) may be necessary.

Lower Cranial Nerve Monitoring

A number of different tumors and congenital anomalies occur below the tentorium requiring monitoring of the cranial nerves at and below Cranial nerve VIII. EMGs of the facial, vagus, accessory and hypoglossal nerves can all be monitored. See placement of electrodes for each specific nerve in an earlier part of this chapter. In addition, median nerve SSEPs are universally monitored while BAEPs can be added in appropriate cases. The type of tumors are the same as previously discussed in other locations, while there are some new ones that have not yet been reviewed. Common congenital anomalies occur in this location, some of which when symptomatic require surgical intervention. Tumors and congenital anomalies in this location may require special positioning of the patient and repeated MN SSEPs are done during positioning making sure that compression of the nervous system does not result from the positioning itself. One common position is the prone position where the neck is held in slight extension. One may have to flex the neck to reduce any slowing in conduction resulting from positioning. **(Fig. 53)** shows such a case.

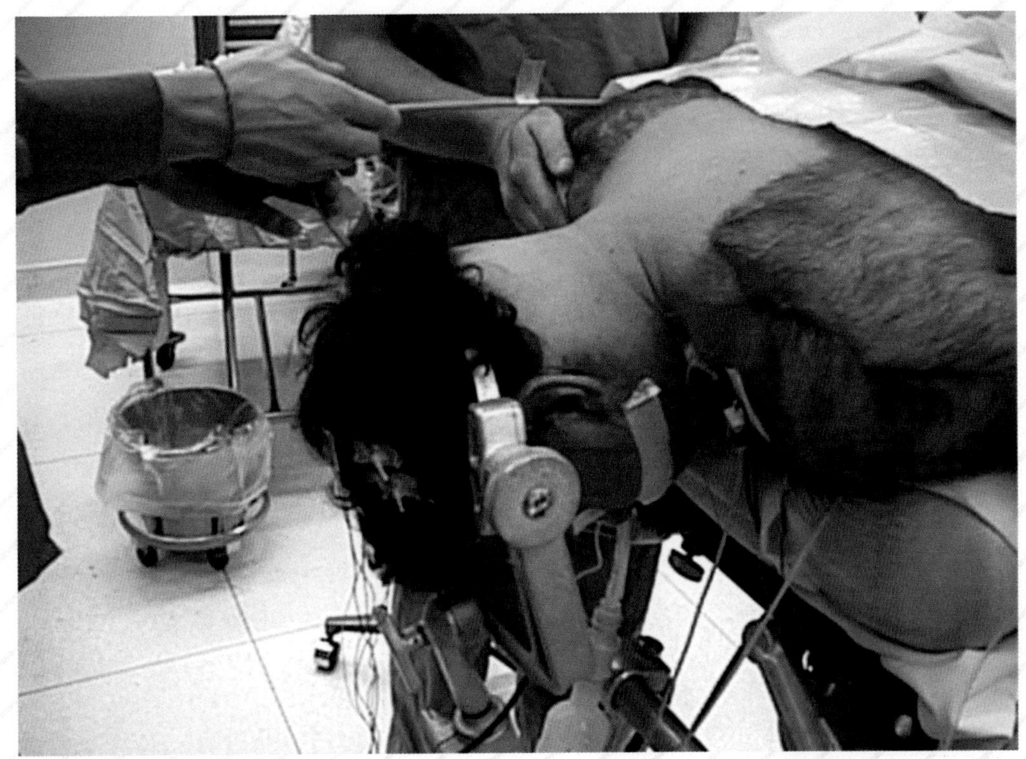

Figure 53. *Patient being positioned for removal of a posterior fossa tumor. Note the slight extension of the neck. Monitoring of brainstem function during position is very important for the patient's welfare.*

A second common position for posterior fossa surgery is the sitting position. Not only is monitoring during positioning critical, anesthesiology has the additional responsibility of monitoring for air emboli that can occur with a patient in the sitting position. **(Fig. 54).**

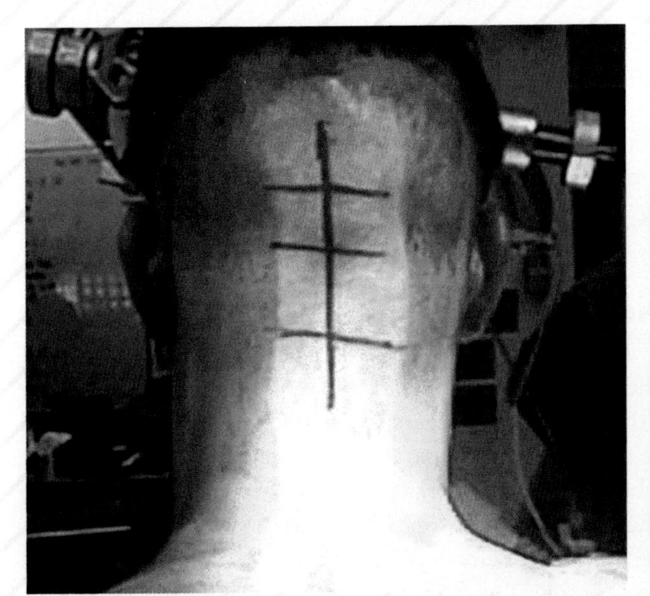

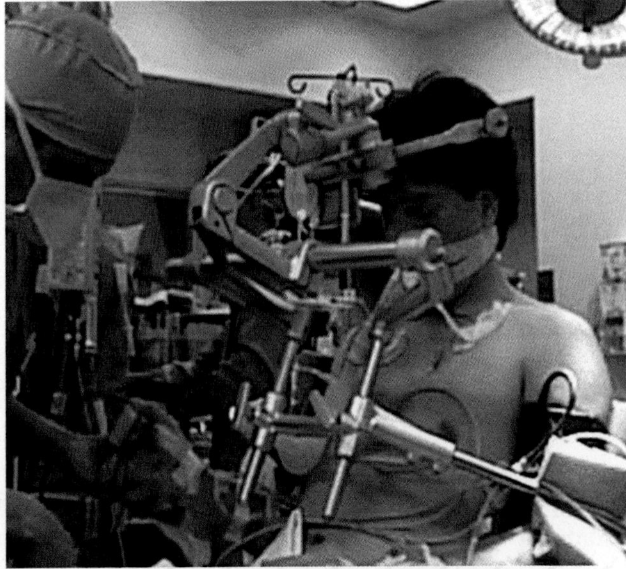

Figure 54. *Patient being prepared for sitting position. This approach is useful for midline lesions affecting the posterior fossa. As patient is positioned median nerve SSEPs should be used to make sure there is no compression of the brainstem with positioning.*

Serial median nerve SSEPs should be used to check positioning of the patient when using the sitting position. It may be necessary to either extend or flex the patient's neck to relieve compression which will cause a delay in the cortical N20 as shown in **(Fig. 55).**.

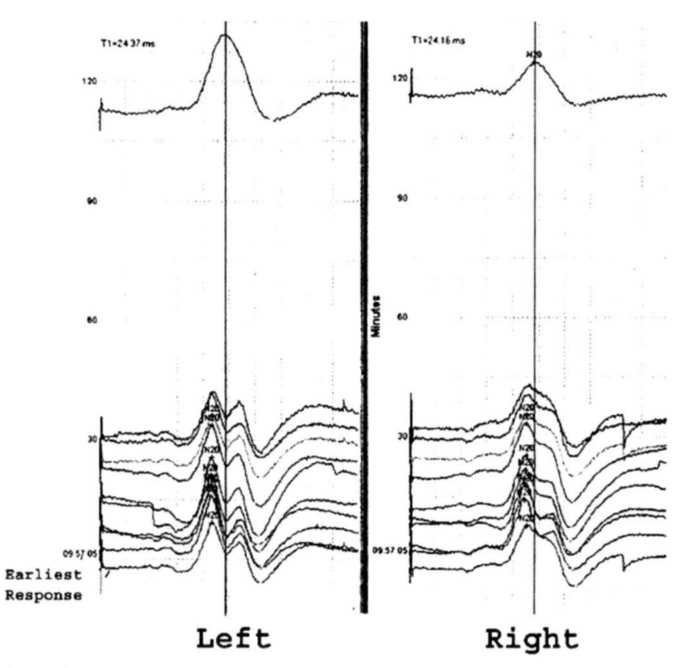

Figure 55. *Example of serial median nerve SSEPs collected while the patient was being prepared for sitting position. Note the slightly prolonged latency of the N20 (Cortical response with the latest position. The patient needs to be repositioned to return to the original baseline.*

Ependymomas

Ependymomas are tumors that are derived from the ependymal lining of the nervous system by neoplastic transformation. They tend to originate in the floor of the 4th ventricle and may extend through the foramen of Luschka and Magendie to enter the subarachnoid space. These tumors are more common in individuals under the age of 20 years, but continue to be present in young adults as well. These tumors can also be found in the lateral ventricles of the hemispheres and in the spinal cord. Once in the subarachnoid space they may ascend up to the cerebello-pontine angle or compromise the lower cranial nerves. Ependymomas vary in their growth characteristics and the more aggressive being identified as ependymoblastomas. These tumors have a strong tendency for recurrence following resection so that post-resection irradiation may be of benefit. Clinical signs include those resulting from obstruction of the ventricular system, nocturnal headaches, vomiting and unsteadiness of gait. Lateralized cranial nerve finding may also be present. **(Fig. 56)** is an example of a large 4th ventricle ependymoma.

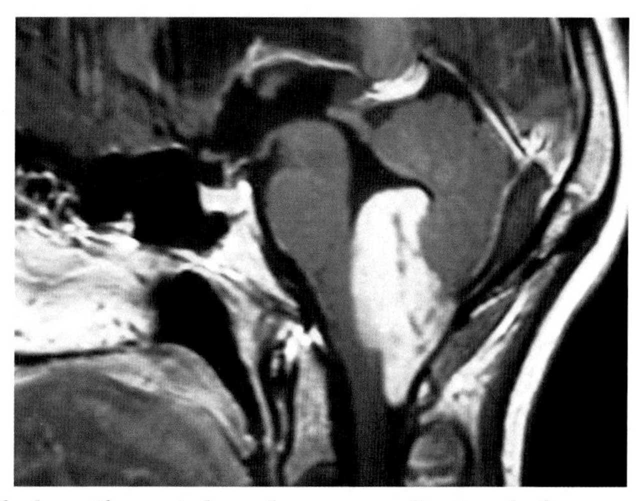

Figure 56. *MRI scan of a large 4th ventrical ependymoma extending into the foramem magnum. Note the enlarged ventricles above the tumor.*

Cerebellar Tumors

Cerebellar tumors are much more common in childhood, but do continue to be present in young adults. In childhood, the two most common cerebellar tumors are the medulloblstoma and the cystic astrocytoma. In the adult, the most common cerebellar tumor is the hemingioblastoma. This tumor generally occurs in one of the cerebellar hemispheres, and may be either solid or cystic. In the cystic tumors an active nodule is generally present. Single tumors are generally sporadic, but multiple tumors associated with hemingioblastoma of the retina are part of "von Hippel-Landau disease". In addition, von Hippel-Landau disease includes the presence of tumors in the kidneys and adrenal glands. About 20 percent of individuals with hemingioblastoma have an increased red cell mass secondary to the secretion of erythropoietin. Clinical symptoms generally result from the enlarging cystic cavity and may lead to unilateral cerebellar ataxia. At times, the cystic lesion compresses the ventricular system leading to increased intracranial pressure and its associated symptoms. Patients with von Hippel-Landau disease may be diagnosed before any symptoms are present and are frequently followed with serial MRI scans and their clinical state. Only when clinically indicated are the tumors removed surgically. The sporadic, but most common form of hemingioblastoma generally presents with clinical features that require surgical removal. **(Fig. 57)** is an example of a cerebellar hemingioblastoma.

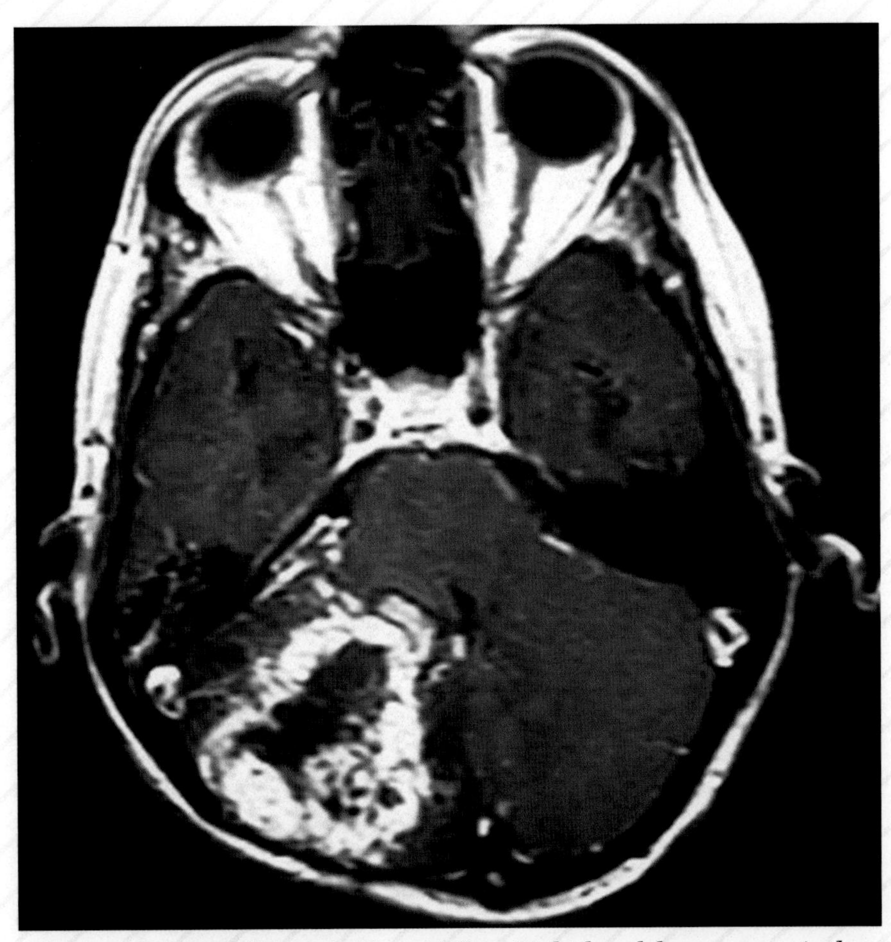

Figure 57. *of a large cystic hemingioblastoma of the cerebellum. Multiple nodules are present in the wall of the tumor associated with small cysts.*

Astrocytomas of the cerebellum are more common in childhood, but may be seen in young adults. Astrocytomas of the cerebellum are generally slow growing tumors that do not extend to other parts of the nervous tissue as do that astrocytoma found in the cerebral hemispheres. These tumors cause symptoms by 1) increasing the intracranial pressure leading to headache (especially at night or in the early morning), nausea and lethargy and 2) unilateral ataxia of the arm and leg. Surgical removal of these tumors is the treatment of choice. These tumors of low grade malignancy generally do not respond well to irradiation. **(Fig. 58)** is an example of a cerebellar astrocytoma.

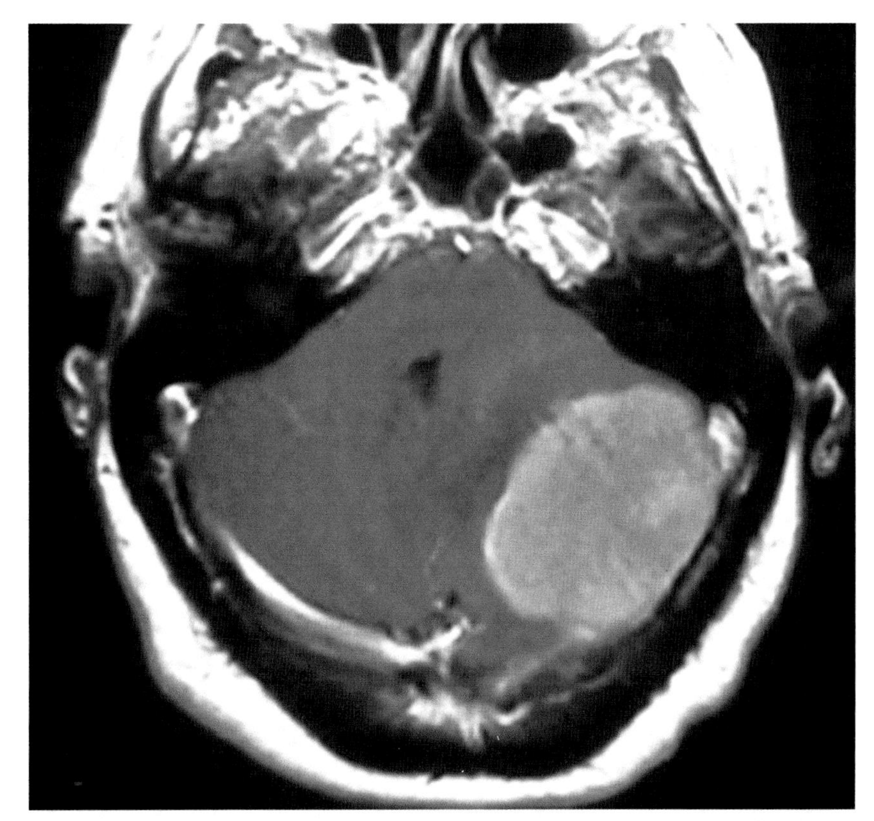

Figure 58. *of a large cerebellar astrocytoma. Note the distortion of the 4th ventricle.*

Meningiomas also occur at the foramen magnum. Meningiomas in this location are like those in other areas. The surgical removal necessitates a posterior approach and like meningiomas in other locations, have a varied blood supply. These tumors are frequently embolized prior to surgical removal. **(Fig. 59 and Fig. 60)** demonstrate a meningioma of the foramen magnum and an angiogram demonstrating a catheter in place for embolization of the blood supply to the tumor.

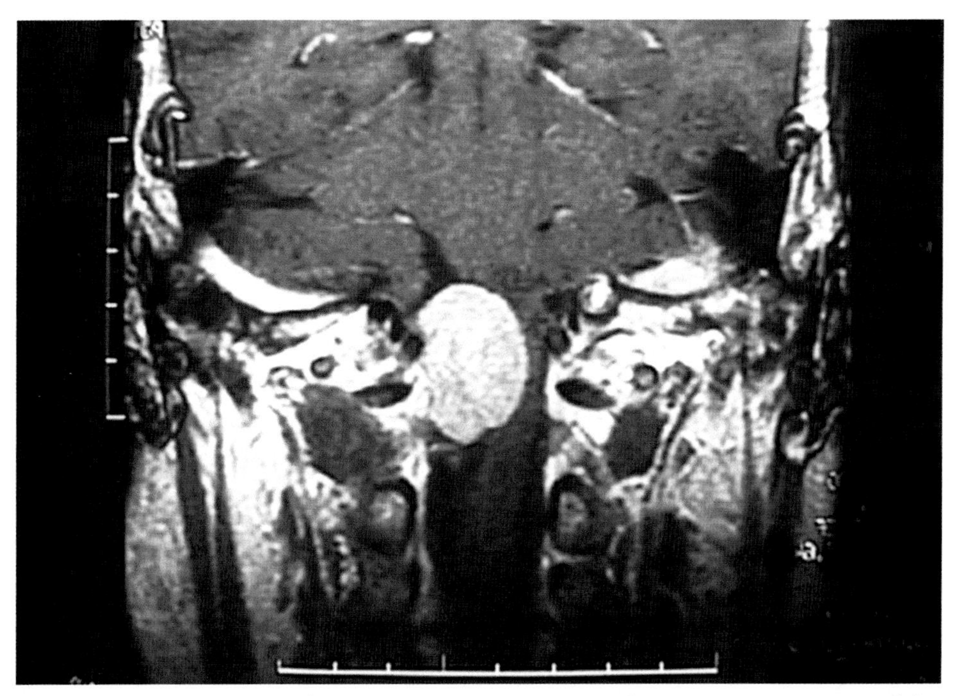

Figure 59. *MRI scan demonstrating a foramen magnum meningioma. There is severe compression of the medulla and upper cervical spinal cord.*

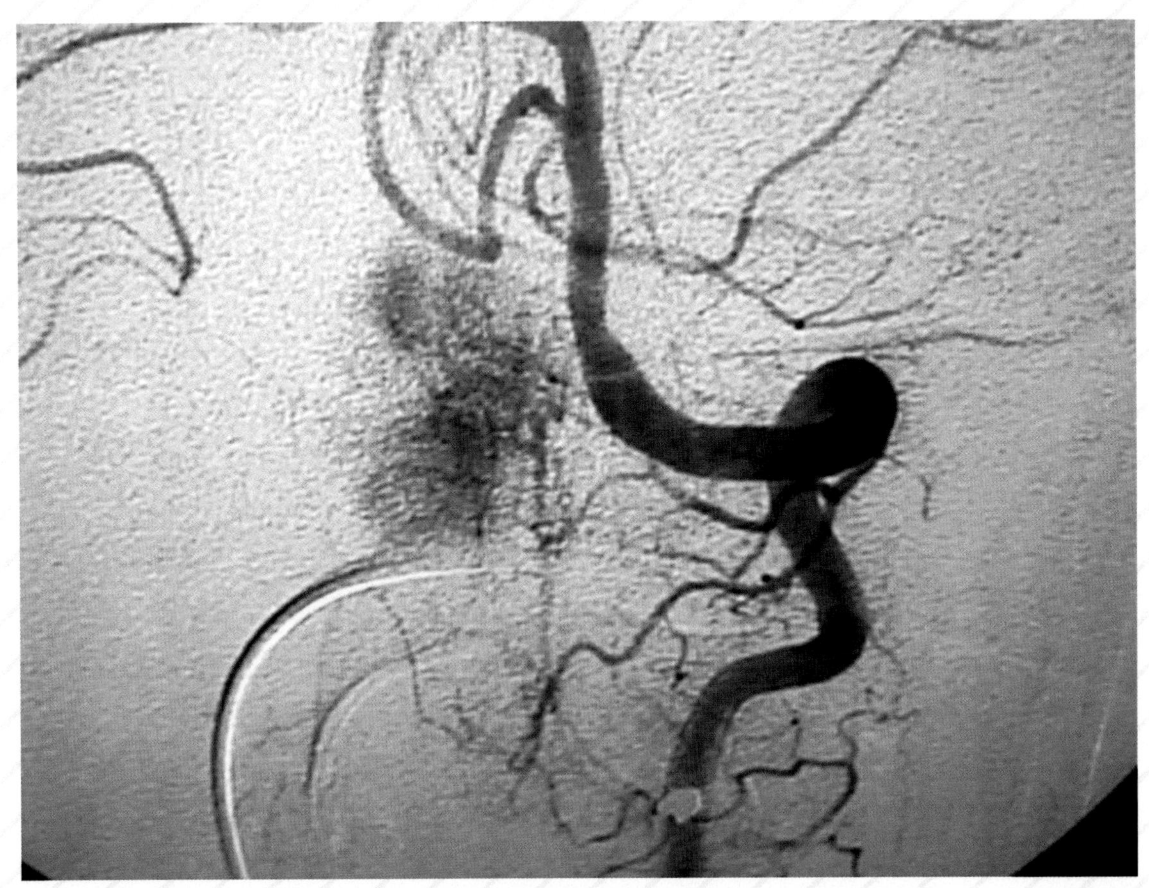

Figure 60. *Angiogram showing the tumor blush and the catheter in place for embolization of the tumor. Reduction of blood supply to the tumor makes surgical removal easier and safer.*

Congenital malformations of the Posterior Fossa

Chiari malformation is a relatively common congenital abnormality of the posterior fossa. Chiari malformations have been divided into types I - IV in order of severity. Type I are for the most part asymptomatic, but when symptomatic leads to occipital headache, made worse with valsalva maneuvers, loss of pain and temperature in the upper trunk and arms (syringomyelia), loss of strength with atrophy of the upper extremities (syrinx), drop attacks, spasticity, balance problems, double or blurred vision (often associated with down-beat nystagmus). Type II is also known as Arnold-Chiari malformation and is associated with a meningomyelocele. Type III usually is accompanied by an encephalocoele at the back of the head with cerebellar or brainstem tissue herniation. Type IV is the most severe with the failure of the cerebellum to develop and other brain malformations are generally present. Syringomyelia and syringobulbia are commonly found in patients with Chiari, Type I. Syringomyelia leads to a typical dissociation of sensation with a loss of pain and temperature over the chest and upper extremities, atrophy of the hands and forearm muscles, long tract signs, loss of bladder and bowel control and chronic pain. As mentioned earlier, many cases of Chiari I are asymptomatic, but when they are symptomatic surgical decompression is used to relieve the pressure from the displaced cerebellar tonsils and to stop progression of the syringomyelia. **(Fig. 61)** is an example of a Chiari I malformation that was symptomatic.

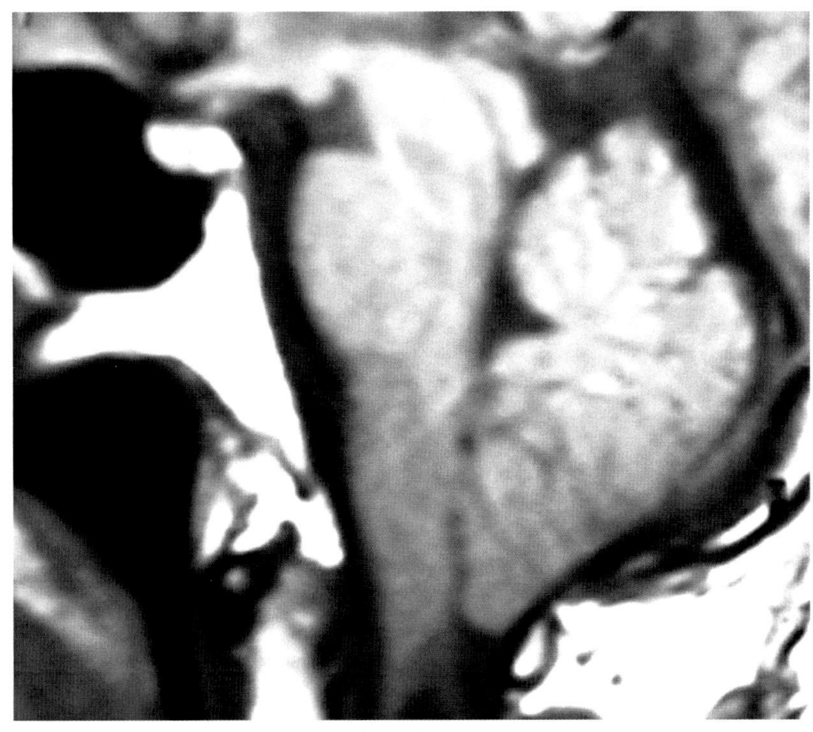

Figure 61. *MRI scan of a symptomatic Chiari I malformation. The cerebellar Tonsils have herniated through the foramen magnum. This patient also had a thoracic syringomyelocoele.*

At the time of surgery to decompress the posterior fossa, the downward herniation of the cerebellar tonsils is clearly seen in **(Fig. 62)**.

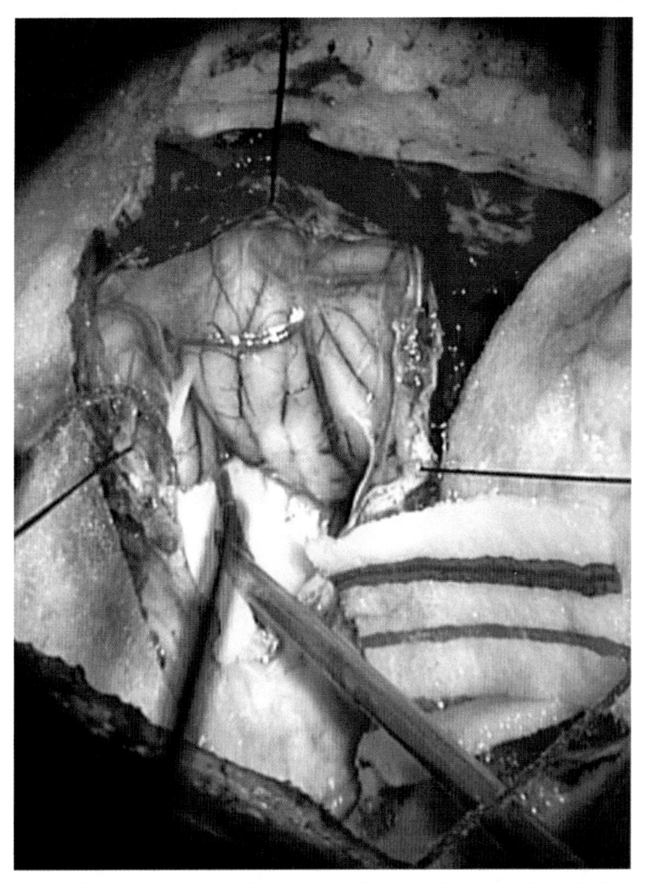

Figure 62. *Intra-operative photograph showing herniated cerebellar tonsils.*

After decompressing the posterior fossa by removing bone, the tips of the cerebellar tonsils are frequently amputated to provide additional space.

(Fig. 63) Demonstrates the amputation of the cerebellar tonsilS.

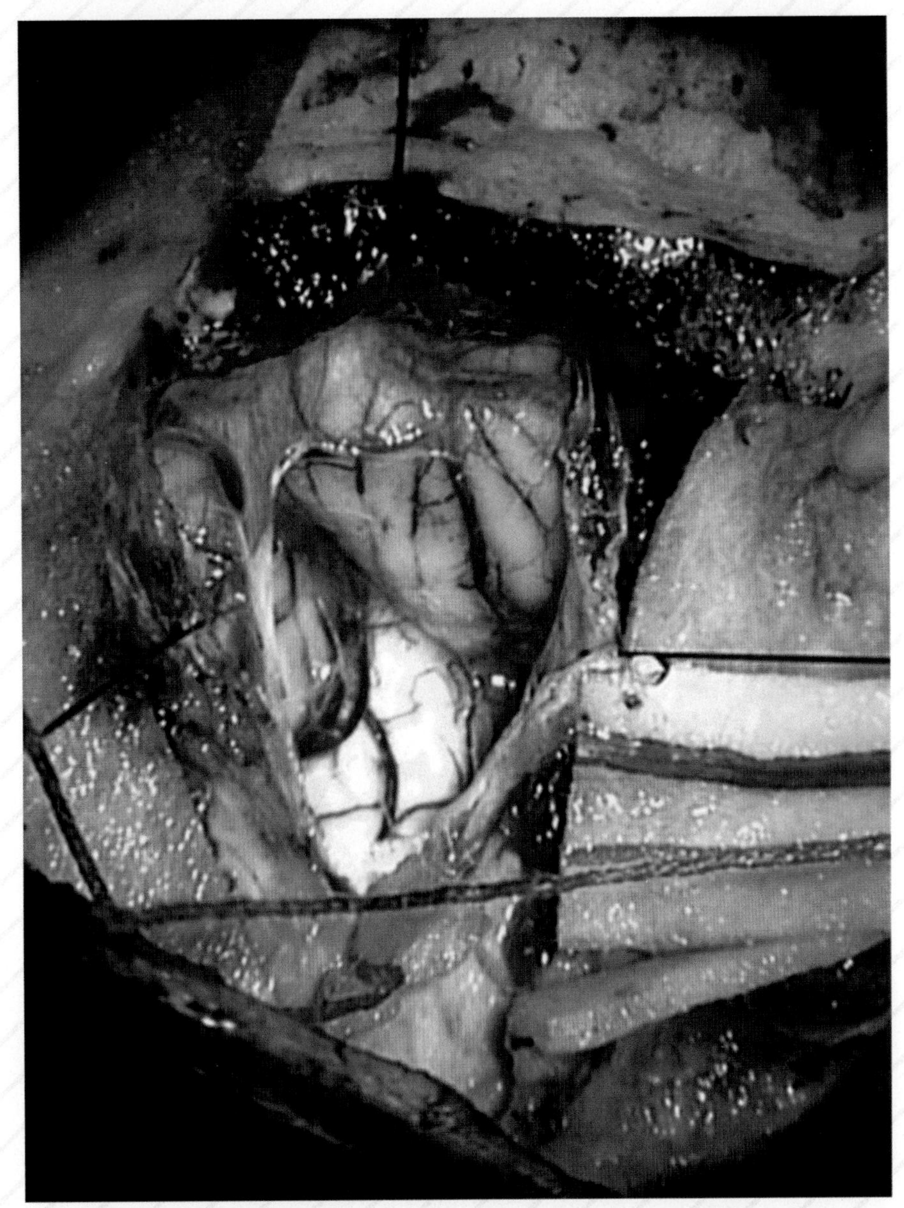

Figure 63. *Intra-operative photo showing amputation of cerebellar tonsils.*

Basilar Invagination.

A number of metabolic bone disorders including After boney decompression and amputation of the cerebellar tonsils a dural patch is frequently used to increase the space for the remaining cerebellum and brainstem.

The final congenital anomaly that may require surgical intervention is that of Basilar Impression. In this condition there is an upper migration of the cervical spine, including the odontoid process through a normal foramen magnum. A similar disorder, although secondary to softening of the bones and ligaments at the base of the skull, is known arheumatoid arthritis, Paget's disease of bone, Osteogenesis imperfecta and Ricketts (Vitamin D deficiency) may lead to this disorder. Symptoms result from the medulla being stretched and angulated over the odontoid process. See **(Fig. 64)** for an example of basilar impression.

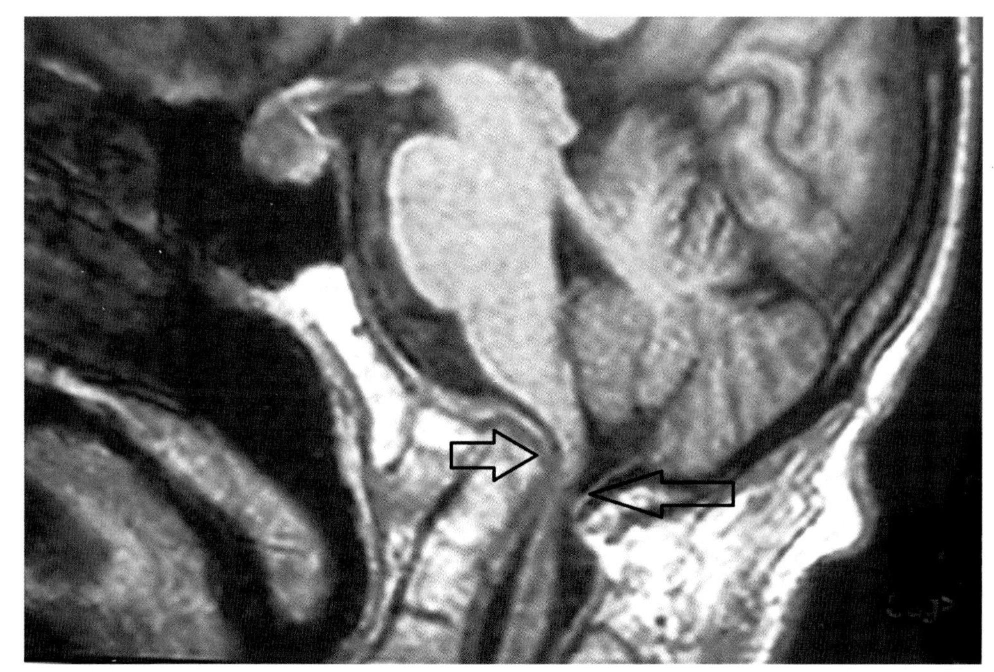

Figure 64. *MRI scan. of a patient with basilar impression. Note the severe compression and angulation of the medulla.*

The goal of surgery is to decompress the posterior fossa and allow the medulla to assume a more normal angle as it enters the spinal canal. **(Fig. 65)** shows the hardware used to stabilize the upper spine and **(Fig. 66)** shows the extension of the hardware to the skull to stabilize the neck.

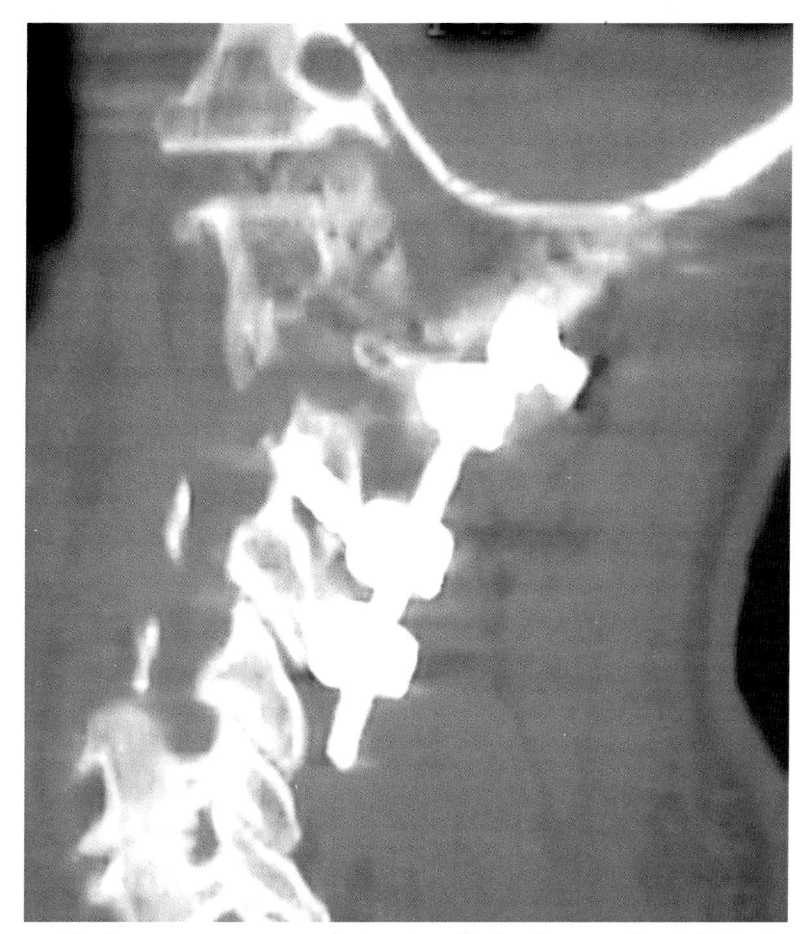

Figure 65. *Intra-operative X-ray showing metallic hardware used to stabilize the spine.*

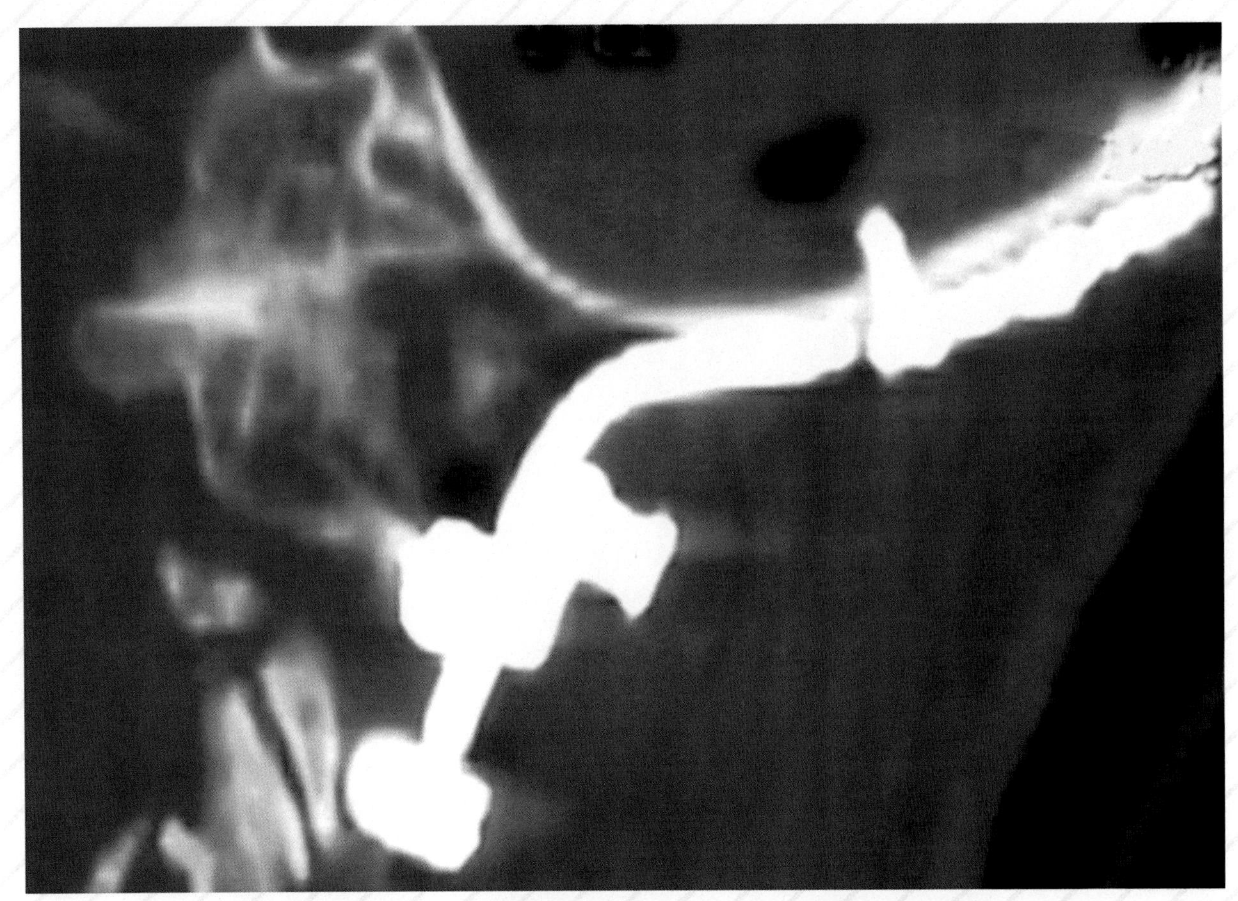

Figure 66. *Intra-operative X-ray showing the hardware used to stabilize the skull to the spine.*

Mobility of the head on the neck is lost with this type of hardware, but it does provide decompression of the problem area and prevents further neurologic damage.

Selected Bibliography

Wirch, A., Farrell, D.F., Grant, F., et. al.. Effect of Papavarine on Intraoperative BAEPs. Am J END Tech 2002; 42: 22-33.

Grant, G.A., Rostomily, R.R., Kim, D.K., et. al.. Delayed Facial Nerve Palsy Following Vestibular Schwannoma Resection. J Neurosurg 2002; 97:93-96.

Sindau, M.P. Microvascular decompression for primary Hemifacial spasm: Importance of intraoperative Neurophysiological monitoring. Acta Neurochir (2005)) 147: 1019-1026.

Ojemann, R.G., Levine, R.A., Montgomery, et. al. Use of Intraoperative auditory evoked potentials to preserve hearing in unilateral acoustic neuroma removal. J. Neurosurg. (1984) 61: 938-948.

Friedman, W.A., Kaplan, B.J., Gravenstein, D., et. al.. Intraoperative brain-stem auditory evoked potentials during posterior fossa microvascular decompression. J. Neurosurg. (1985) 62: 552-557.

Grundy, B.L., Jannetta, P.J., Lina, A., et al. Intraoperative monitoring of brainstem auditory evoked potentials. J. Neurosurg. (1981) 57: 674-677.

Harner, S.G., Daube, J.R., Ebersold, M.J., et. al. Improved preservation of facial nerve function with use of electrical monitoring during removal of acoustic neuromas. Mayo Clin. Proc. (1987) 69: 92-102.

Harner, S.G., Daube, J.R., Beatty, C.W., et. al. Intra-operative monitoring of the facial nerve. Laryngoscope (1988) 98: 209-212.

Moller, A.R., and Jannetta, P.J. Monitoring auditory function during cranial nerve microvascular decompression operations by direct recording from the eighth nerve. J. Neurosurg. (1983) 59: 493-499.

Moller, A.R., and Jennetta, P.J. Preservation of facial function during removal of acoustic neuromas. Use of monopolar constant-voltage stimulation and EMG. J. Neurosurg. (1984) 61: 757-760.

Moller, A.R., and Jannetta, P.J. Microvascular decompression in Hemifacial spasm. Neurosurgery (1985) 16: 612-618.

Radke, R.A., Erwin, C.W., Wilkins, R., et. al. Intraoperative brain-stem auditory evoked potentials: Significant decrease in postoperative deficit. Neurology (1987) 37 (Suppl. 1) :219.

Tator, C.H., and Nedzelski, J.M. Preservation of hearing in patients undergoing excision of acoustic neuromas and other cerebello-pontine angle tumors. J. Neurosurg. (1985) 63: 168-174.

Acioly, A.A., Lisbsch, M., Carvalho, C.H., et. al. Transcranial Electrocortical Stimulation to Monitor the Facial Nerve Motor Function During Cerebellopontine Angle Surgery. Operative Neurosurgery 2. (2010) 66: 354-362.

CHAPTER 6
Spinal Cord Intra-operative Monitoring

SPINAL CORD MONITORING during surgery on the spine is more commonly accomplished than for skull base surgery. Continuous posterior tibial somatosensory evoked potentials (PTN SSEPs) in conjunction with electromyography (EMG) of muscles innervated by segmental nerves in the surgical field are the principle methods used with spinal cord monitoring. This type of monitoring is carried out continuously during the course of the surgery. A wide variety of clinical conditions are monitored including intrinsic and extrinsic tumors, spinal spondylosis, herniated intervertebral discs, descending aortic aneurisms and for the correction of curvatures of the spine. Unmonitored, the relative risks by the procedures include: scoliosis surgery: 0.72 to 4 percent descending aortic artery aneurisms as high as 40 percent and anterior approach to cervical spine : 0.3 percent, but for nerve root 10 percent. The goal of monitoring is to reduce these risks by warning the surgeon of changes in measured functions. Fine wire electrodes are placed in skeletal muscles whose nerve roots are at risk during the surgical procedure. It must be realized that SSEPs are dependent upon the large ascending fiber system confined to the posterior columns of the spinal cord. The posterior columns do not provide any information related to either the lateral small fibe system (pain) or the descending motor system. The PTNSSEPs methods used for intra-operative monitoring are virtually the same as those used in the diagnostic set-up. See Chapter 3 for these methods.

Virtually all spinal surgery is carried out in the prone position. This requires monitoring for positioning of the patient in preparation for surgery.

Intra-operative X-ray is of great value to help the surgeon locate the correct level for surgery. The C-Arm X-ray machine is designed to be moved in and out of the operative field. Its shape allows the X-ray generator to be oriented to the patient to take films of the spine. Frequently, surgical clamps are placed on the patient's operative drapes and the X-ray not only shows the spine, but the placement of the clamps, helping to identify the operative site. Special surgical frames (Stryker or Jackson Tables) are used in spinal surgery. The patient is placed on his or her back and the frame is flipped to make the patient prone.

PTN SSEPs should be carried out with the patient lying on his or her back as a baseline. Once the patient has been placed in the prone position repeated PTN SSEPs should be carried out to make sure that no changes have occurred as a result of positioning. If changes have occurred then they must be corrected by re-positioning of the patient.

A normal PTN SSEPs is demonstrated in **(Figure 1).**

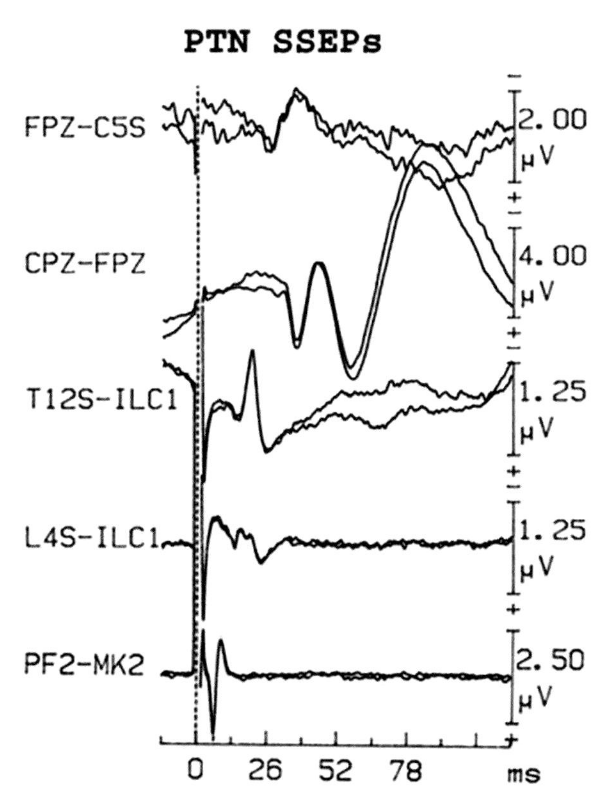

Figure 1. *Normal PTN SSEPs. Responses bottom to top. Popliteal response is a peripheral response obtained at the popliteal fossa. Two examples of the lumbar response are seen at the L4S and T12S levels. The lumbar response is a stationary negative wave that has the highest amplitude at the T12S electrode. At the L4S level a bifid peak is generally seen. The first component is the traveling wave and the second a stationary, small lumbar response. A bipolar derivative, CPZ-FPZ shows the cortical response (P37). Finally the scalp electrode (FPZ) to off head electrode (C5S) provides the subcortical responses, P31 and N34 that originate at the level of the medulla.*

(Fig. 2) represents a direct cortical recording taken at the time of epilepsy surgery. Note the very large difference in amplitude of the P37 when recorded directly from the brain and compared to scalp responses.

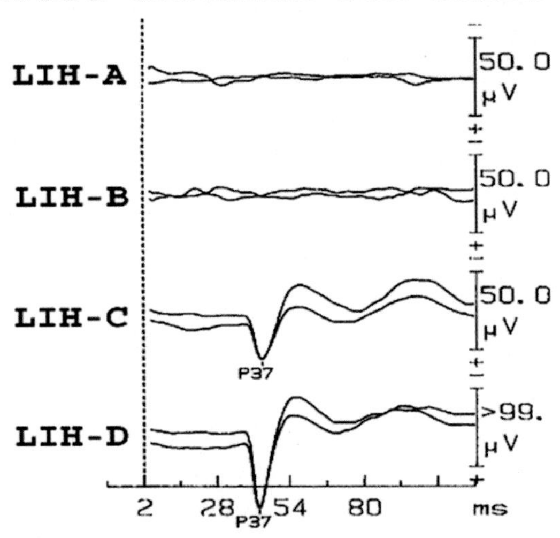

Figure 2. *Recording of direct cortical response (P37) from an Inter-hemispheric electrode strip. LIH D shows the maximal response followed by LIH-C.*

Continuous PTN SSEPs monitoring should be accomplished for positioning and during opening through closing. One must also realize that during surgery if a reduction in wave amplitude or a prolongation in latency of the wave occurs, one need not complete the current test cycle. Once the wave forms are resolved that test can be completed and a new test immediately started. This allows one to rapidly resolve the question if the change noted is due to a surgical cause or if technical factors such as sample contamination by an electrical artifact have occurred. This is by far the most rapid way of collecting the data that might provide the surgeon with a correct reason for the changes in the signal. During critical parts of the surgery this same technique speeds up the collection of data and does not require the time necessary to collect an "n "of 500 to 1000 samples.

Electromyogram (EMG) is of value in identifying nerve root irritation or the spontaneous development of fibrillation potentials when the anterior horn cells are damaged.

Intra-operative Monitoring for Spinal Cord Tumors

By far the most common extrinsic tumor compressing the spinal cord is the solitary schwannoma followed by the meningioma. These two tumors are followed by a wide variety of tumors. The most common intrinsic tumor is the ependymoma followed by different grades of astrocytoma. Intrinsic tumors are approached by splitting the posterior columns in the midline then removing the tumor which is frequently encapsulated. With a proper midline incision no changes in the PTN SSEPs are seen.

During opening there is a great deal of surgical electrical artifacts generated and the monitoring will require a pause of data collection during these periods. During opening, the monitoring team develops a series of baseline studies which are used for comparative data for the remainder of the surgery. In many conditions affecting the spinal cord the PTN SSEPs cortical latency (P37) will be abnormally prolonged at the beginning of surgery. Any additional changes must be reported as they occur.

An example of an abnormal PTN SSEPs is shown in **(Fig. 3)**.

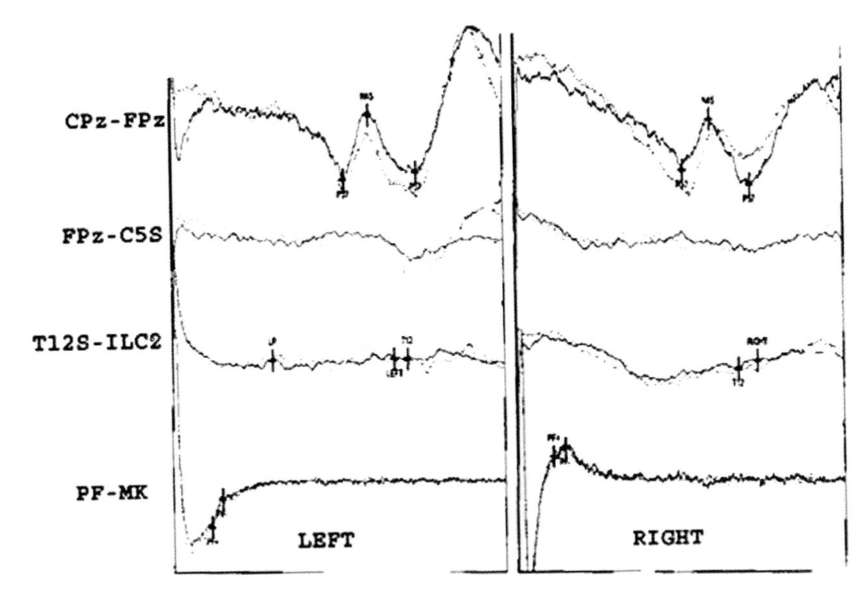

Figure 3. *Bilateral prolongation of the latency for the cortical response (P37) resulting from an intrinsic ependymoma of the spinal cord. About 1/3 of normal individuals have an absent lumbar response. Subcortical responses are also missing in this individual. This is likely due to temporal dispersion of the ascending volley.*

Spinal cord ependymomas are rare tumors that can be found anywhere in the spinal cord. They are intrinsic tumors that originate from the transformation of ependymal cell rests (once the central canal). **(Fig. 4 and Fig. 5)** are examples of a spinal cord ependymoma.

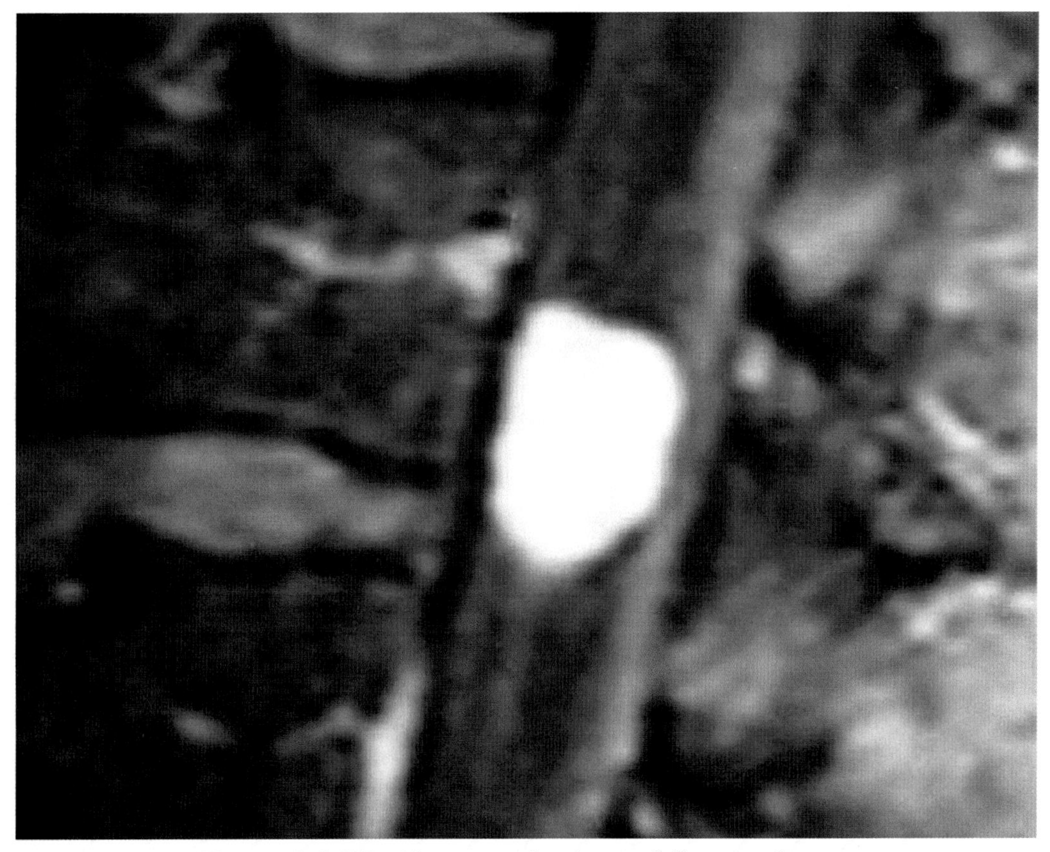

Figure 4. *MRI with contrast of an intramedullary ependymoma.*

Figure 5. *Intra-operative photograph of intramedullary ependymoma. The spinal cord has been entered through a midline incision of the Posterior columns to reveal the tumor. A midline incision did not alter the PTN SSEPs. There is a tendency for this type of tumor to recur.*

Extrinsic tumors compressing the spinal cord include schwannomas and meningiomas. These tumors are benign and their removal should be curative. **(Fig. 6, Fig. 7 and Fig. 8)** are examples of an extrinsic tumor.

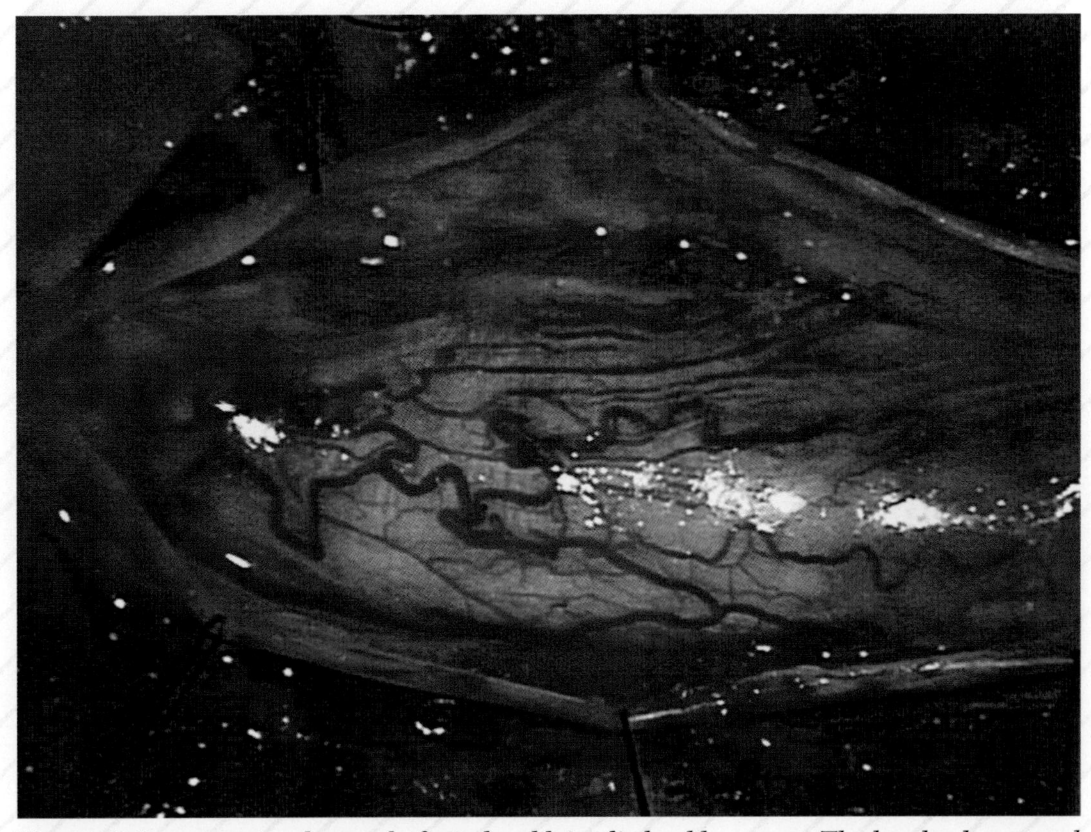

Figure 6. *Intra-operative photograph of spinal cord being displaced by a tumor. The dura has been opened.*

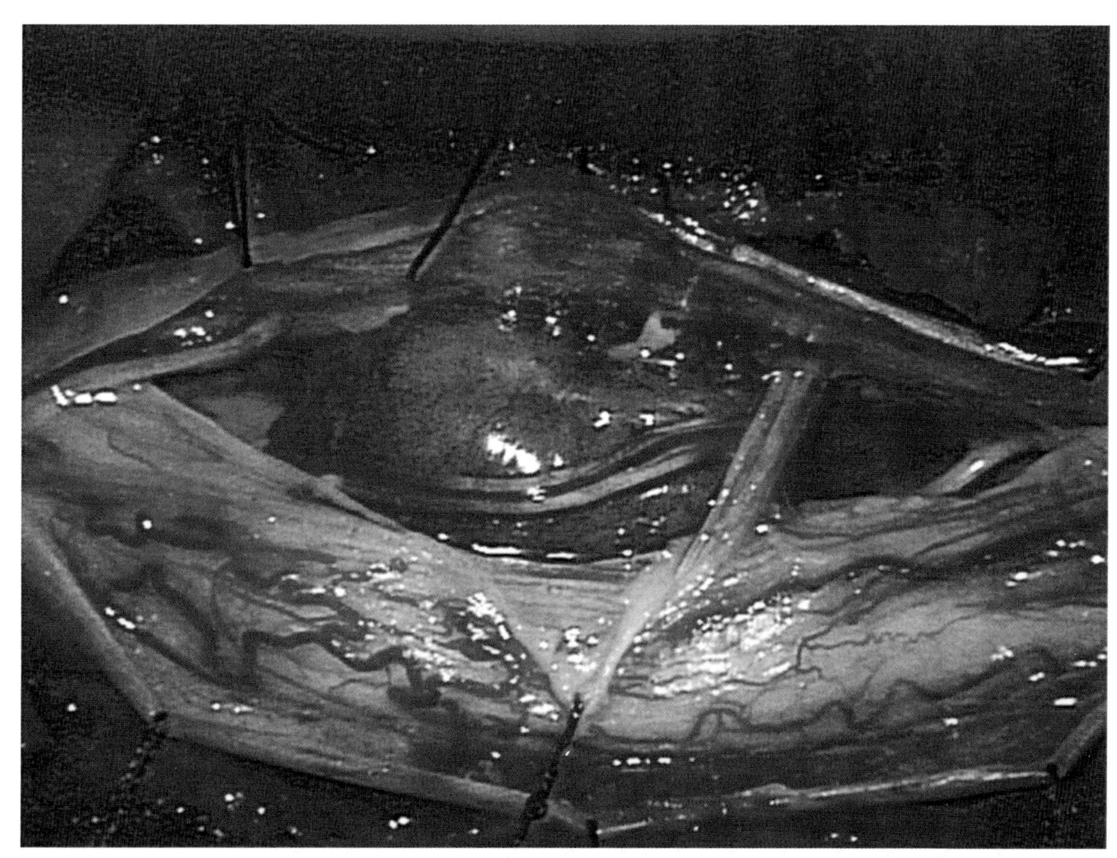

Figure 7. *Intra-operative photograph showing the meningioma after the spinal cord has been rotated to gain access to the tumor. Note the nerve root leaving the spinal cord and crossing the body of the tumor.*

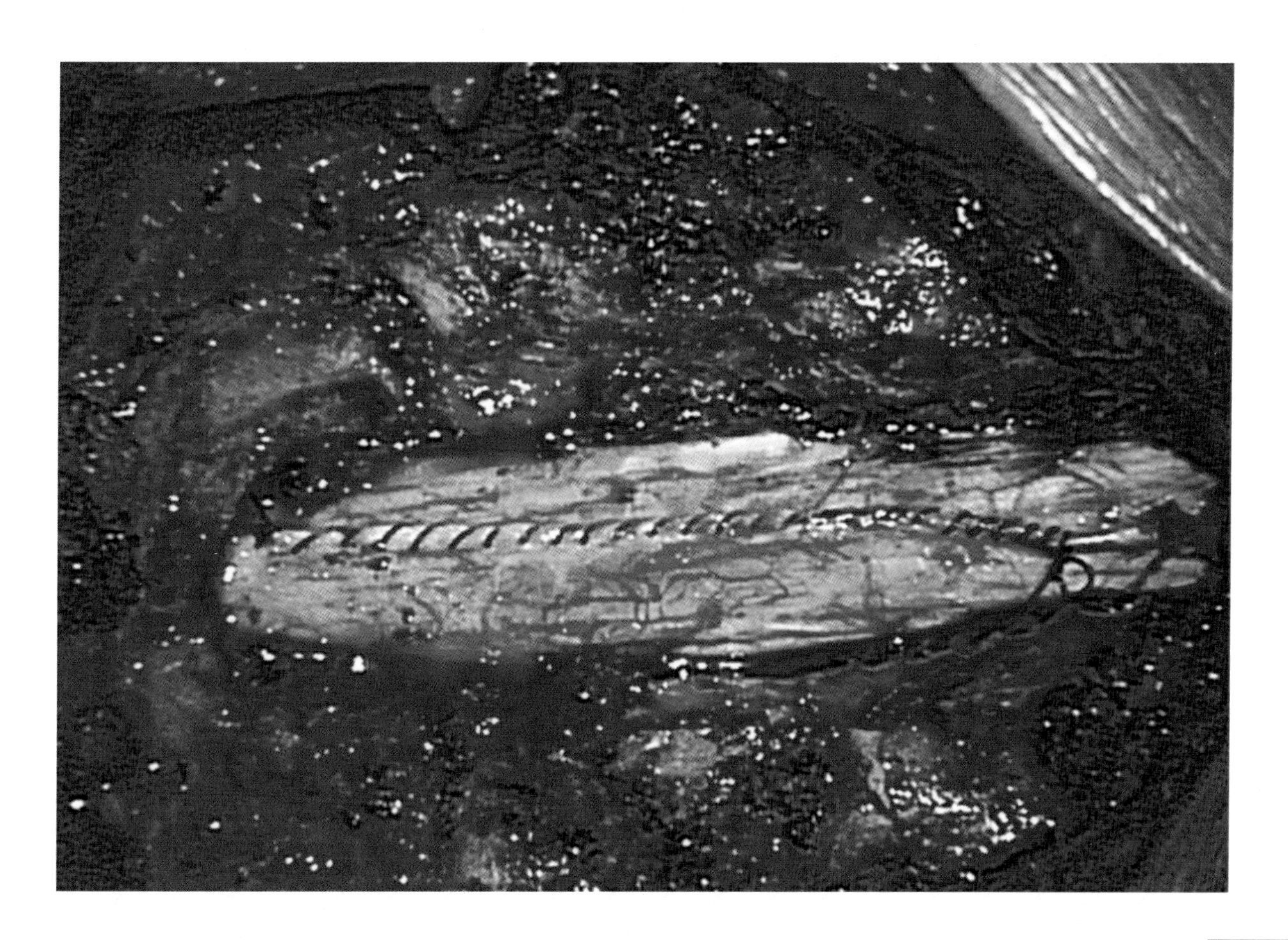

Figure 8. *Intra-operative photograph after the tumor has been removed and the dura closed. Preparations are being made to close the surgical site.*

Tumors of the filum terminale and other tumors affecting the region of the cauda equina pose special problems not seen with tumors of the spinal cord itself. Additional monitoring is needed when the surgery involves this portion of the nervous system. The external anal sphincter is made up of skeletal muscle innervated by nerve fibers from the sacral cord (S2, 3, 4). The normal functioning of this voluntary muscle is critical and preservation of this function allows the patient to remain continent of feces. Damage to these nerves causes incontinence. Two fine wire electrodes are inserted into the external anal sphincter, one on each side. At rest skeletal muscle is electrically silent; any irritation of the nerves to the anal sphincter result is spontaneous EMG activity. When this occurs the surgeon is warned that he is working in a very vital region and great care must be exercised not to damage these critical nerves. **(Fig. 9, Fig. 10, Fig. 11, and Fig. 12)** are examples of a tumor of the filum terminale.

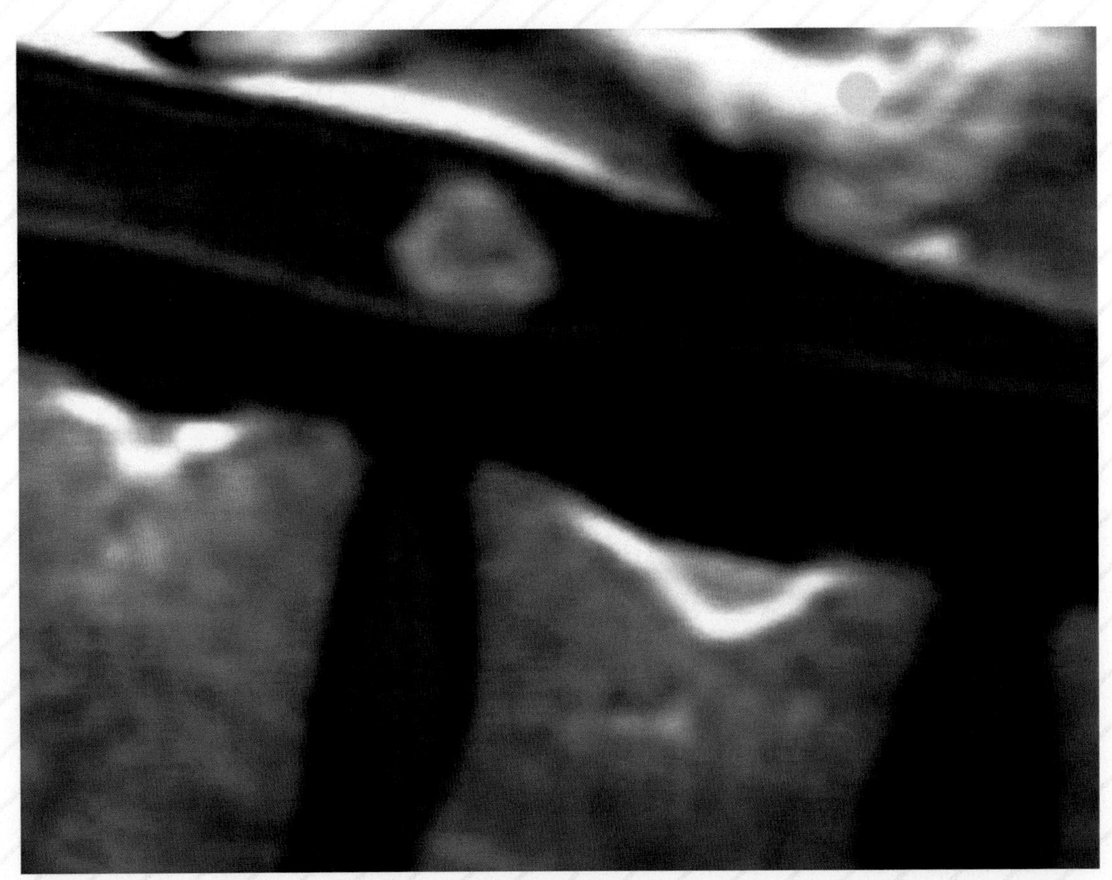

Figure 9. *MRI scan demonstrating a tumor of the filum terminale.*

Figure 10. *Intra-operative photograph showing the beginning of the mobilization of a tumor of the filum. Note that the tumor is surrounded by the nerve Roots that make up the cauda equina.*

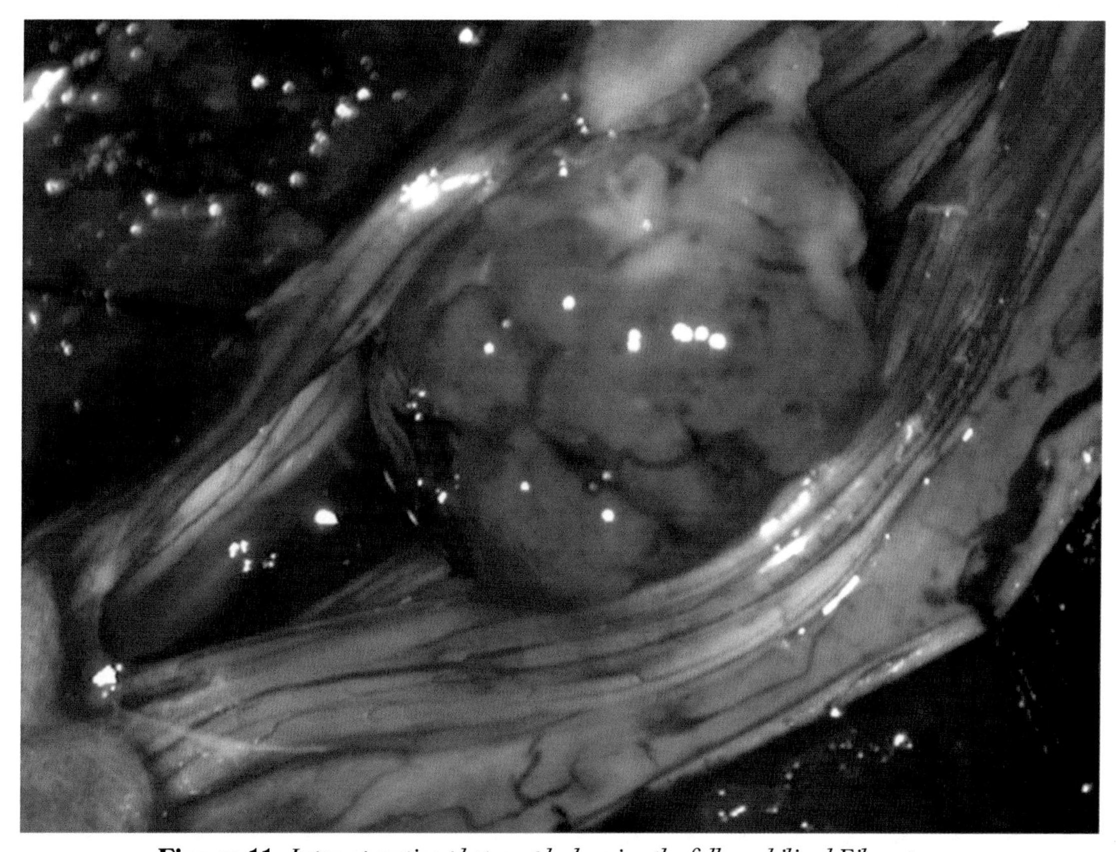

Figure 11. *Intra-operative photograph showing the fully mobilized Filum tumor.*

Figure 12. *Intra-operative photograph showing the operative site after removal of the filum tumor. The nerve roots making up the cauda equine have returned to a more normal position.*

Monitoring during surgery for Intervertebral Disc Disease

Intervertebral Disc Disease

Lumbar Intervertebral disc herniation is a common problem. These disc herniations occur most commonly at the L4-5 or L5-S1 level. This is the area of the lumbar spine that has the most flexibility. Herniated discs occur at higher spinal levels, but are much less common. Disc herniation may occur as an isolated event alone or as part of a more complex degenerative process, lumbar Stenosis. There may be a hereditary component for disc herniation as familial examples are very common. This condition may also be known as sciatica, as the nerve pain radiates down the back of the leg, from the buttocks to the knee... The onset is frequently sudden, especially with certain activities, usually of a minor nature. Heavy lifting generally does not cause herniation as the back and abdominal muscles are prepared for the lift. It is more likely to be some little movement of the back with a twisting motion that leads to acute disc herniation. The pain is aggravated by sitting and may be relieved by standing and lying (especially with the knees elevated to take pressure off the nerve roots. The normal lumbar curvature is frequently lost due to intense muscle spasm of the paraspinal muscles. Either the knee or ankle reflex is diminished or absent depending on the level of herniation. Sensory abnormalities may also occur. Straight leg raising causes increased pain at the site of injury. Conservative management with rest, muscle relaxants and mild analgesics for up to a month is the primary treatment. About 70-80 percent of patients will respond to this treatment and not require surgery. If there is early motor weakness or if the person has not responded well to conservative therapy, then surgery may be necessary. It should be remembered that a herniated lumbo-sacral disc does not cause spinal cord compression; therefore no upper motor neuron findings should be present. The spinal cord ends at about the spinal level of L2 and herniated discs at lower levels impinge upon the nerve roots as they head for their spinal exit foramina. **(Fig. 13)** is an example of a herniated intervertebral disc.

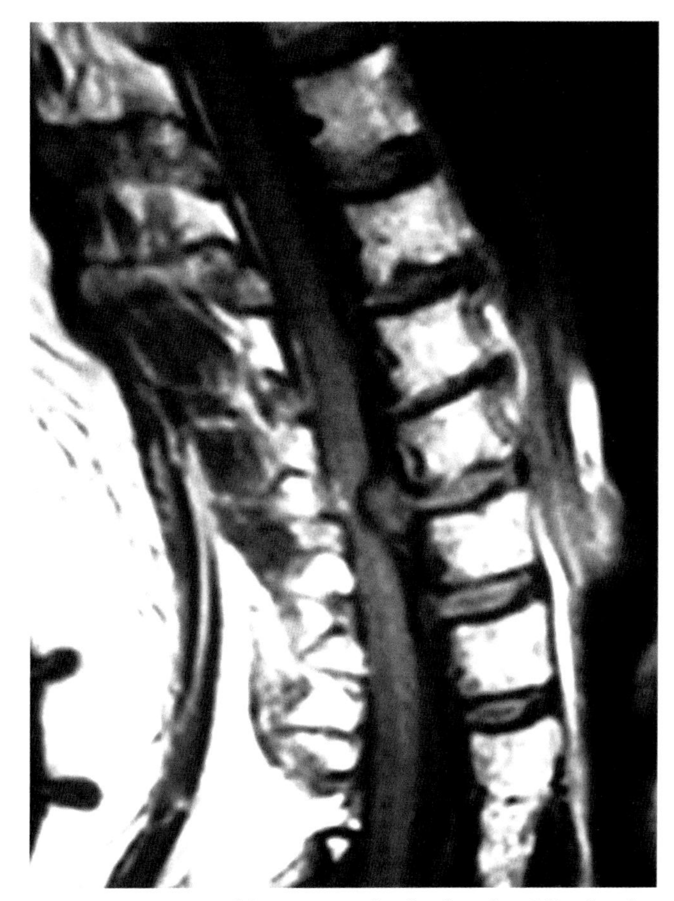

Figure 13. *MRI Scan demonstrating a central herniation of a lumbar disc. The thecal sac which contains the cauda Equina is compressed. Note, the potential space between the vertebral bodies and the dural sac (theca).*

Surgical removal of a symptomatic herniated disc generally results in a cure. With a simple disc removal spinal instrumentation is generally not needed.

However, there may be a predisposition for future disc herniation as the removal at one level puts additional stresses on the disc spaces above and below the operated level. Bulging of a disc seen on a MRI scan does not generally cause symptoms and should not be a reason for surgical intervention in a patient with chronic low back pain. These unfortunate patients will continue to have pain even when they have undergone back surgery. Only those patients that have an anatomic reason for the low back pain respond to surgical correction.

Cervical Disc Disease

The most common cervical disc herniation occurs at the C5-6 level followed by the C6-7 level. Like the lumbar spine herniation occur at the level of greatest spine flexibility. Unlike the lumbar spinal area, cervical disc herniation may impact the spinal cord. Central disc herniation compresses the cervical cord, whereas, lateral herniation affect the cervical roots. Cervical disc disease is less likely to resolve spontaneously than lumbar disc disease. Cervical disc herniation is more likely to have early surgery than lumbar disc disease. The operative approach for single level and possibly 2 level disc herniation is from an anterior approach. This type of surgery has less morbidity than a posterior approach. Recovery is more rapid and there is much less post-operative pain. Surgery from the anterior approach means that the disc removal occurs through the disc space. Following removal of the disc, the disc space is maintained by either a bone plug obtained from the iliac crest or by an external plate. **(Fig. 14)** demonstrates a metal plate inserted to stabilize the operative site.

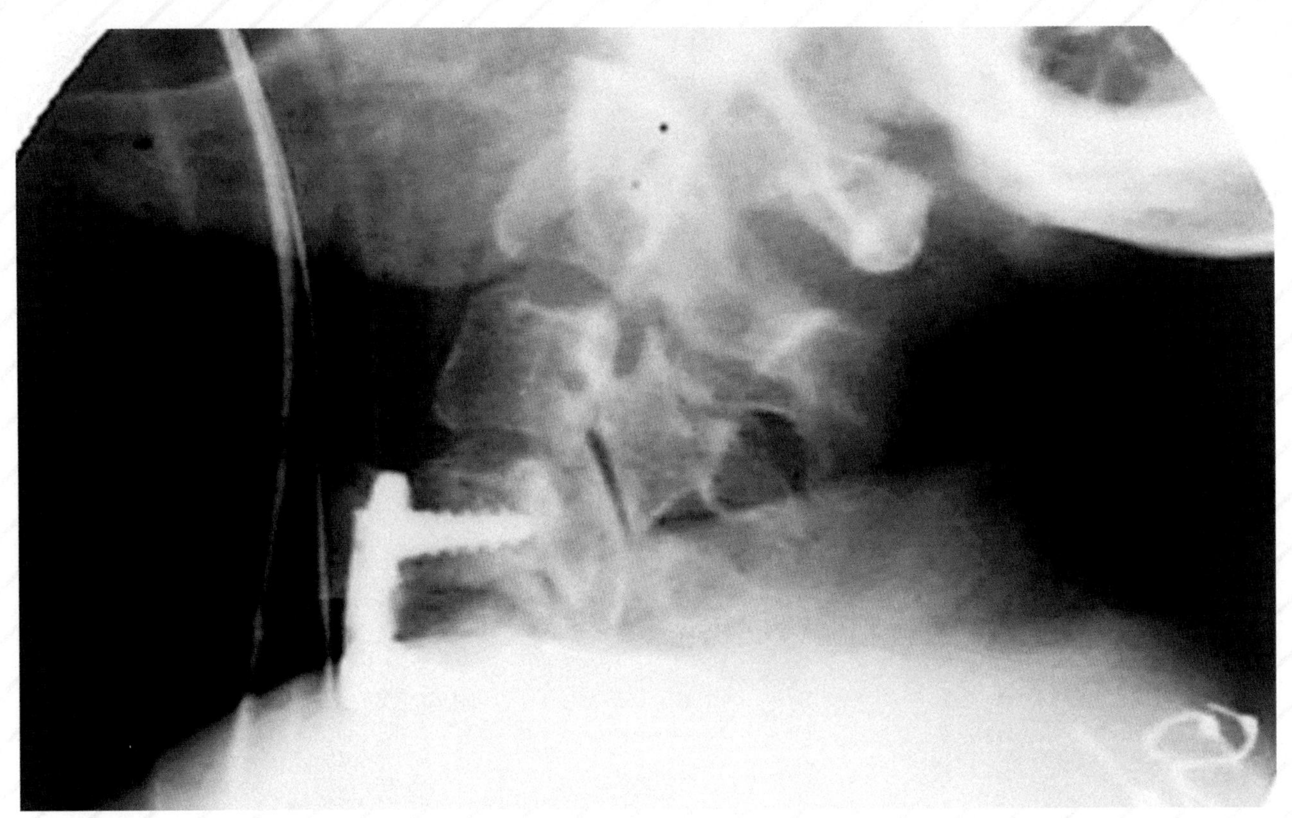

Figure 14. *Lateral X-ray demonstrating plate position following cervical disc removal from an anterior approach.*

Fusion of the operative site precludes any subluxation of the spinal canal following surgery.

Cervical and Lumbar Spondylosis and other Vertebral Disorders

Degenerative disorders of the spine are many times more common than tumors in and around the spinal canal. These disorders take on a number of forms and the surgical therapy varies depending on the nature of the abnormality.

Cervical Spondylosis

Cervical Spondylosis is a very common disorder. This disorder results from arthritis of the cervical spine with the development of boneyspurs and bars developing around the intervertebral disc spaces. The spurs and bars are formed from new bone formation and calcification of the anterior spinal ligament. The pathophysiology is more complicated than just compression, as spinal vascular insufficiency tends to occur at the site.

Cervical spondylosis is by far the most common cause of a "gait disturbance" in middle aged and older individuals. Patients generally have a spastic gait and associated neurologic findings are present. Dermatomal SEPs are of value in diagnosis of the level of involvement, but no value during surgery. Dermatomal SEPs are labile to levels of consciousness and unreliable under anesthesia so are not used during the surgical procedure. PTN SSEPs generally show a prolonged P37 latency and are used in conjunction with MN SSEPs and Trans-cranial electrical motor evoked potentials (TceMEOs) for monitoring during decompressive surgery. If the cervical spine is unstable, subluxation of the vertebral bodies occurs with flexion and extension, and then spine stabilization is necessary. Cervical stenosis occurs in individuals who have a congenitally narrowed spinal canal. The most common form or surgical therapy is to remove the posterior spine structures (laminectomy). This is frequently carried out at multiple levels. Laminectomy effectively decompresses the cervical spinal cord and prevents additional damage. At times, some neurologic improvement occurs, but cannot be relied upon. If the cervical spine is unstable spinal instrumentation and fusion may be necessary. If nerve roots are compressed, decompression of the lateral recess may be necessary. The combination of spinal cord compression and lateral radicular disease in the cervical region is uncommon, estimated to be about 1 in 20 individuals.

Lumbar Stenosis

Stenosis of the lumbo-sacral spine is moderately common in the middle- aged to elderly patient. The compression is of nerve roots of the cauda equine and the symptoms are quite different than seen in cervical stenosis. By far the most common presenting symptom is that of "spinal claudication". After walking a short distance the legs become heavy and weak requiring rest. Only a very short rest is necessary to restore function, but the symptoms return after additional walking. Claudication of the lower extremities has two major causes, 1) lumbar stenosis and 2) vascular insufficiency to the legs. The later is usually identified by the absence of pulses at the ankle. If necessary, vascular insufficiency can be confirmed by flow Doppler examination of the arteries of the legs. Again, there is a very strong congenital component to the development of lumbar stenosis in those individuals who become symptomatic. These individuals are generally born with a small spinal canal or have minor congenital anomalies of the spine. With a small spinal canal the amount of arthritic change necessary to cause symptoms is less than those who have a large spinal canal. Unlike, cervical stenosis the combination of central compression of the cauda equina and lateral radiculopathy is common. Pre-operative dermatomal SEPs are of value in the identification of which lateral recesses will need to be decompressed to eliminate nerve root compression and eliminate the resulting pain syndrome. At times lumbar stenosis leads to a very unstable spine which can be identified with flexion and extension lateral X-rays. Treatment consists of surgical laminectomy, frequently at multiple levels, lateral recess decompression and at times spine stabilization with instrumentation and spinal fusion. **(Fig. 15).**

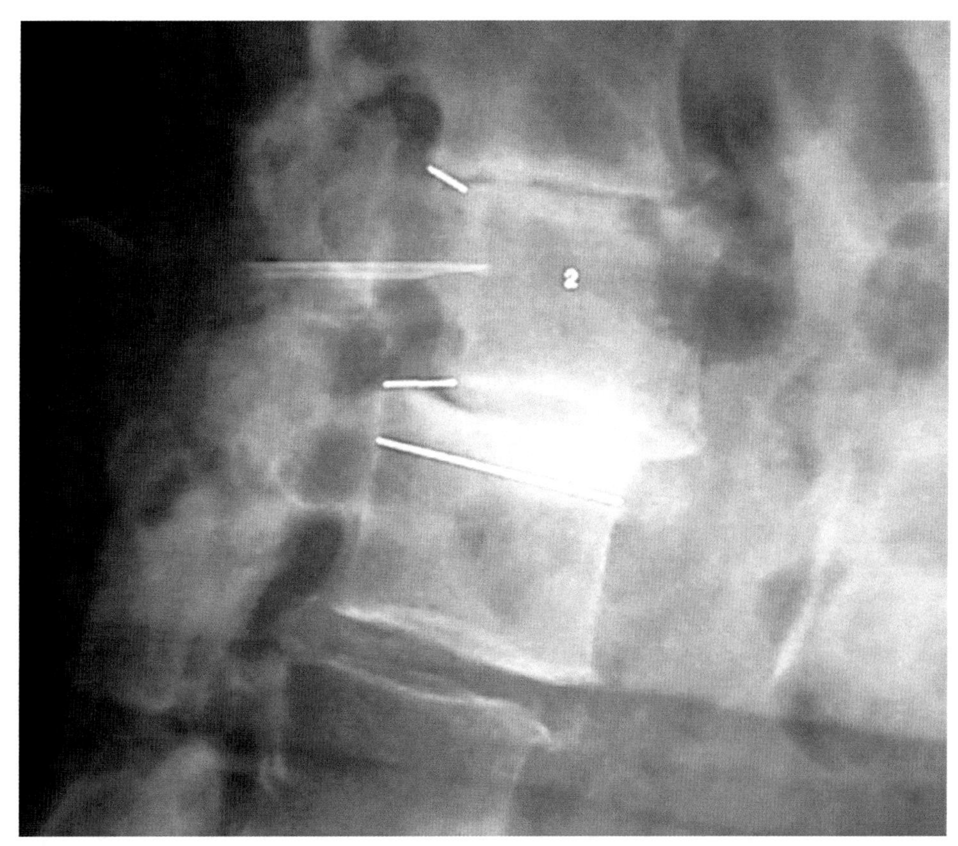

Figure 15. *Lateral X-ray showing anterior subluxation of vertebral body. Changes like this frequently will require spinal fusion to prevent additional Subluxation.*

An additional example of lumbar stenosis with anterior subluxation is shown in **(Fig. 16).**

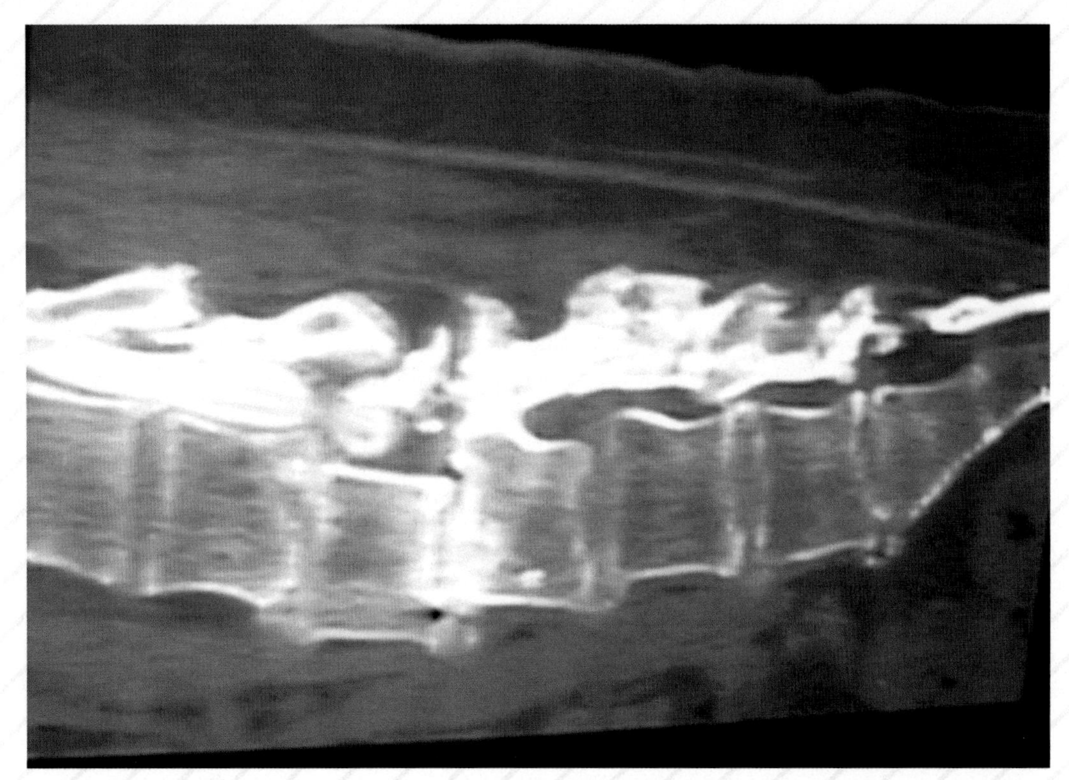

Figure 16. *Lateral X-ray showing an anterior subluxation of the lumbar Vertebrae. This subluxation is at about ¼ of a vertebral body. This spine is unstable and will require spinal instrumentation and probable fusion.*

Spinal instrumentation is a relativly new development and is rapidly advancing. New titaniam, titanian alloy, or stainless steel devises such as screws, plates, rods, woven mesh, hooks and threaded interbody cages allow the surgeon to create individual theraputic devises for each individual case. These devises are frequently combined with bone grafts of various sorts to stabilize the spine. In earlier times, bone grafts were used exclusively. The patient would then spend prolonged periods in a plaster spica giving the graft time to fuse the spine. Currently, the vast majority of patients undergoing this type of surgery will not require prolonged periods of external immobilization as the internal hardware provides the stability. It is now technically possible to surgically remove a complete vertebral body and processes in a patient with an osteosarcoma of the vertebral body and replace the vertebra with a prosthetic metal cage and bone graft.

Intraoperative Monitoring during Scoliosis Surgery

Scoliosis or curvature of the spine remains a relatively common medical problem. In the past, poliomyelitis with its assymetric muscle paralysis was a common cause of this condition. In fact, much of the surgery to correct severe scoliosis was devised during the later stages of the existence of polio. Following the development of polio vaccines (killed vaccine and the live modified polio vaccine) polio virtually disappeared from the United States. Currently, the vast majority of patient's with scoliosis do not have a known cause, i.e., "ideopathic". Cases of scoliosis are frequently identified during school screenings for scoliosis. Once identified, these children and adolescents need to be followed on a regular basis to determine if the condition is stable, improving, or worsening. Many cases will stablize and require no special therapy, others will progress slowly and may require bracing to slow down or prevent additional worsening of the curvature. These braces are generally worn 23 hour a day until growth of the spine stops. In that group of patients where the curvature increases rapidly and does not respond to brace therapy or if the curvature exceeds 45 degrees then surgical intervention is likely necessary to prevent further progression. A patient with a curvature of 45 degrees or more is likely to have an acceleration of the deformity. Severe curvatures are likely to compromise the shape of the chest wall as the vertebral bodies rotate with advancing curvature leading to impaired respiration. Excessive respiratory insufficiency can lead to right heart failure (cor pulmonale) in these individuals.

Symptomatic cases of scoliosis, those resulting from a known neuromuscular disorder such as X-linked muscular dystrophy, juvenile spinal muscular atrophy, cerebral palsy and other inherited neuromuscular disorders tend to have a more rapid course than the typical ideopathic form.

Surgical spinal instrumentation started in the 1950's with the introduction of the Harrington rod. This rod was introduced to partially straighten the spine and to prevent further advances in the curvature. It is the oldest and most proven form of spinal instrumentation. Harrington rods were used when the curvature of the spine is 60 degrees or greater. In general, the lower lumbar spine is not included in the field of the Harrington rod allowing some flexibility of that part of the spine. The Harrington rod is no longer used, nor is the Luque rod. The Luque rod was a custom length rod applied to the most affected part of the spine. This rod was fixed to each segment of the spine giving this rod a higher risk for neurologic damage than the Harrington rod. The Luque rod was used less frequently than the Harrington rod in the surgical management of scoliosis. In certain individual cases combinations of a single Harrington rod on the concave side of the curvature and a Laque rod on the convex have been used. Newer types of instrumentation include the use of rods cross-linked with hooks to realign the spine and redistribute the physical forces of the curved spine. These newer rods are generally named after the manufacturor of the rod such as "Synthes" rods. A combination of hooks and screws are used with the rod to stabilize and reduce the curvature of the spine.

Surgical correction of scoliosis requires continuous monitoring, especially when traction to straighten the spine is accomplished. Any change in the PTN latency (> 10 percent) or reduction in the amplitude of the cortical response (P37) of >greater than 50 percent requires warning the surgeon of the change. This warning allows the surgeon to make changes in the surgical procedure to correct what caused the changes in the evoked responses. The somatosensory system which can be continuously monitored during surgery is used as a substitute for monitoring the motor system which cannot be done continuously **(Fig. 17).**

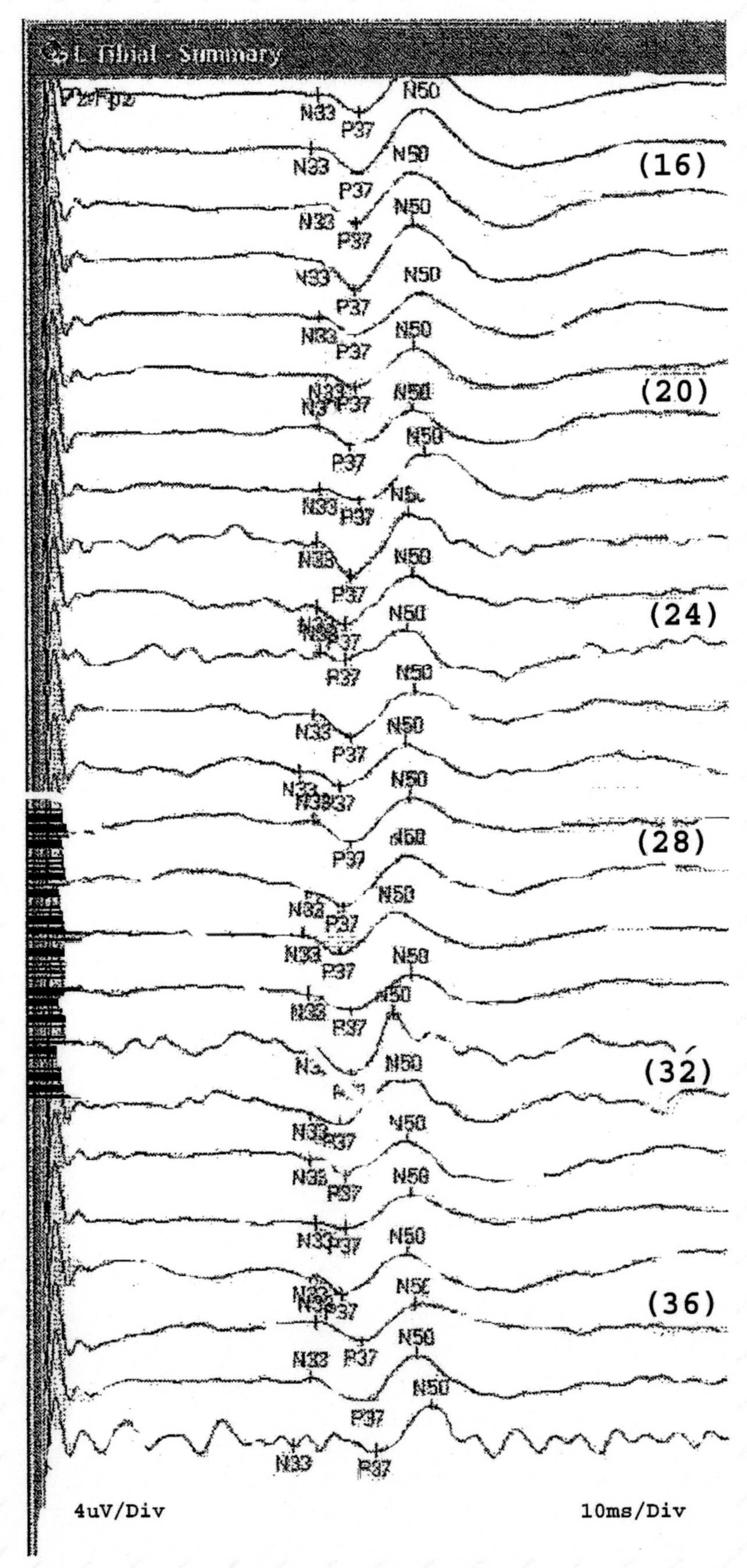

Figure 17. *Series of PTN SSEPs taken during spinal surgery. Note there has been only minor changes in latency (2-3 msec) during the course of the surgery.*

Before the advent of trans-scalp cranial electrical motor evoked potential (TceMEP) stimulation, to determine if the motor system was intact a "wake-up test" which was the "gold standard" for testing motor function had to be performed. This procedure had a significant risk as the anesthesia had to be stopped, the patient awakened in the midst of surgery and the surgical team would have the patient voluntarily move their lower extremities. This is no longer necessary in the vast majority of cases as the integrety of the cortical spinal tracts can be obtained by electrical stimulation of the cortex to induce motor activity. Currently,trans scalp motor stimulation is more successful for the upper extremities than the lower. This is because of the location of the hand region of the Rolandic cortex is more accessible than the foot, which may be deep in the interhemispheric fissure. This has been overcome by stimulating between C3 and C4 electrode positions. This produces stimulation of the upper and lower extremities simultaneously. (See Section on Transcranial Motor Stimulation).

Transcranial Motor Stimulation

Over the past few years transcranial motor stimulation (TcMEP) has been added to the regimen of monitoring spinal surgery. Theoretically, this type of stimulation should be very helpful in preventing damage to the cortical spinal tracts and reduce the incidence of paralysis. This procedure has not yet been standardized, so there are no Guidelines for its use. Different laboratories use various parameters to determine the presnece of an abnormality. In some laboratories a reduction of 50 percent of the recorded CMAP amplitude, after increasing the stimulus intensity by 20 percent is considered abnormal, while other laboratories use a reduction of 80 percent. Recording methods have not yet been standardized. Care must be taken not to induce seizures, bites or burns with this procedure. Contraindications include: patients with known seizure disorders, implanted pacemekers and other cardiac implants. The two major recording types are 1) surface electrodes placed over the muscles of interest (TcMEP). The second is the use of two needle electrodes (bipolar) inserted directly into the muscle of interest, recording a Campound Motor Action Potential (TcCMAPs). Neither of these recording systems have been shown to be superior to the other and both have certain shortcomings. These shortcomings are primarily based on the fact that supra-threshold stimulation cannot used for fear of causing an epileptic seizure. Peripheral nerve stimulation and nerve stimulation for somatosensory evoked potentials uses suprathreshold levels of current as the stimulus. This means that every nerve fiber is contributing to the response and that an increase in stimulus intensity will not alter the amplitude of the response. By using subthreshold stimuli, variation in response is likely as a variable number of cortical neurons and their fibers are stimulated with each trial, resulting in variable responses. The recorded evoked waveforms from an entire muscle is likely to vary greatly and make it difficult to determine an abnormality. With TcCMAPs, there should be less variation as each motor action potential (motor unit) is an all or none phenomen, although some variation might be seen if the recording electrodes (bipolar) are too far apart so that differenct components are being included in the compound motor action potentials. All of this means that reproducible responses are unlikley and that serial responses may be of different amplitudes. TcMEPs suffer from a very high false positive rate of approximately 12 percent. One group reported that a reduction of greater that 80 percent in the response occurred and that in 2 patients a wake-up test showed no neurologic abnormality and in 3 additional cases surgery was completed and the patients did not have a post-operation neurologic deficit. Somatosensory evoked potentials did not change in this same patient population. This group found that patient obesity and prolonged surgery accounted for these false positives. If a response is obtained one assumes the cortical spinal tracts are intact. In children under 7 years of age it is nearly impossible to obtain motor responses even when directly stimulating the cerebral cortex. Complex ramping schedules have been developed to stimulate the Rolandic cortex of Children, but is not routinely available. Adults, who are under anesthesia, are also very difficult to obtain a motor response even with direct cortical stimulation. A current of up to 20 milliamps applied directly to the cortex may not elicit a response in the anesthesized patient. In those surgeries for epilepsy and brain tumor removal when the patient is awakened for cortical mapping procedures, motor responses are generally elicited with a low current of 2-5 milliamps. The other major problem with transcranial motor stimulation is that it cannot be used continuously because of the potential movement of the patient.

Originally, this test was only used when the surgeon asked for a motor check therefore, did not provide anticipatory information that might allow the surgeon to intervene and prevent injury to the spinal cord. However, more recently a regular schedule is set up with the surgeon so that TcCMAPs are carried out every 10 minutes during the surgery allowing adequate time for the surgeon to correct any potential problems. A sample of TcMEPs is shown in **(Fig. 18)**. While the PTN SSEPs carried out during the same surgery is shown in **(Fig. 19)**.

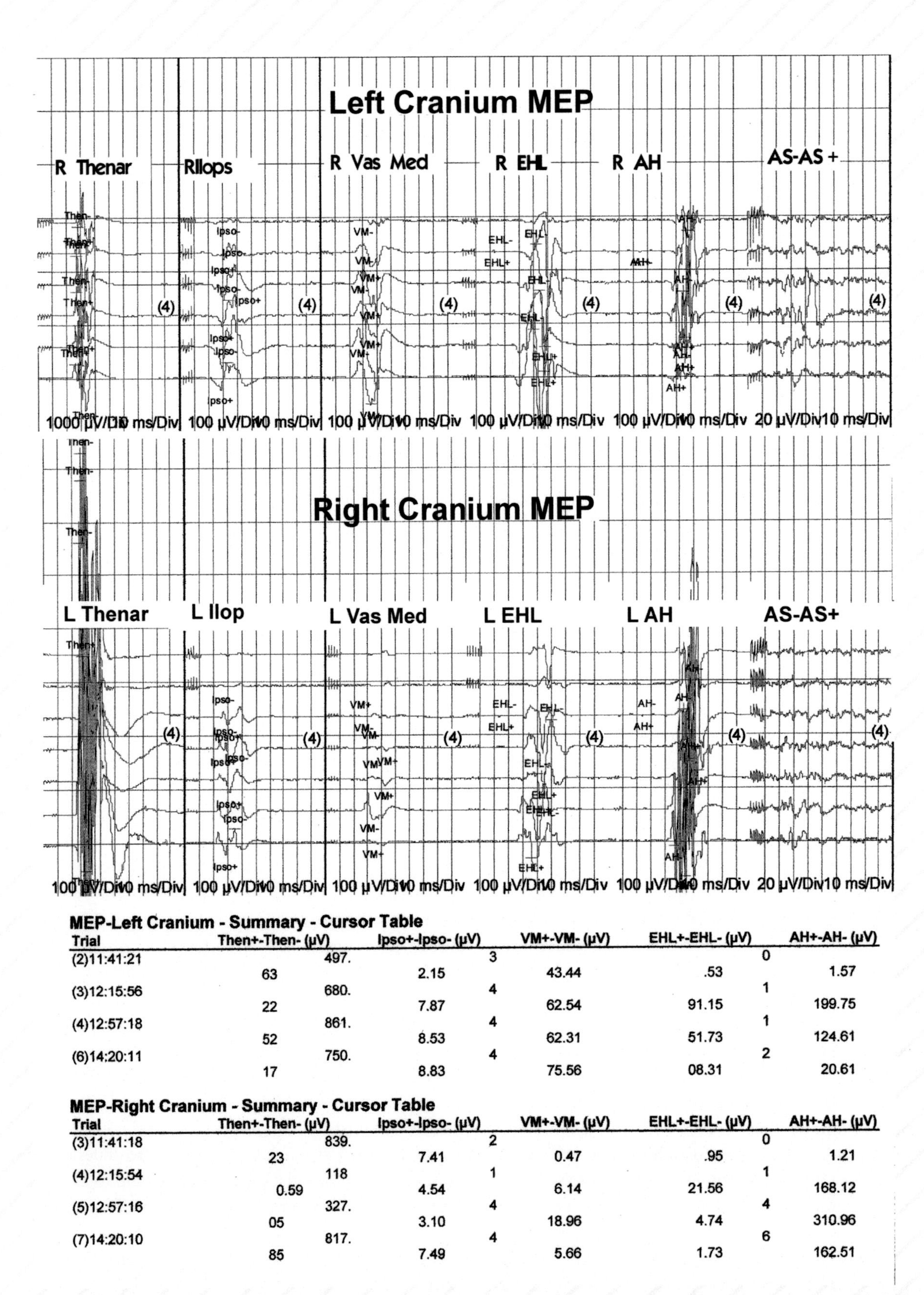

MEP-Left Cranium - Summary - Cursor Table

Trial	Then+-Then- (µV)	Ipso+-Ipso- (µV)	VM+-VM- (µV)	EHL+-EHL- (µV)	AH+-AH- (µV)
(2)11:41:21	497.	3		0	
	63	2.15	43.44	.53	1.57
(3)12:15:56	680.	4		1	
	22	7.87	62.54	91.15	199.75
(4)12:57:18	861.	4		1	
	52	8.53	62.31	51.73	124.61
(6)14:20:11	750.	4		2	
	17	8.83	75.56	08.31	20.61

MEP-Right Cranium - Summary - Cursor Table

Trial	Then+-Then- (µV)	Ipso+-Ipso- (µV)	VM+-VM- (µV)	EHL+-EHL- (µV)	AH+-AH- (µV)
(3)11:41:18	839.	2		0	
	23	7.41	0.47	.95	1.21
(4)12:15:54	118	1		1	
	0.59	4.54	6.14	21.56	168.12
(5)12:57:16	327.	4		4	
	05	3.10	18.96	4.74	310.96
(7)14:20:10	817.	4		6	
	85	7.49	5.66	1.73	162.51

Figure 18. *Summary of MEP stimulation during the course of spinal surgery. Bipolar recordings are accomplished in the following muscles: Thenor, Iliopsoas, Vastuc Medialis, Extensor Hallicus longus, Abductor hallicus, and Anal Sphincter. Note the wide fluctions in the amplitude of the responses.*

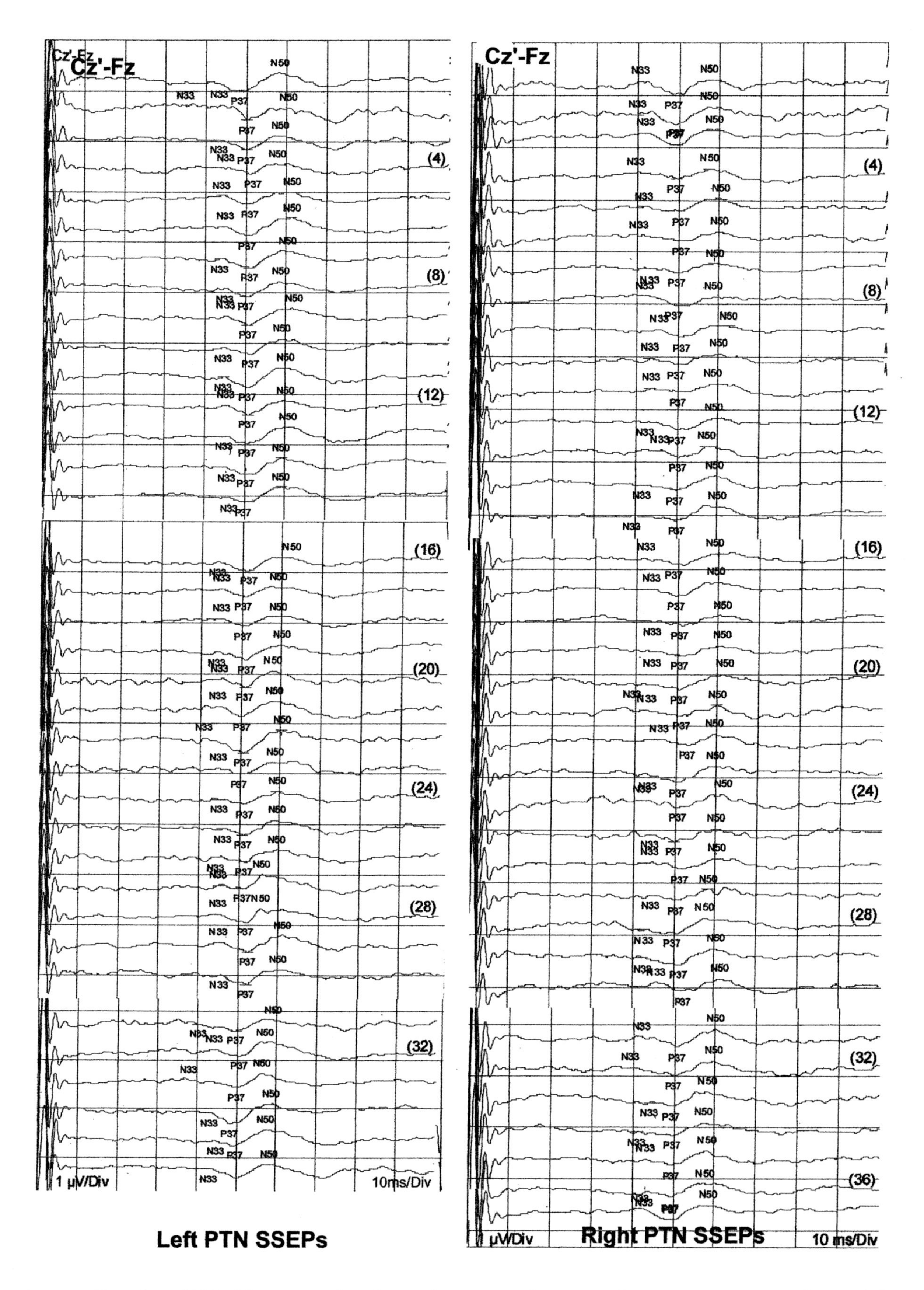

Figure 19. *Bilateral PTN SSEPs carried out during the same spinal surgery as shown in Figure 18. Note the recorded waveforms are very consistent, both in latency and amplitude.*

Currently, the available cortical stimulators provide constant voltage stimulation to the cortex. There is a great need for the development of a constant current stimulator which would be much safer to use than a constant voltage stimulator. With a constant voltage stimulator large changes in impedence could lead to excessive current being delivered to the tissues including the cerebral cortex with resulting burns. Inspite of the short-comings of the current type of stimulator noted above, its use continues to expand. Fine needle stimulating electrodes (commercially available) are placed in the subdermal tissues of the scalp to reduce the variable impedence caused at the skin of the scalp. The motor cortex is stimulated through the skull, and dura mater using the direct voltage stimulator. Transcranial stimulation is performed with a high frequency (500 Hz) electrical train stimulator. The voltage is generally set at 400 V with a square wave pulse width of 0.2-0.7 milliseconds and provided either as a single pulse or as a train of 5, stimulation rate of 0.1 Hz, but up to 1200V is sometimes used. The fine needle electrodes are inserted subdermally at C3 and C4 for the upper extremities and at C1 and C2 for the lower extremeties. It has been reported that responses are obtained by recording compound motor actiion potentials from the muscles of interest, generally the abductor hallicus for the upper extremity and the tibialis anterior for the lower extremity. Intermittenly, on a pre-arranged schedule following notification of the surgeon, the cortex is stimulated and the integrety of the motor system tested. Because of the sub-threshold stimulation some variation in the amplitude of the motor response is to be expected, but any significant response provides evidence that the cortical spinal tracts are functioning. Variation in response may vary greatly as one group reported a temporary reduction of 50 percent in 44 percent of their cases and 37 percent needed an increase in stimulation intensity of more than 20 percent to correct for the reduced amplitudes. This group reported a sensitivity of 88 percent and a specificity of only 56 percent in predicting postoperative deterioration of motor function. In England and other parts of the world transcranial magnetic stimulation (ŕTMS) is now being used for stimulating Rolandic cortex (not approved by FDA for use in the United States). Magnetic transcranial stimulation uses a transcranial stimulator that has a maximal output of 2.5 Tesla (T). Either a single pulse or a train of pulses with an interpulse interval of 2 ms is used. The magnetic field is created in the stimulator with a high-energy capacitor bank. Triggering the stimulator causes a large pulsed current (5000 A, 150 µS) through a circular copper coil to induce a magnetic field. The hand held coil is placed over the appropriate cortical motor area. The magnetic field induces electric current in the cerebral cortex causing the cortical neurons to fire. Stimulation is thought to occur through the activation of cortical interneurons. Typically, the CMAP generated by magnetic stimulation is of a lower amplitude that that caused by transcranial electrical stimulation. Transcranial magnetic stimulation is also more sensitive to anesthetic agents than is transcranial electrical stimulation. At this time it is not clear whether one form of stimulation is superior to the other, but magnetic stimulation appears to be safer from the standpoint of potential scalp burns that could be induced by electrical stimulation. One major advantage of magnetic stimulation is that the patient can undergo pre-operative studies in the laboratory to detect whether the MEP is present and to obtain CMAP amplitutes to compare during surgery. This is impossible with electrical stimulation because of the intense pain generated by the procedure. Transcranial electrical motor stimulation can only be used in an anesthesized patient.

Transcranial motor stimulation is now being used to prevent spinal cord damage during Thoracoabdominal Aortic Surgery. A small percentage of patients (4-13 percent) undergoing surgery for abdominal aneurism are at risk of spinal cord infarction because of the anatomic variation in radicular arteries supplying the anterior spinal artery. A few individuals do not have adquate intercostal radicular arteries to provide blood if the Great Artery of Adamkiewicz is sacrificed (generally present at the L2 level) resulting in infarction of the anterior portion of the spinal cord. The infarctions are frequently in the mid-thoracic region of the spinal cord where the descending components of the anterior spinal artery and the lack of intercostal radicular arteries fail to provide adequate collateral circulation. In one large study, 8 patients out of 72 had a CMAP amplitude drop of greater than 75 percent with all having neurologic impairment after completion of surgery. An additional 21 patients showed transient ischemic changes leading to intervention by repurfusion of intercostal or lumbar arteries, aortic declampling and correction of hypotension. Like somatosensory evoked potentials, TcMEPs disappeared with deep hypothermia but returned to normal with rewarming of the patients. It is likely that newer surgical techniques such as reattachment of intercostal arteries and the great lumbar artery of Adamkiewicz and in reducing severe hypotension will reduce the incidence of spinal cord infarction in Thoracoabdominal aortic aneurism surgery.

Selected Bibliography

Celesia, G.G., Allison, T., Bodis-Wollner, I., et. al. American electroencephalgraphic society committee on guidelines for intraoperative monitoring of sensory evoked potentials. Guideline 11: guidelines for intraoperative monitoring of sensory evoked potentials. J. Clinc Neurophysiol (1994) 11: 77-87

Hall, J.E., Levine, C.R., and Sudhir, K.G. Intraoperative awakening to monitor spinal cord function. Description of procedure and report of three cases. J. Bone Joint Surgery (1978) 60 (4): 533-536

Balzar, J.R., Rose, R.D., Welch, W.C., e. al. Simultaneous somatosensory evoked potential and electromyographic recordings during lumbosacral decompression and instrumentation. Neurosurgery (1998) 42 (6): 1318-1325

Lesser, R. P., Raudzens, P. Luders, H., et. al.. Postoperative neurological deficits may occur despite unchanged intraoperative somatosensory evoked potentials. Ann Neurol (1986) 19:22-25

Chatrian, G.E., Berger, M.S. and Wirch, A.L. Discrepancy between intraoperative SSEP's and postoperative function. Case Report. J. Neurosurg (1988) 69:450-454

Nuwer, M.R. and Dawson, E. Intraoperative evoked potential monitoring of the spinal cord: enhanced stability of cortical recordings. Electroencephalogr Clin Neurophysiol (1984) 59: 318-327

Nuwer, M.R. Spinal Cord monitoring with somatosensory techniques. J Clin Neurophysiol (1998) 15 (3): 183-193

Macri, S., De Monte, A., Greggi, T, et. al. Intra-operative spinal cord monitoring in orthopedics. Spinal Cord (2000) 38 (3): 133-139

Weiss, D.S. Spinal cord and nerve root monitoring during surgical treatment of lumbar stenosis. Clin Orthop (2001) 384: 82-100

Thornton, C., and Sharpe, R.M. Evoked responses in anesthesia. Br J Anesth (1998) 81 (5): 771-781.

Lubitz, S.E., Keith, R.W., and Crawford, A.H. Intraoperative experience with neuromotor evoked potentials. A review of 60 consecutive cases. Spine (1999) 24 (19): 2030-2034

Hsu, B., Cree, A.K., lagopoulos, J., et. al. Transcranial Motor-Evoked Potentials combined with response recording through compound muscle action potential as the sole modality of spinal cord monitoring in spinal deformity surgery. Spine (2008) 33 (10) 1100-1106

Kim, D.H., Zaremski, J., Kwon, B., et. al. Risk factors for false positive transcranial motor evoked potential monitoring during surgical treatment of cervical myelopathy. Spine (2007) 32 (26) 3041-3046

Langeloo, D.D., Lelivelt, A., Journee, H., et. al. Transcranial electrical motor-evoked potential monitoring during surgery for spinal deformity: A study of 145 patients. Spine (2003) 28 (10) 1043-1050

Ubags, L.H., Kalkman, C.J., Been, H.D., et. al. A Comparison of myogenic motor evoked responses to electrical and Magnetic transcranial stimulation during nitrous oxide/opioid anesthesia. Anesth. Analg. (1999) 88: 568-572

Hargreaves, S.J., and Watt, J.W.H. Intravenous anesthesia and repetitive transcranial magnetic stimulation monitoring in spinal column surgery. British Journal of Anaesthesia (2004) 94: 70-73

Kawanishi, Y., Munakata, H., Matsumori, M., et. al. Usefulness of transcranial motor evoked potentials during thoracoabdominal aortic surgery. Ann Thorac Surg. (2007) 83: 456-561

CHAPTER 7
Peripheral Nerve Monitoring

NEUROPHYSIOLOGICAL MONITORING DURING peripheral nerve surgery can provide essential information and significantly augment the surgical treatment of peripheral nerve injuries and disease. In addition to the conventional use of neurophysiological techniques to protect functional nerves from injury during surgery, electrophysiological techniques may be used as diagnostic tools to assist and guide the surgeon's actions and to determine the most effective course of surgical treatment. Peripheral nerve monitoring may prevent inadvertent injury, guide dissection of affected nerves, identify peripheral nerves, and evaluate the function of peripheral nerve.

The electrophysiological tools used include SEPs, MEPs, spontaneous EMG, triggered EMG, and nerve conduction techniques. Several or all of these techniques may be interwoven together to address specific aspects of peripheral nerve surgery. Because these techniques are used not only to monitor nerve function but also to provide the surgeon with diagnostic information, it is imperative there is a continuous dialogue among the members of the surgical team, particularly among the neurophysiologist, the surgeon, and the anesthesiologist, in order to obtain optimal results. All members of the team should understand the anatomy, the nature of the pathology, and the intent and implication of neurophysiological data.

Pre surgical investigations should detail the functional and anatomical status of the peripheral nerves in question. Needless to say, a thorough clinical examination is paramount in defining function. Electrodiagnostic studies, such as EMG, SEPs, nerve conduction studies, F waves and H reflexes, can further elucidate the functional status of nerves. Anatomical information from CT myelogram, MRI, magnetic resonance neurography (MRN), and angiograms complete the diagnostic picture and provide the information needed to determine when surgery should be done, what the surgical approach will be, and what form neuromonitoring will take.

In general, peripheral nerve procedures that may be monitored include:

- Neurolysis or nerve graft following traumatic injury. Because of the abnormal tissue due to regrowth and scarring, nerves may be displaced or absent. Intraoperative electrodiagnostics may be used to locate and define functional nervous tissue. A classic application, pioneered by Kline and colleagues is the electrodiagnostic analysis of a neuroma in continuity with nerve conduction techniques to determine if nerve conduction exists through the neuroma. If so, a neurolysis would be advisable; if not, then a nerve graft could be considered.

- Excision of tumors. Neuromonitoring is used to protect functional nerve fibers from injury, usually by monitoring SEPs and MEPs. Additionally and most importantly, neuromonitoring, using recording of CMAPs to intrafield stimulation, can identify the location of nerve fibers in tissue in the surgical field. For encapsulated tumors, such as schwannomas or neurofibromas, the nerve fibers are generally spread over the capsule leaving little space free of nerve fibers. Intrafield stimulation can be used to map the course of nerve fibers across the tumor capsule and define an area free of nerve fibers for the incision of the capsule. Similarly, intrafield stimulation can be used to identify and track the course of nerve bundles through tissue so the nerve bundle may be dissected away from tissue and preserved.
- Resection of cysts: Neuromonitoring to protect nerve and to identify and locate nerve during cyst resection.
- Nerve entrapment: In cases of extreme entrapment with lots of scarring, Intrafield stimulation techniques may be used to locate a nerve in the scarred regions then together with SEPs or MEPs monitor the nerve to protect it from injury.

Most procedures involve the brachial plexus, lumbar plexus, or the major nerves of the arm and leg (median, ulnar, radial, peroneal, and tibial) but neuromonitoring techniques can also be applied to less commonly considered nerves such as facial, axillary, femoral, and posterior interosseus. In any type of case, the preoperative information should guide the surgeon's and the neurophysiologists planning and execution of the surgery and its monitoring.

Electrodiagnostic Techniques:

SSEPs, MEPs, triggered EMG, spontaneous EMG, and nerve conduction studies are techniques that may be used interchangeably or continuously throughout a procedure in either a monitoring function to protect from injury or a diagnostic function to provide the surgeon with guiding information.

Somatosensory Evoked Potentials (SSEPs):

SSEPs are an electrodiagnostic test of sensory function. Most peripheral nerves are mixed sensory and motor nerves. There are a few exceptions that are pure sensory nerves, such as sural and saphenous nerves, and some that are pure motor nerves, such as the posterior interosseus nerve. For those with sensory fibers, SSEPs provide information about the integrity of the sensory components of these nerves. Specifically, SSEPs reflect the activity of the larger diameter fibers involving the sensations of proprioception, vibration, and fine touch and do not reflect activity in smaller diameter fibers, associated with pain and temperature sensations. For the spinal cord this may be an issue because the posterior column pathway is separate from the anterolateral column pathway as well as separate from the motor pathways. With peripheral nerves, however, fascicles of large and small diameter sensory fibers are intermingled and both are intermingled with fascicles of motor fibers and so SSEPs are well suited to test the functionality of these nerves. It should be noted that SSEPs to more pure sensory nerves such as sural, saphenous, and radial nerves are usually smaller amplitude than those to the major mixed nerves and may be more difficult to record due to their small size. The reason for this difference may be that the sensory nerves contain proportionally smaller diameter Group II fibers rather than larger diameter Group I fibers as are found in mixed nerves, and as a result project a smaller electrical field around the nerves when activated. This is likely the reason for smaller peripheral nerve responses and possibly for smaller spinal cord responses with pure sensory nerve stimulation. For cortical level responses, the reason for smaller responses may also be that the smaller fibers project to a different neuronal pool that forms a smaller overall electrical field than that formed from larger fibers.

SSEPs to Nerve Stimulation Outside of the Surgical Field:

SSEPs to stimulation of nerves outside of the surgical field done in a conventional manner may serve two purposes. One is to monitor any nerves that may be at risk during the surgery and a second is to provide diagnostic information about functionality of nerves in question.

Stimulation parameters for conventional SSEPs are similar to clinical EP technique, i.e. stimulus duration of 0.1 to 0.2 msec and a stimulation intensity of 1.5 times motor twitch threshold, approximately 25-30 mA for upper extremities and 35-50 mA for lower extremities. For sensory nerves an intensity of 25-30 mA will usually suffice. Stimulation rates should be commensurate with the type of recording done. For SEPs, cortical level recordings must be done with low frequency stimulation rates of 3-5/sec, whereas if the recording is at subcortical or peripheral nerve level, then stimulation rates may be increased to 10-15/sec.

SSEPs, both clinically and intra-operatively, are typically recorded from peripheral, spinal cord, brainstem, and cortical levels. Because the surgical field is peripheral, proximal recordings at spinal cord, brainstem, or cortical levels are "after" the surgical field, reflect neural activity passing through the surgical field, and serve to assess and monitor nerve function in the surgical field. It can be preferable to use spinal cord and brainstem level recordings (as opposed to cortical level recordings) because these signals will follow higher rates of stimulation more faithfully, allowing stimulation rates of 10 to 15/sec and significantly shortening the averaging time. However, they can be more difficult to record due to lower signal to noise ratio, and in that case cortical level recordings are more appropriate.

Stimulating electrodes placed outside of the surgical field may be either surface or 0.5 inch bare needle electrodes affixed with tape or even stapled in position. For convenience and ease of application, needle electrodes are most commonly used. In some cases, the entire limb is sterile and within the surgical field. The technologist then must mark the placement sites prior to skin preparation, direct the sterile needle insertion by the surgeon, ensure placement of leads off the surgical field, and connect the leads to connector pods under the surgical table.

The selection of what nerves to stimulate for SSEPs depends on the focus of the surgery. As a rule, SSEPs should be done for any major nerve that is under potential surgical manipulation. Sometimes, nerves that could suffer collateral injury should also be considered. For example, a tumor on the median nerve would obviously require monitoring of median nerve SSEPs, but if the ulnar nerve is in close proximity and could receive pressure from retraction or other manipulation, then ulnar nerve SSEPs should also be part of the protocol. So for upper extremities, SSEPs typically can be done for median, ulnar, and radial nerve stimulation at the wrist. On occasion, stimulation of the musculocutaneous nerve at the elbow may also be considered. For the lower extremities, SSEPs are typically done to tibial nerve stimulation at the ankle, peroneal nerve stimulation at the fibular head (common peroneal) or peroneal nerve stimulation at the ankle (superficial peroneal nerve stimulation). On occasion, stimulation of the saphenous nerve at the medial knee or the sural nerve at the lateral ankle may be considered.

An example showing the use of cortically recorded SSEPs to monitor function of the median and ulnar nerve during resection of a schwannoma of the median nerve in the forearm is found in Figure 1. Cortical level recordings were used because subcortical level signals were too small and not easily discerned. This case illustrates the use of SSEPs in detecting a compromise of the median nerve due to positioning of the arm in preparation for tumor resection. The patient's arm was positioned on an arm board at 90 degrees to the body with the elbow fully extended and the arm flat on the arm board. The baseline runs showed good median nerve SSEPs to stimulation at the wrist. As seen in Figure 1, after 10-15 minutes, the median nerve SSEPs clearly became absent at a time prior to incision. The control ulnar nerve SSEPs were unchanged pointing to a focal change in the median nerve. The surgeon was alerted and a towel roll was placed under the wrist to flex the elbow a few degrees. Within minutes the median nerve SSEPs recovered fully and remained without incident for the remainder of the procedure. The case illustrates ability of SSEPs to detect inadvertent compromise to the median nerve related to arm positioning. Most likely, the insult was caused by interruption of circulation in the forearm due to the fully extended elbow. **(Fig. 1).**

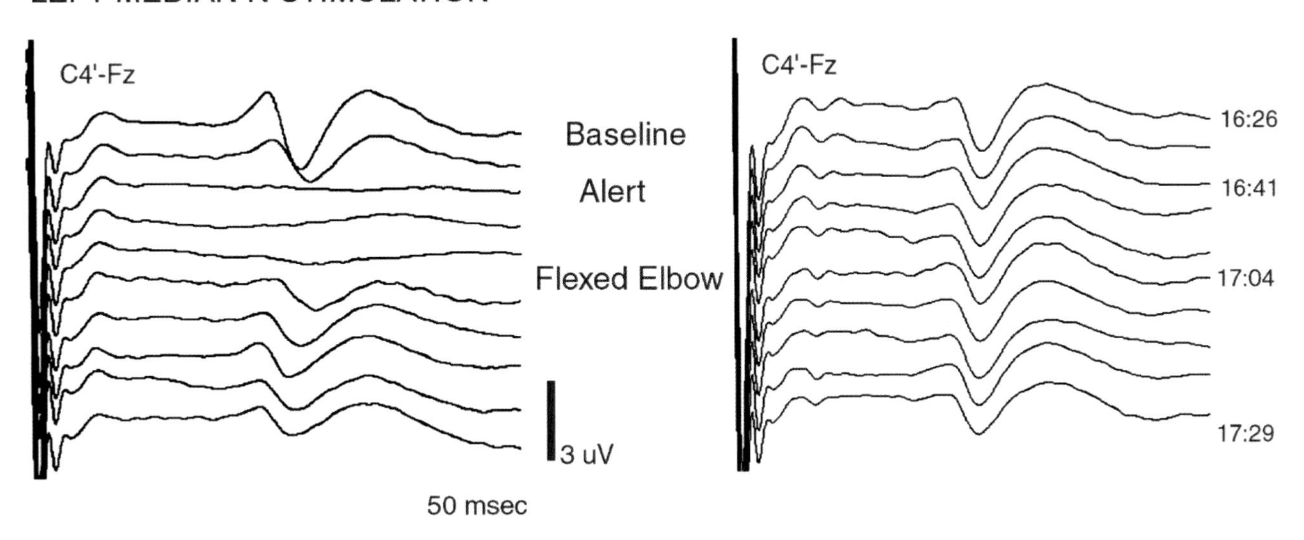

Figure 1. *Contralateral scalp recorded SSEPs to median nerve stimulation (left) and ulnar nerve stimulation (right). Median nerve SSEPs show a loss of signal shortly following baseline recordings prior to incision for resection of a forearm median nerve schwannoma. The insult was considered related to arm position and was resolved by flexing the elbow. Ulnar nerve SSEPs were unchanged demonstrating the focal nature of the insult to the median nerve.*

Mixed nerve SSEPs as shown in the previous example are the most common choice for stimulation sites. There can be, however, certain situations when dermatomal stimulation may be applied. Since radial, median and ulnar nerves may enter the spinal cord over multiple roots, dermatomal stimulation may be used when greater specificity of a nerve root is required. For example, dermatomal stimulation of the fingers can be done with needles inserted on each side of the digit, one the cathode and the other the anode so current passes across the digit and activates the digital cutaneous nerves. The disadvantage of the dermatomal technique, as noted elsewhere in this book, is its small response amplitudes and potential variability under operative conditions (see Figure 2 for comparison of relative amplitude of mixed nerve and associated dermatomal responses). In our experience, stimulation of the C6 (thumb), C7 (middle finger) and C8 (little finger) for the upper extremity and of the L5 and S1 dermatomes for the lower extremity have proven the most successful over stimulation of other dermatomes.

Figure 2 illustrates a case that utilized dermatomal stimulation as part of the evaluation of a right brachial plexus traumatically injured by a stab wound. Clinically, the 21 year old patient showed a lack of shoulder abduction and poor elbow flexion specifically with reduced supraspinatus, no infraspinatus, no deltoid, and no bicep muscle function. Additionally, the patient had nearly absent sensation over the cap of the shoulder. All other muscle function and sensation were normal. These clinical findings as well as a high resolution MRI suggested an injury of the upper trunk near the junction of the C5 and C6 nerve roots. Intraoperatively, SSEPs to stimulation of the radial, median, and ulnar nerves were obtained (top two traces of each set of SSEPs in Figure 2). To confirm specificity of root function, SSEPs to C6 (thumb), C7 (middle finger), and C8 (little finger) dermatomes were recorded, which confirmed the integrity of the C6, C7 and C8 nerve roots. The dermatomal SSEPs, while smaller than mixed nerve responses as noted previously, were stable and were monitored throughout the case (lower 4-5 traces) to protect the function of these roots. After dissection of the brachial plexus, further intraoperative evaluation with intrafield stimulation of the C5 nerve showed responses in serratus anterior, rhomboid, and supraspinatus muscle but no responses in deltoid or bicep muscles consistent with the clinical examination and consistent with a scarred region in the upper trunk at the junction of the C5 and C6 nerve roots. A nerve graft was performed and at two years there was a near full recovery of elbow flexion and substantial recovery of shoulder abduction. **(Fig. 2).**

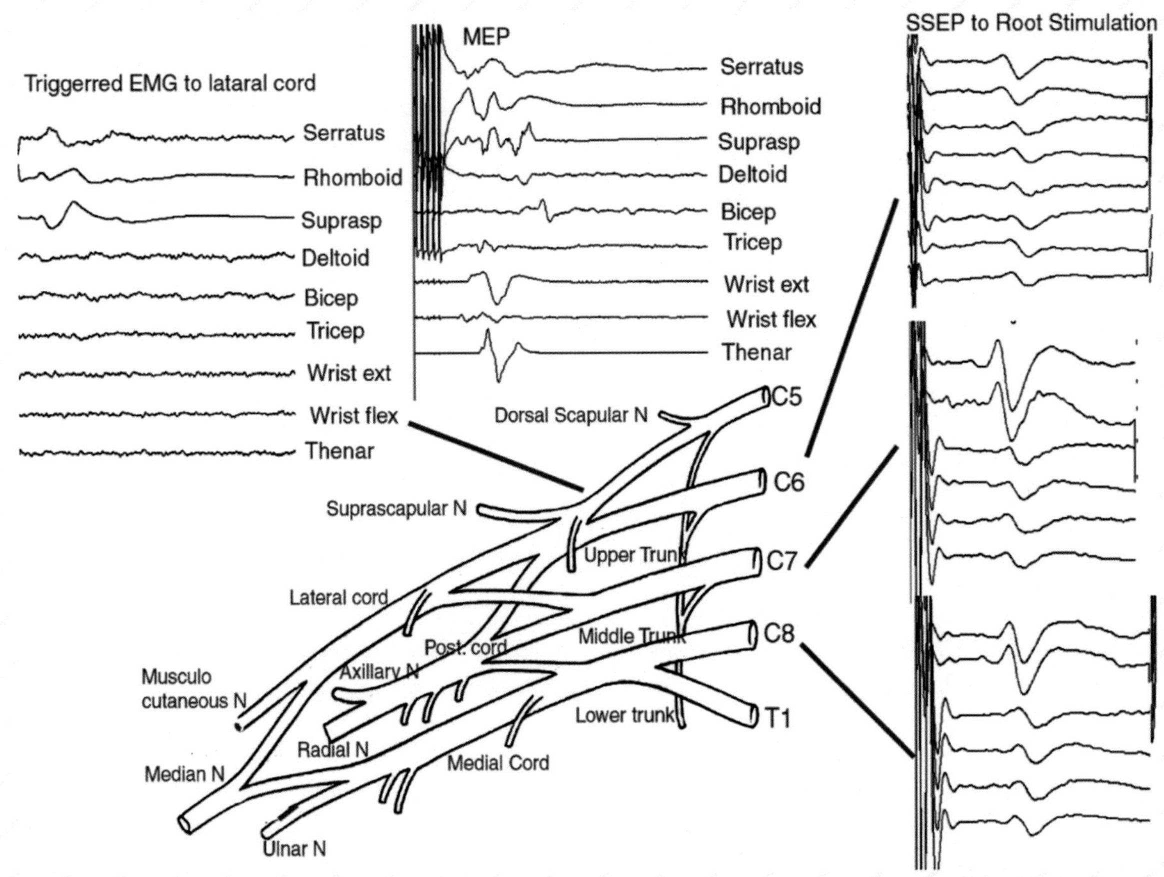

Figure 2. *Case of a stab wound injury affecting the upper trunk of the brachial plexus. Right side panel shows SSEPs to mixed nerve (top two traces in each set) and dermatomal stimulation (bottom 4-5 traces in each set). The intact responses confirm functionality of the C6, C7 and C8 nerve roots. Intrafield stimulation of the C5 nerve root near its junction with the C6 nerve root yielded responses in proximal shoulder muscles and no responses in deltoid or bicep muscles.*

As with SSEP protocols for spinal cord or brain procedures, peripheral nerve recordings serve an important role in assessing the neural activity distal to the surgical field and affirming appropriate functioning of the averaging equipment and establish the integrity of the nerve stimulated. For mixed nerves, it must be recognized that the response recorded from a peripheral site to distal nerve stimulation is a combined response of orthodromic sensory activity and antidromic motor activity. While it may be recorded with SSEP protocols, the response may be more precisely referred to as a nerve action potential (NAP). This author prefers to use the term compound nerve action potential (CNAP) to reflect the activity of mixed sensory nerves, where compound refers to the different types of fibers. Some authors use the term NAP when referring to nerve recordings to nerve stimulation within the surgical field (to be discussed forthwith). When only sensory fibers are stimulated, for example digital nerve stimulation, the response recorded from a nerve reflects only sensory nerve activation and in clinical electrodiagnostics is called a sensory nerve action potential (SNAP).

Peripheral level recordings, which are an essential component of any SSEP protocol, may theoretically be done anywhere along the peripheral nerve there is good access. Typically for upper extremity SSEPs, Erb's point is the recording site of choice, such as for median, ulnar, or radial nerve stimulation. However, because Erb's point is likely in the surgical field for brachial plexus surgeries and also because Erb's point recording may yield insufficiently small signals even for median nerve, an alternative recording location is at the axilla in the crease formed by the bicep and tricep muscles. This location yields well developed responses, although occasionally radial nerve responses can be small or absent. Additionally, recordings at the elbow may also be considered with electrodes in the antecubital fossa for median nerve and the ulnar groove for ulnar nerve. For the lower extremities, the popliteal fossa is a site of choice for tibial nerve recordings and posterior to the fibular head for superficial peroneal nerve recordings. As mentioned, these responses can assure the monitorist that peripheral nerves are functional.

However, in cases in which the surgical field is in the arm or the leg, then peripheral nerve recordings may, in fact, be proximal to the surgical field and may serve to monitor the function of the nerve through the surgical field in the classical monitoring role. The advantage of using a peripheral response for monitoring over a more central recording is the robust nature of the response. It can be recorded quickly because the average requires fewer sweeps per average and a higher stimulation may be used. The response itself is usually much larger amplitude than any central recording.

An example of the utility of monitoring SSEPs with peripheral nerve recording is shown in Figure 3. The axilla level recordings to median and ulnar nerve stimulation were used in addition to intrafield stimulation and recording of CMAPs to monitor function of the median nerve during resection of a forearm schwannoma. Median nerve SSEPs, which were clearly present at baseline, became significantly reduced about 13 minutes after start of tumor resection. The responses remained reduced throughout the remainder of the procedure. The focal nature of the insult to the median nerve was confirmed by the stable ulnar nerve SSEPs. In addition, intrafield stimulation of the median nerve proximal to the tumor showed severely reduced CMAPs in the thenar muscle coincident with loss of SSEPs. After resection, the CMAP to proximal stimulation were about one half that of the CMAP to distal stimulation, confirming potential compromise of the median nerve through the tumor site. The surgeon confirmed that a 1-2 mm diameter bundle of fibers had to be transected to remove the tumor and likely contributed to the signal loss. Postoperatively, the patient has a slight weakness of the abductor pollicis and opponens muscles and diminished sensation in the first and second fingers. **(Fig. 3).**

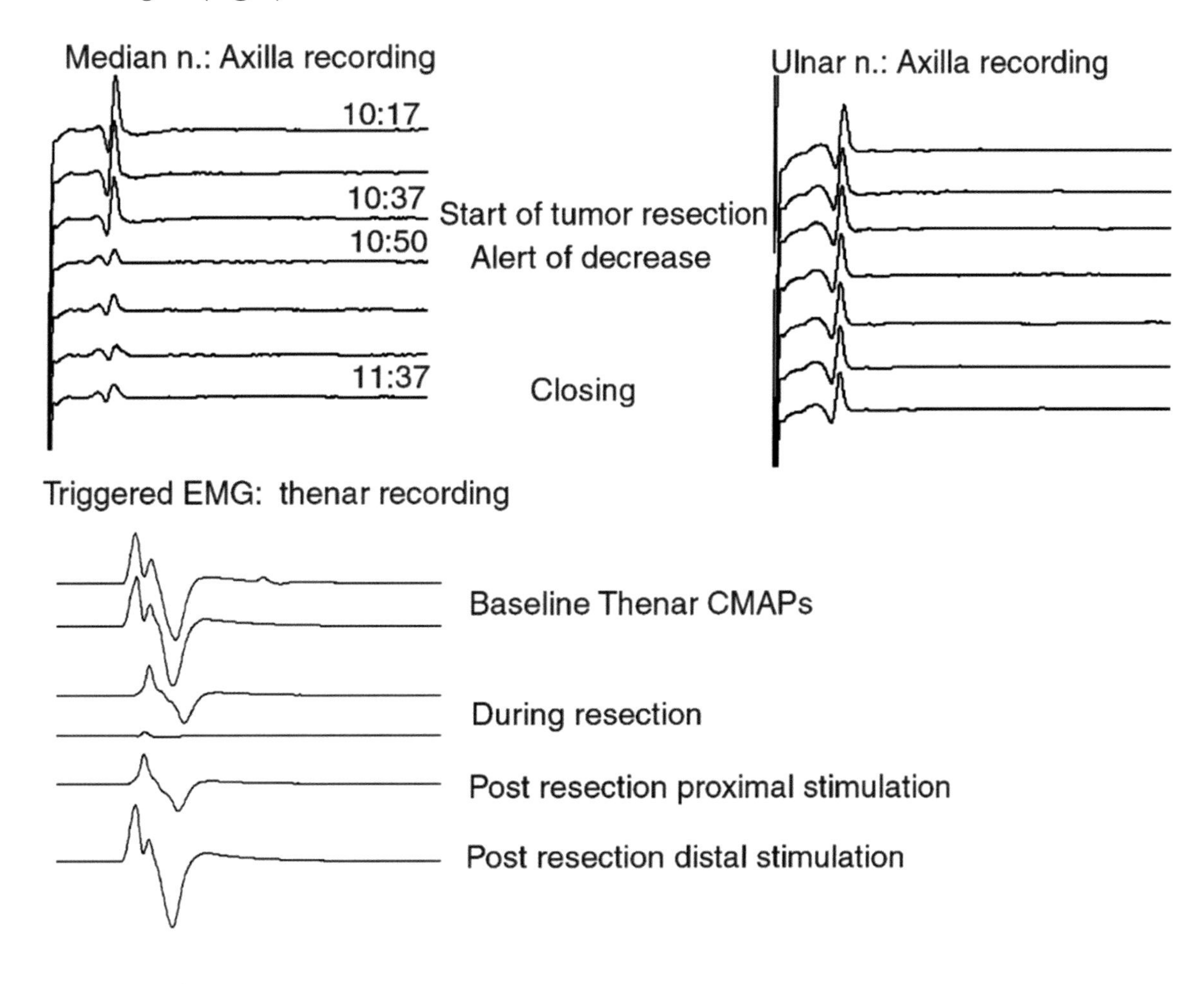

Figure 3. Median and ulnar nerve SSEPs recorded at the axilla during resection of a median nerve schwannoma in the forearm. Reduction of the median nerve SSEP about 13 minutes into tumor resection together with severe reduction of thenar CMAP to intrafield stimulation of proximal median nerve suggested nerve compromise and correlated with transection of a bundle of nerve fibers in order to remove the tumor. Postoperative weakness and decreased sensation ensued.

SSEPs to Stimulation within the Surgical Field

A quite useful variant of the SSEP technique is SSEPs to stimulation of nerves or nerve roots within the surgical field. Such stimulation can be used to determine if a sensory nerve root is intact or may define a point along a nerve that contains viable sensory fibers.

Stimulation of nerves within the surgical field is done with handheld electrodes applied to the nerve by the surgeon. These electrodes may be monopolar electrodes or bipolar electrodes. Monopolar electrodes may be of a flush tip or ball tip configuration. Flush tip electrodes provide for a restricted, focal application of current that is useful for location of nerve fibers. Ball tip probes deliver a somewhat broader electric current and are useful when localization of a nerve deep in tissue is needed and focal delineation is not necessary. Bipolar electrodes, either straight tip or hooked electrodes are typically used when focal stimulation of a whole nerve is needed such as when nerve conduction studies are done. A bipolar electrode has the advantage of focal stimulation of a nerve and a smaller stimulus artifact compared to monopolar stimulation. Some have advocated tripolar hook electrodes with a center cathode and two anodes to each side as a means to minimize stimulus artifact. With any electrode the nerve should be lifted away from adjacent tissue and fluid suctioned away to prevent spread of current to adjacent nerve or shunting of the current, respectively.

For direct nerve stimulation, stimulus duration may be 0.05 to 0.2 msec. and intensities carefully increased until a response is seen, usually between 0.05 and 1.0 mA, although higher values may be needed in injured nerve or if nerve is deep in the tissue. A word of caution about stimulus output is that one must know the stimulus waveform shape of one's equipment to be sure the shape is rectangular and delivering an appropriate shape and intensity expected. Bioengineering help may be needed if one is not able to establish stimulator output by oneself.

Again stimulation rates should be commensurate with the type of recording done. For SSEPs, cortical level recordings must be done with low frequency stimulation rates of 3-5/sec, whereas if the recording is at subcortical or peripheral nerve level, then stimulation rates may be increased to 10-15/sec.

If stimulation is within the surgical field and is only a few centimeters from the recording electrodes, then the stimulus artifact may prevent clear signal assessment. The short latency due to the short conduction distance from brachial plexus or nerve roots to spinal cord or brainstem means the response may be buried in the artifact. Then, cortical level recordings become the recording site of choice because, even though the latencies will be shorter than those expected with distal stimulation, the response should be late enough to avoid the stimulus artifact. One word of caution with this stimulation and recording scenario is that the stimulation may often cause muscle twitch artifacts that may mimic the intended evoked potential. To resolve, stimulation intensity may be reduced to lessen the twitch or, preferably, a short acting muscle relaxant may be used to eliminate any motor contributions to the signals leaving the neural component.

Nerve root stimulation:

SSEPs to stimulation of nerve roots can determine the functional integrity of the sensory portion of nerve roots suspected of being avulsed. Following exposure of nerve roots, they can be stimulated directly with bipolar stimulating electrodes. If a centrally recorded response is recorded, then that nerve root is functional. Figure 4 illustrates such a case. These intraoperative data are from a person who sustained a severe brachial plexus injury in a motorcycle accident that left him with a painful and flail left upper extremity and a great deal of scarred tissue both above and below the clavicle. The upper left panel of the figure shows SSEPs to radial, median, and ulnar nerve stimulation. No responses are noted at the central levels. Most importantly, no responses are noted at the axilla, suggesting a postganglionic injury involving the C6-C8 nerve roots. (A postganglionic injury is located distal to the dorsal root ganglion, which means the sensory fibers are no longer nourished by their cell bodies in the dorsal root ganglion and have atrophied, become nonfunctional, and fail to produce action potentials. In contrast, a preganglionic injury is located proximal to the dorsal root ganglion and involves an avulsion of the roots from the spinal cord. With a preganglionic injury peripheral sensory response may be recorded because the dorsal root ganglion is connected to and nourishes the distal sensory fibers.) Baseline MEPs revealed responses in the serratus anterior, rhomboid, and infra/supraspinatus and, possibly, the deltoid muscles. Once, the initial exposure was done above the clavicle, intrafield stimulation of the upper trunk elicited responses in the same muscles. Together, the data from MEPs and intrafield stimulation indicated intact function of the upper trunk and possibly its posterior division. Because of the postganglionic injury, SSEPs to nerves at the wrist provided no assessment of nerve roots. To assess nerve root function, SSEPs were done to direct nerve root stimulation. In this example, following exposure of the nerve roots, intrafield stimulation elicited central SSEPs to stimulation of the C6 and C8 nerve roots but not the C7 nerve root (upper right panel in figure 4), indicating that only the C7 nerve root was avulsed from the spinal cord and not the C6 or C8 nerve roots. Not shown were the absent central SSEPs and absent CMAPs in biceps to stimulation of the musculocutaneous nerve. These findings suggest conduction impairments above the clavicle of the lateral trunk between the branching of the posterior division and the musculocutaneous nerve and below the clavicle involving the posterior and medial cords. Since the lateral cord and musculocutaneous nerve did not show severe intraneural fibrosis, the nerves were left intact. At two years, some sensation had returned to the level of the elbow but no motor function had been restored. **(Fig. 4).**

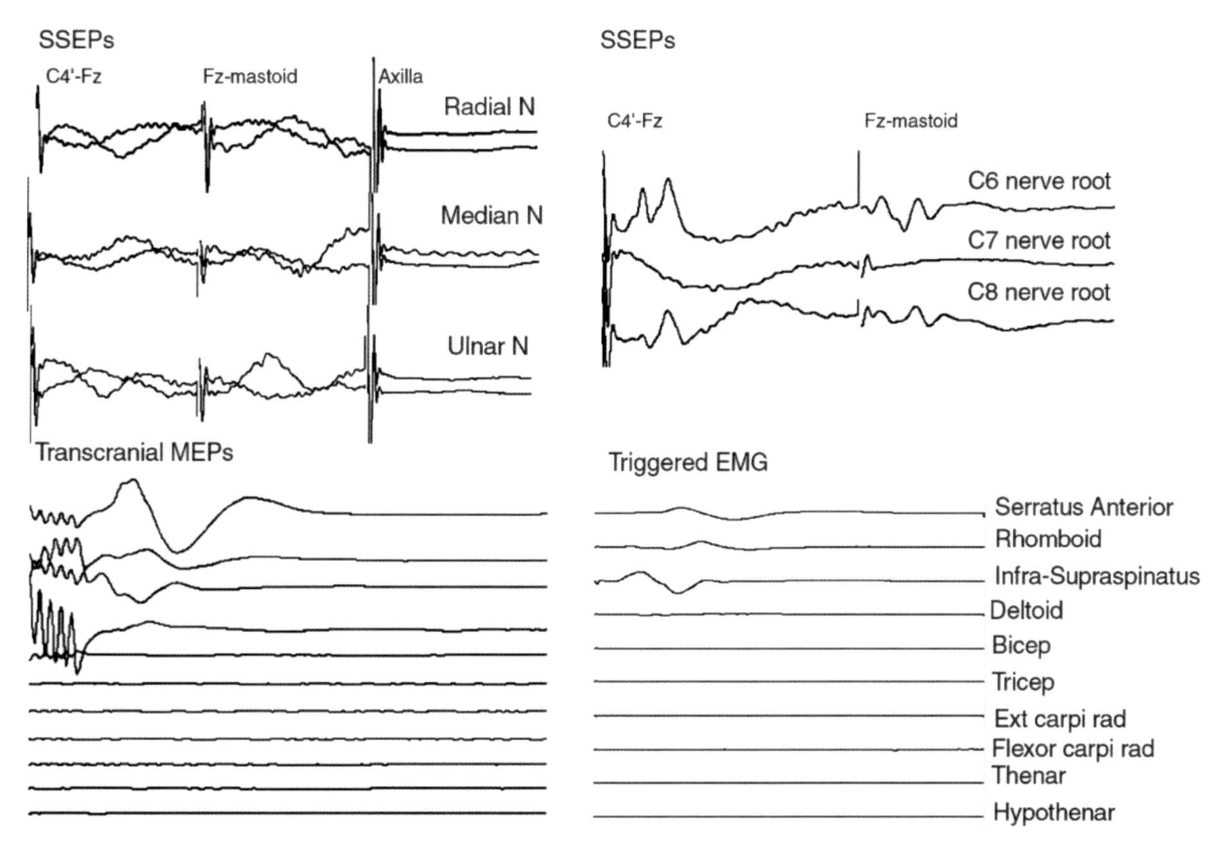

Figure 4. *Data from a person involved in motorcycle accident who sustained a profound brachial plexus injury resulting in a painful and flail arm. SSEPs (upper left) to mixed nerves were absent at all levels, including axilla recordings, indicating postganglionic injuries at C6-C8 levels. MEPs (lower left) and intrafield stimulation (lower right) confirmed an intact upper trunk and possible posterior division to the deltoid. SSEPs to stimulation of nerve roots (upper right) showed the C7 root to be avulsed but not the C6 and C8 roots.*

Nerve action potentials (NAPs) recorded in the Surgical Field to Nerve Stimulation:

The intrafield recording of NAPs to intrafield stimulation of a nerve is a technique originated by Kline and colleagues for the identification of functional nerve fibers in continuity across a neurofibroma (see application section of this chapter). During the post-traumatic period, traumatized nerves may develop a mass of scar tissue that contains the attempts of nerve fibers to regenerate through the scarred tissue. Some may successfully do so, while other fibers become a tangled part of the neurofibroma. An analysis of conduction across a neurofibroma can help determine the course of treatment.

For NAP stimulation and recording, handheld bipolar hook electrodes or cuff electrodes are the main choice. Monopolar electrodes may be used but stimulation is not precisely focused on the nerve and will cause a larger stimulus artifact. For bipolar stimulating electrodes a separation of 2-4 mm is ideal with a larger separation used for larger diameter nerves. Since the depolarization of the nerve occurs at the cathode of a stimulating electrode, the cathode of a bipolar pair should ideally be oriented in the direction of axonal travel. For monopolar stimulation the cathode is placed on the nerve and the anode at a distal point. In order to avoid the recorded signal occurring during the stimulus artifact, the distance between stimulating and recording electrodes should be sufficient large, typically 4 cm or more. Newer equipment, which uses floating grounds, by its design, has a larger stimulus artifact than artifacts in older equipment that directly grounded the equipment to earth ground and, thus, necessitates a longer stimulus and recording electrode separation. A tripolar electrode with a center cathode and two anodes on each side was designed to help minimize stimulus artifact. In all cases, the electrodes should be lifted out of the surgical field as carefully as possible to avoid stretching but to isolate the nerve from tissue and fluid that would shunt current away from the nerve and degrade the stimulation or the recorded signal. Stimulus duration should be 0.05 – 1.0 msec with single or a low repetition rate. Stimulus intensity is usually on the other of 1-5 mA but higher intensities may be required. For recording bipolar hook electrodes, the interelectrode distance should be about 3-5 mm. This distance is based on the length constant of the wave of depolarization. It has to be far enough apart for the action potential to pass beneath the active electrode before passing beneath the reference electrode. If the electrodes are close together the action potential appears on both the active and reference electrodes. Because of differential recording, the signals can be reduced in size and can likely show a complicated multicomponent waveform. As with stimulation electrodes, recording electrodes should be clear of tissue and fluid to avoid shunting of the recorded current. The response is on the order of microvolts with frequencies in 10-1000 Hz range. Amplifier settings should be set commensurate with these observations. Although, in the case of a large stimulus artifact, the low frequency filter may be brought up from 10 Hz to 100 Hz with the effect of reducing the slower components of the artifact and "collapsing" it sooner to baseline. The latency of the response will be short and depend on the distance between the stimulating and recording electrodes, on the order of 1 msec for a separation of 4-5 cm.

In addition to assessing neural integrity over short segments of nerves, NAPs may be used to analyze nerve conduction related to nerve entrapments such as the ulnar nerve at the elbow, the median nerve at the wrist, or the peroneal nerve at the fibular head. The nerve's conduction velocity may be derived to determine if slowing is occurring. Here, the distance from the stimulating cathode to the active recording electrode must be measured as well as the onset latency of the response. The distance in millimeters divided by the latency in millisecond will yield the conduction velocity in meters/sec. Another method is to simulate in an "inching" manner at 1 cm intervals along the suspected injured area of the entrapment. The latencies of the MCAP to stimulation at each centimeter can be examined for focal slowing across one or several segments indicating the location of the severest entrapment. An example of such a study may be found in Figure 4 of Slimp, 2000.

Finally, NAPs can serve to locate the position of a nerve in the surgical field when the nerve is embedded in scar tissue or displaced by scarring or a tumor. The surgeon my carefully dissect the tissue and look for the nerve but this can be laborious and time consuming. In this application, recordings are made in the surgical field with ball tip probes or needle electrode to stimulation of the nerve distally. Responses are recorded from successive locations orthogonally to the path of the nerve, usually at intervals of 2-3 mm. The amplitude of the response should show a maximum at some point with lower amplitudes to each side of the maximum. The site of maximum amplitude will indicate the nerve location under that recording site. An example may be seen in Figure 21-2 in Slimp and Kliot, 1995. Using this technique, the location of a nerve underneath scarred tissue or tumor can be made and focus the dissection in a directed and timely manner.

Motor potentials:

Recordings from muscle may reflect single/multiple motor unit activity or compound motor action potentials (CMAPs). Motor unit activity, recorded as spontaneous or free running activity is used to indicate irritation or potential injury to nerves innervating a muscle. CMAPs to stimulation of nerve, sometimes called triggered EMG, can be used as an endpoint to indicate functionality of motor fibers, to measure conduction velocity, and to indicate normal response patterns by measuring size of the CMAP. Muscle potentials are also the basis of motor evoked potentials (MEPs) which are elicited most typically by transcranial electrical stimulation.

Free Running Spontaneous EMG

As with spinal root or cranial nerves, a continuous recording of EMG activity may be analyzed for spontaneous activity. The amplification gain (and display gain) should be sufficiently high (10-20uV/div) to reveal single motor unit activity. Spontaneous activity in the form of short bursts or longer trains of motor unit potentials may indicate iatrogenic, mechanical irritation to the nerves. These responses are useful feedback to guide the surgeon's actions around the nerves, but it must be stressed that the activity is not a reliable predictor of outcome. It should also be pointed out that not all spontaneous activity is manipulation driven. Some activity, usually at a low level of sporadic single or small collection of motor units, may simply be spontaneous or caused low level activity of the patient's nervous system. If this activity occurs in several muscles or becomes more prevalent in a particular muscle, especially if surgeon activity is minimal, then the activity may reflect an increasing level of arousal and can be beneficial for anesthetic management of the patient.

Triggered EMG:

Nerve stimulation may elicit CMAPs in the target muscles appropriate for the nerves under investigation. Because these CMAPs are evoked or triggered by stimulation they are also sometimes referred to as triggered EMG. The selection of muscles to record CMAPs should be the most prominent target muscles for nerve(s) that may be stimulated in the surgical field. Selection and analysis for a single nerve is straightforward e.g. the musculocutaneous nerve would necessitate recording from biceps muscles, whereas for proximal stimulation sites, such as the trunks and cords of the brachial plexus or of nerve roots, several muscles must be selected. Stimulation of proximal sites elicits responses in an array of muscles. The constellation of responses may used to identify the likely site of stimulation. Tables 1-3 give the innervation schemes for the muscles of the upper **(Table 1)** and lower **(Table 3)** extremities and for the trunks and cords of the brachial plexus **(Table 2).** As one can see, the more proximal the stimulation site is generally the more muscles that will be activated.

Table 1

		Nerve Root				
Nerve	Muscle	C5	C6	C7	C8	T1
Dorsal scapular	Rhomboid					
Suprascapular	Supraspinatus/Infraspinatus					
Axillary	Deltoid					
Musculocutaneous	Biceps					
Radial	Triceps					
	Brachioradialis					
	Extensor carpi radialis longus					
(posterior interosseous)	Extensor pollicis brevis					
(posterior interosseous)	Extensor indicis proprius					
Median	Pronator teres					
	Flexor carpi radialis					
(anterior interosseus)	Flexor pollicis longus					
(anterior interosseus)	Pronator quadratus					
	Abductor pollicis brevis					
Ulnar	Flexor carpi ulnaris					
	Flexor digitorum profundus					
	Adductor pollicis					
	First dorsal interosseous					

Table 2

	SA	Rh	IS/SS	Delt	Bic	BR	PT	ECRL	Tri	FCR	APB	FPL	FDI	FCU	EIP
C5/Prox Upper Trunk															
Upper Trunk															
Middle trunk															
Lower trunk															
Lateral cord															
Posterior cord															
Medial cord															

Table 3

		Nerve Root					
Nerve	Muscle	L2	L3	L4	L5	S1	S2
Psoas, Iliacus	Iliopsoas						
Obturator	Adductor longus						
Femoral	Vastus medialis/lateralis						
Femoral	Rectus femoris						
Superior gluteal	Tensor fascia lata						
Superior Gluteal	Gluteus medias						
Inferior gluteal	Gluteus maximus						
Sciatic (tibial)	Semitendinosus/membranosus						
Sciatic (tibial)	Long head Biceps femoris						
Sciatic (peroneal)	Short head biceps femoris						
Deep peroneal	Tibialis anterior						
Deep peroneal	Extensor hallucis longus						
Superficial peroneal	Peroneus longus						
Deep peroneal	Extensor digitorum brevis						
Tibial	Tibialis posterior						
Tibial	Flexor digitorum longus						
Tibial	Lateral gastrocnemius						
Tibial	Medial gastrocnemius						
Tibial	Soleus						
Tibial (medial plantar)	Abductor hallucis						

For muscle recordings, 0.5 inch bare needles, insulated needles, or fine wire electrodes may be used. Half inch bare needles are essentially "surface" recordings and reflect not only the activity generated beneath the electrode but likely reflect activity generated at some distance from the electrode. In most arm and leg locations, several muscles may be within the recording distance of these electrodes and, therefore, do not connote specificity of muscle origin. To obtain recordings specific to a particular muscle, insulated electrodes, preferably of a bipolar variety or two closely spaced monopolar electrodes such as fine wires or insulated needles, must be placed within the muscle of choice. If the nerve is known and stimulation is to determine functionality, then specificity of the CMAP is not as important, since any response will reflect the innervation of that nerve. However, if the nerve is unknown and the object is to identify the nerve by the CMAPs elicited, then the specificity of recording offered by insulated needles is an advantage.

For triggered EMG stimulation, a single stimulus or a slow rate of 1-3/sec is recommended to reduce the interference caused by contraction of the muscle. Nerves may be stimulated with either bipolar hook electrodes or with monopolar electrodes. Monopolar flush tip type probes deliver a smaller electrical field than ball tip type probes. Flush tip probes have the advantage of selective stimulation of nerve fibers and are useful for discreet localization and mapping of motor fibers. Ball tip probes with their larger stimulation fields are useful for testing tissue for motor fibers prior to dissection of that tissue.

Motor Evoked Potentials:

MEPs have an application in peripheral nerve surgery. MEPs may be recorded from the nerve root, a nerve segment, or from muscle. Nerve root or nerve recordings are done with bipolar hook electrodes. Responses are small and may be contaminated by volume conduction from surrounding muscles. Deep neuromuscular blockade must be employed to be absolutely sure the response is from the nerve and not from adjacent muscle. A present response, which at the nerve root would occur at about 8 msec, would indicate intact conduction of the ventral root.

MEPs more typically are recorded from muscle. To be sure, MEPs recorded from muscle may serve as a monitor of function in the same way as SSEPs can and in the same way they do for spinal cord or brain surgery. An MEP in a particular muscle will indicate intact conduction of motor fibers to that muscle but not provide any information about a specific root or nerve pathway. Perusal of tables 1-3 will reveal that no muscle with the exception of perhaps rhomboid is innervated by a single nerve root.

Another complication with MEPs recorded with "surface" electrodes, as is commonly done, is volume conduction of the response. Because muscles of the arm and leg are several and are in close approximation, a surface electrode or subcutaneous needle will "see" several muscles. MEP stimulation is not selective but activates many, if not all, muscles in the arms and legs (and trunk). This raises the possibility that an electrode, which is intended to record from a muscle directly below it, may in fact "see" a response generated at some distance and conveyed to the electrode by volume conduction. This possibility is illustrated in Figure 5. The patient had a complete traumatic transection of the peroneal nerve as demonstrated visually and by lack of CMAP to direct stimulation. However, as can be seen in the upper traces of Figure 5, MEP responses recorded with subcutaneous needles were discernible over tibial nerve innervated muscles (gastrocnemius/soleus) and peroneal innervated muscle (tibialis anterior, extensor hallucis longus). The responses in the peroneal leads are actually volume conducted activity from the gastrocnemius/soleus muscles. When MEPs are recorded with paired insulated needles (2-4 mm separation of active and reference), peroneal muscles show essentially no responses. Interpretation of the location of origin of signals recorded with subcutaneous needles must be done with great caution. **(Fig. 5).**

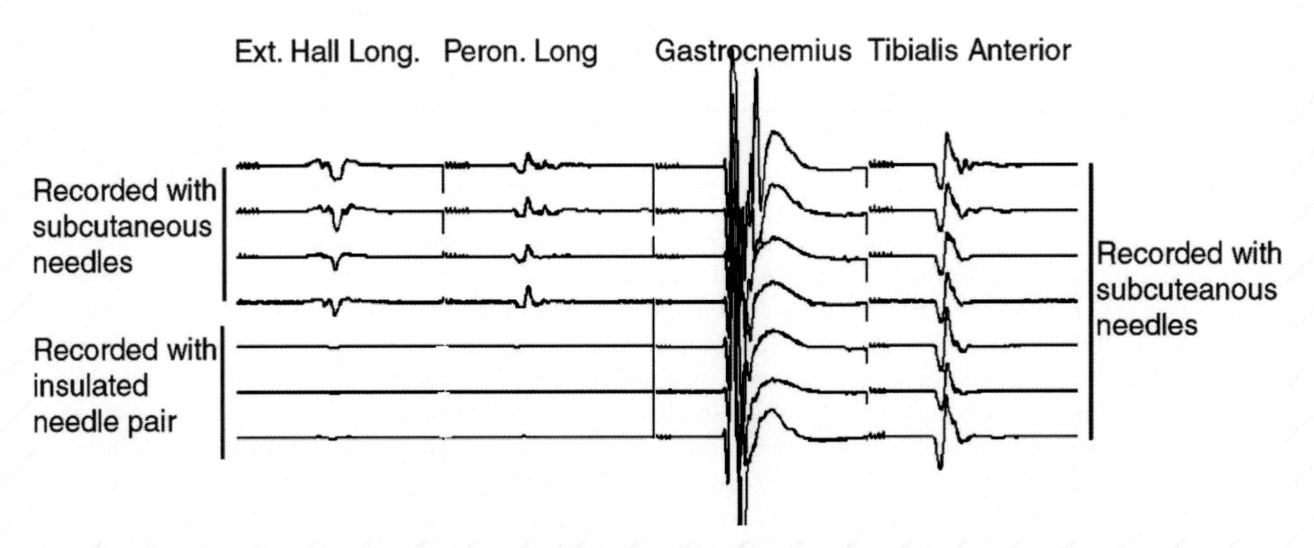

Figure 5. *MEPs recorded from tibial (gastrocnemius) and peroneal (extensor hallucis longus, peroneus longus, and tibialis anterior) innervated muscle using subcutaneous needles or insulated needles. The insulated needle recordings suggest that the extensor hallucis longus and peroneus longus responses recorded with subcutaneous needles are likely volume conducted from the gastrocnemius and soleus muscles.*

As mentioned in previous chapters, anesthesia can be an issue when using MEPs. If MEPs are used, preferred anesthesia is total intravenous anesthesia (TIVA) or low dose gas anesthesia. However, if only SSEPs and EMG are done, then inhalational agents are acceptable at 1.0 MAC or less. In any case that a motor response is recorded it is advisable to refrain from muscle relaxants at all as this will insure a maximal response. Some advocate a partial blockade and monitor twitches to a train of four to maintain appropriate partial blockade, but partial blockade should be used cautiously because reduction of motor responses with partial blockade may make threshold determination difficult and erratic and may block small responses such as reinnervated motor units.

The temperature of the limbs generally is less than trunk temperatures and can be a factor in peripheral nerve analysis, especially when nerves are exposed as they are during peripheral nerve surgery. When doing nerve conduction velocity measurements one can expect a 10m/sec slowing in exposed nerve due to cooling. Latencies of SSEPs may be also be prolonged as a result of cooling. CMAPs may be delayed slightly but amplitudes are generally not affected unless cooling is quite severe.

Use of a tourniquet has the effect of blocking conduction about 10-20 minutes after application and as such should not be used when monitoring peripheral nerves. If the tourniquet is applied only briefly then responses will recover within seconds to minutes, but if the tourniquet is applied to longer periods up to an hour or more, then recovery of nerve function is much slower.

Application of Electrophysiological Tools:

This section outlines some of the applications of the electrophysiological tools heretofore described. These tools may be orchestrated to augment the surgical treatment of peripheral nerves. The first objective is to monitor the relevant function nerves associated with the surgical site and prevent inadvertent injury to these nerves. A second objective that is outside "standard" monitoring is to guide dissection. Intrafield stimulation and recording motor responses can be used to guide dissection by identifying if any motor fibers exist in tissue considered for resection. A third objective is to evaluate the nerves under surgical examination for their identity and their functionality. Adequate knowledge of peripheral nervous system anatomy is necessary to identify the appropriate nerves associated with a particular surgical field. These nerves must then be appropriately monitoring with SEPs, MEPs, and/or EMG to achieve the goals of monitoring.

An appropriate protocol requires a full understanding of the patient's condition, the surgeon's surgical plan, and the goals for surgical intervention. Not only should nerves be monitored that are directly involved in the surgical plan but attention should be paid to possible surrounding nerves that may be potentially put at risk.

Neuroma in continuity:

The analysis of neuromas in continuity with NAPs as described by Kline and colleagues is considered by many to be an essential component of this type of peripheral nerve surgery. Gross appearance of a neuroma in continuity may suggest intact axons, but it is only by electrophysiological analysis can a functional assessment be made.

Injuries may be classified functionally as neurapraxic, axonotmetic, or neurotmetic. Neurapraxic injuries show a functional block of conduction along a nerve. The axon and endoneurium are intact but traumatized sufficiently to block conduction. These fibers continue to provide metabolic support for the axons distal to the neuroma and so the axons remain functional. The prognosis in these cases in quite favorable since recovery of nerve structures is quite likely. An axonotmetic lesion is a disruption of the axon with preservation of the endoneurium. Functional recovery requires regrowth of the axons to their target structures, which requires the endoneurial tubes to act as guides for the growing axon. Axonal regrowth occurs at approximately one millimeter per day. The most severe injury is a neurotmetic injury, which involves disruption of both the axon and the endoneurium, effectively blocking any regrowth of axons for lack of guiding endoneurial tubes. Axonal regeneration, if it occurs, is a slow process of about 1 mm per day.

Electrodiagnostically, none of these types of injury will show conduction across the injury site. In the short term, before degeneration of the axons occur, stimulation of the nerve distal to the injury site will show electrical excitability as measured by nerve potentials recorded distal to the injury to stimulation distally (CNAPs or SNAPs) or by motor potentials recording in target muscles to stimulation distal to the lesion. When the axon is disrupted, in either axonotmesis or neurotmesis, the axon will continue to be electrically excitable for up to a week after injury then wallerian degeneration occurs. Wallerian degeneration of the axon will occur in the nerve distal to the injury back to the injury site. There is some evidence that degeneration or dysfunction of the nerve may also occur proximally. If after sufficient time for wallerian degeneration to occur, one sees distal nerve or muscle responses to nerve stimulation distal to the injury, then the injury is a neurapraxic type and recovery is likely.

More difficult is the distinction between neurotmetic and axonotmetic lesions. Neurotmetic lesion will never generate any electrical signals distal to the lesion because both axons and surrounding neural structures are absent and regeneration is incomplete. Axonotmetic lesions, on the other hand, may lead to functional recovery if the axons can effectively regrow; however, length of the growth, health of the endoneurial tubes, degree of scarring, and other factors may prevent successful regrowth of axons. Patients with no clinical improvement 2-6 months after injury may become surgical candidates for neurolysis or nerve graft repair and their regrowth is not likely to be functional. The decision to perform surgery is a complicated process, requiring careful understanding and planning.

For neuromas containing axonotmetic injuries, NAPs may identify those axons that are in the process of regeneration. Since regenerating axons may take weeks or even months to complete their journey to their target muscle, functional recovery may take longer than the optimal 3-6 month window after injury for nerve grafting. A NAP can appear weeks before reinnervation results in functional recovery in muscle or in identification of voluntary motor units by EMG. Demonstrating regeneration by NAP would avoid unnecessary surgery. A sufficiently sized NAP recorded proximal to the neuroma to stimulation just distal to the neuroma would indicate functional continuity of axons through the neuroma with the size of the potential giving an indication of the number of functional axons. The extent of regeneration may be gathered by stimulating at successive points more distally in and "inching" fashion. The transition of a positive response to no response would indicate the leading edge of regeneration. Treatment may be nothing or a neurolysis. If there is no NAP, then there is no regeneration, probably due to neurotmetic type injury or regeneration is stunted. In this case, a nerve grafted would be considered.

Localizing Viable Nerve Function:

When nerve grafting, the proximal stump should be transected at a point where there are viable nerve fibers to optimize regeneration. For example, a traumatic transection may require nerve graft across the transection site in order to offer the possibility or functional regrowth and recovery. To ensure success of the graft, it is best to graft at a point along the proximal where the nerve fibers are intact and functional. While this may be determined by visual inspection or by cutting the nerve and inspecting the end for evidence of fascicles, SSEP to stimulation of a sequence of points along the proximal segment can establish electrophysiologically a point on the nerve that has functional sensory fibers. Often this point of functionality is quite proximal to the transection sites, suggesting that a wallerian degeneration may occur quite proximal to the lesion.

An example of SSEPs to stimulation of a peroneal nerve proximal to its injured site is given in Figure 6. Seven months prior to surgery, this person sustained multiple stab wounds to the right leg involving the left peroneal nerve. Clinically, the patient had a foot drop and electromyographically showed complete denervation of peroneal supplied muscles. The short head of the biceps were muscle was functional as was tibial nerve innervated muscles, indicating the peroneal nerve was traumatized distal to the junction of the nerve to the short head of the biceps. Intraoperatively, the peroneal nerve was completely transected and electrophysiologically showed no motor or sensory conduction across the injured area. To better establish a proximal location on the peroneal nerve for nerve grafting SSEPs to intrafield stimulation at points approximately 1, 2, and 3 cm proximal to the injury were done. The results shown in Figure 5 show large amplitude SSEPs to stimulation at points 2 and 3 cm proximal with a slightly smaller response at the 1 cm point. Distal stimulation elicited no responses as expected. The proximal neuroma was resected and normal appearing fibers occurred at approximately 1 cm proximal to the neuroma. A nerve graft was performed. **(Fig. 6).**

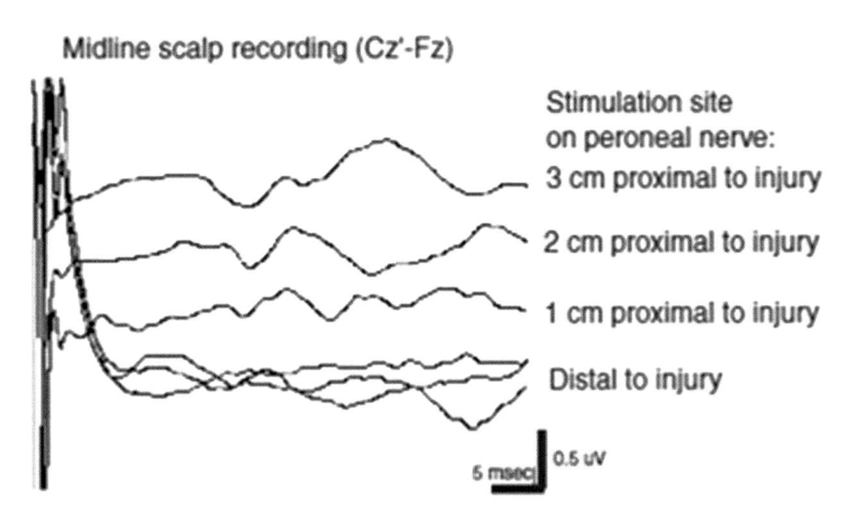

Figure 6. *SSEPs to intrafield stimulation of the peroneal nerve at points 1, 2 and 3 cm proximal to traumatically transected peroneal nerve.*

Brachial plexus injuries:

While neuromas in continuity are certainly part of brachial plexus injuries, overall the neurophysiological analysis of brachial plexus injuries is more involved. Preoperatively, a good EMG examination should be done to identify affected muscles and by implication affected components of the brachial plexus and to differentiate root (avulsion), plexus injuries, and multiple peripheral nerve injuries. Differentiation between root and plexus injuries is made with paraspinal muscle EMG and sensory nerve action potential amplitudes. A root lesion may be associated with abnormal paraspinal muscle EMG due to injury of the motor fibers to the paraspinal muscle which branch off at the root level. Also, because a root injury is proximal to the dorsal root ganglion, root lesions tend to show normal SNAPs. On the other hand, normal paraspinal muscle EMG with absent SNAPs would imply a plexus lesion.

Intraoperatively, preoperative observations may be confirmed and the analysis extended. SSEPs can assess the function of the sensory pathways through the brachial plexus. With a diagram of the brachial plexus in mind (see Figure 2), radial, median, and ulnar nerve SSEPs can reveal information about conduction through the brachial plexus. Radial nerve fibers travel in the radial nerve, posterior cord, posterior division, upper trunk, and C5/C6 nerve roots. Median nerves travel in the median nerve, lateral cord, anterior and posterior divisions, upper trunk, and C6/7 nerve roots. There is a possibility that median sensory fibers also travel in the medial cord, anterior division, lower trunk and C8 nerve root, but this does not seem to be the case based on empirical observation. The ulnar nerve travels in the medial cord, anterior division, lower trunk, and C8/T1 nerve roots. The presence or absence of central SSEPs to these major nerves would imply the presence or absence of viable sensory fibers in the pathways described.

Additionally, SSEPs can be used to differentiate root from plexus lesions by peripheral responses recorded to the radial, median, or ulnar nerve stimulation. The presence of peripheral responses (best recorded at the axilla to optimize chances of seeing responses) and the absence of central responses would imply a root level lesion (preganglionic injury), whereas absent responses at both the peripheral and central level would imply a plexus lesion (postganglionic injury). SSEPs to direct stimulation of roots may also be used to assess nerve root function and determine its viability for nerve grafting **(see Figure 4).**

After an initial assessment of nerve function, SSEPs are used to monitor the approach and the manipulation of nerve during a brachial plexus to prevent inadvertent of the brachial plexus. In addition to using SSEPs to assess nerve function in the brachial plexus, spontaneous EMG should be used throughout to serve as a warning of mechanical irritation sufficient to elicit action potentials in motor nerves. The presence of EMG activity may, but does not necessarily imply actual injury to the fibers.

Triggered EMG performs a valuable function in both guiding dissection and evaluating and identifying nerve components. Oftentimes, there is immense scarring and displacement of nerve in the injured brachial plexus. By judicious use of intrafield stimulation and recording CMAPs and testing tissue prior to dissection for functional motor fibers, inadvertent transection of nerves hidden in the scarred tissue can be avoided. Moreover, by analyzing the pattern of responses in muscle (see tables 1-3) the identity of nerve in question may be revealed. Figure 7 shows example of CMAPs to stimulation of the upper, middle and lower trunks. If the pattern of CMAPs includes serratus anterior, rhomboid, or infraspinatus/supraspinatus muscles, as seen in the left panel, then stimulation is at the upper trunk (refer to brachial plexus drawing in Figure 2). If serratus anterior is present then the stimulation is proximal to the dorsal scapular nerve. If not, but infraspinatus/supraspinatus CMAPs are present then stimulation is between dorsal scapular nerve and suprascapular nerve. If the pattern does not include these proximal muscles and does not include the tricep but does include bicep and median nerve innervated muscles then stimulation is more distal in the lateral cord. If the pattern includes tricep, as seen in the middle panel, then stimulation is at the middle trunk. If the pattern of responses includes the forearm muscles, thenar, and first dorsal interosseus but not proximal shoulder muscles then the stimulation site is at the lower trunk, as seen in the right panel. It should be noted that it is not practical to sample all of the muscles listed in tables 1-3. Selection of muscles should be tailored to the surgical procedure and the nerves that are likely to be encountered. Furthermore, if using surface leads, discrimination of closely placed muscles does not likely occur. So for example, if leads are placed over the extensor carpi radialis muscle, it should be realized that all the extensor muscles of the forearm may contribute to a response, and it may be more practical to consider such leads as reflections of "wrist extensors". **(Fig. 7)**

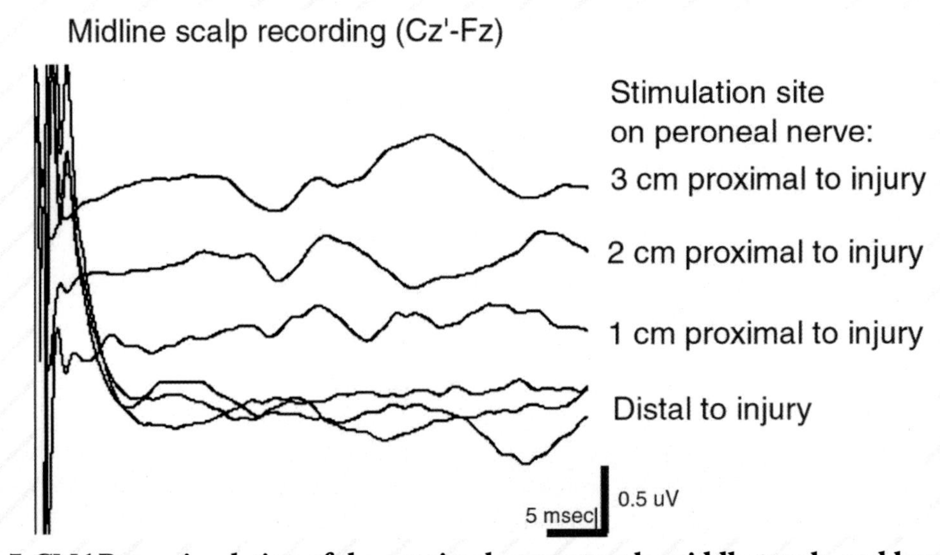

Figure 7. CMAPs to stimulation of the proximal upper trunk, middle trunk, and lower trunk.

Tumors of peripheral nerves

Tumor resections:

Resection of tumors or cysts of peripheral nerves is not necessarily a simple operation. Most peripheral nerve tumors are schwannomas, which start internally in a nerve bundle, grow slowly and force the nerve fascicles to splay out over the capsule of the tumor. Safe resection requires identification of an area of the capsule that is devoid of nerve fibers, incision at this point, and hulling out of the tumor leaving the capsule and nerve fibers intact.

Mapping of motor fibers on the tumor capsule can be done with intrafield stimulation. A flush tip probe is a good choice, since its spread of current is small and allows a focal stimulation. Stimulation can be done in 1-2 mm steps across the tumor capsule, noting the points on the capsule with an ink pen where stimulation elicits a CMAP in the nerve's target muscle. An example may be seen in Figure 8. It is often found that fibers are present across nearly the entire capsule but there is usually one area that no or little response is elicited and may be used for entry to the tumor. In Figure 8 strong responses were seen at the sides of the tumor, show by solid lines, whereas in the middle of the tumor the dots indicated smaller responses. Stimulation of the capsule and tumor internally should continue to be done to verify no motor fibers in tissue to be removed. Following resection, responses to stimulation of the nerve proximal to the tumor can be compared to responses to distal stimulation to verify that they are similar and, hence, give no evidence for any impairment. **(Fig. 8).**

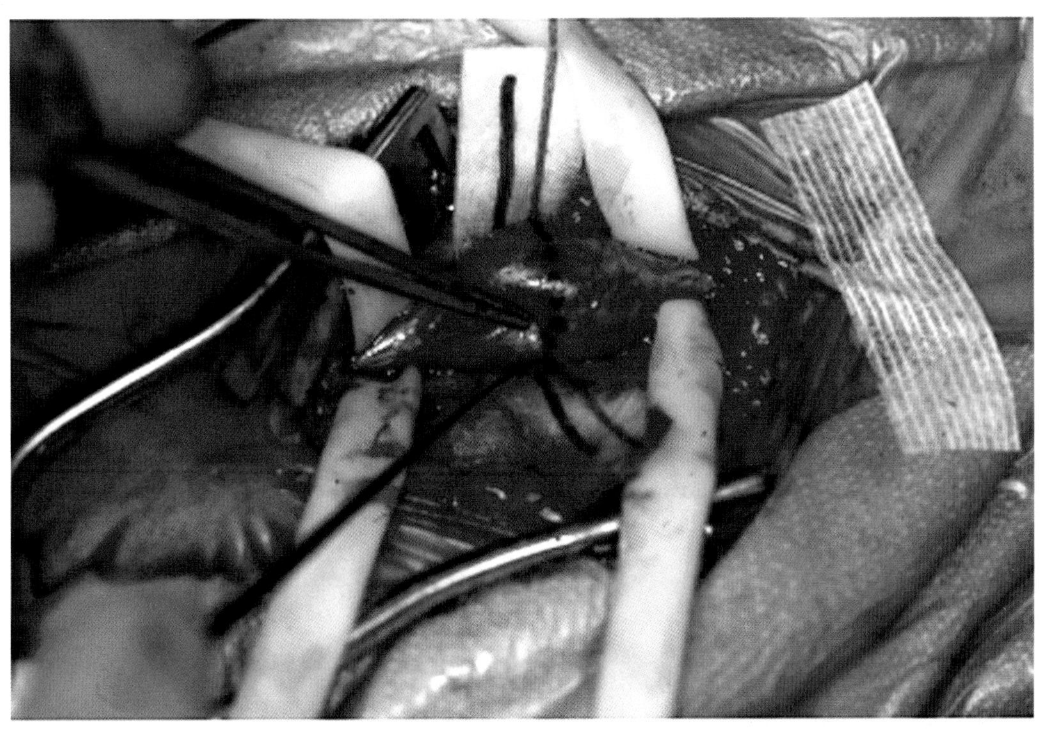

Figure 8. *The results of intrafield stimulation of a median nerve tumor and recording of CMAPs in thenar muscles shows prominent response indicated by the inked lines and lesser responses showed by the ink dots. A longitudinal incision was made in the middle of the dotted area.*

Some tumors show a nerve entering one side and seemingly exiting the other side of the tumor. Often when stimulated these nerves show no response, suggesting they are either sensory nerves only or, if mixed nerves, the motor fibers are not functional. In these cases the nerves are transected and the tumor removed. Figure 1 showed the consequences of having to transect a functional bundle of fibers in order to remove the tumor.

Cysts present a similar situation but cysts are usually outside the nerve bundle. Intrafield stimulation helps identify the location of the nerve as opposed to the cyst. Repeat stimulation during cyst resection can be used to monitor the nerve and give indication of any impending impairment.

Tumors of the Head and Neck:

Resection of tumors of the face and neck are complicated by the uncertainty of the location of the nerves of the face and neck in the presence of large and sometimes widely spread tumors. Depending on the location, the facial nerve, spinal accessory nerve and even nerves of the brachial plexus may be displaced or encased by tumors of the face and neck such as hemangiomas, parotid tumor, neurofibroma, sarcomas. Many surgeons have accepted that loss of nerve function is a given consequence of these resections. The nerves are small, have multiple branches, and are easily altered from typical anatomical location by the growth of the tumors. Visual identification of their location is difficult and tedious and their integrity is threatened by dissection, retraction, cautery, and compression.

At the University of Washington, the use of neurophysiological techniques has provided surgeons with information to plan the surgical approach, identify the location of nerves of face and neck in relation to the tumor, and assist in the dissection and preservation of these nerves. The techniques utilize recording CMAPs as well as spontaneous, free running EMG. The choice of muscles sampled depends on the tumor location and the nerves involved. For facial tumors, facial nerve muscles are sampled for the branches of the facial nerve:

Zygomatic branch:	frontalis, orbicularis oculi
Buccal branch:	levator nasii, orbicularis oris
Mandibular branch:	depressor labii

For tumor of the neck, the spinal accessory nerve target muscles are sampled:

Trapezius

Sternocleidomastoid

For deep neck tumors, the spinal accessory nerve is monitored, and, in addition, the brachial plexus muscles are sampled. The number of muscles used on the following list depends on the extent of the tumor:

Serratus anterior

Rhomboid

Infraspinatus/supraspinatus complex

Deltoid

Biceps

Triceps

Extensor carpi radialis (wrist extensors)

Flexor carpi radialis (wrist flexors)

Thenar

Hypothenar

A most effective component of the University of Washington's protocol is percutaneous mapping of the course of the nerves performed after anesthetic induction and prior to skin incision. A ball tip probe is used in a monopolar stimulation mode with the reference on the neck or shoulder. CMAPs to the stimulation are displayed and viewed by the person doing the stimulation. With the probe placed at a point likely to be over the nerve usually a proximal location along the nerve path, stimulation intensity is increased gradually until a CMAP is seen in the target muscle(s). The probe is then moved back and forth in a direction presumed to be approximately 90 degrees to the path of the nerve. The amplitudes of the CMAPs at each point are compared. The point with the largest amplitude will indicate the closest position to the nerve, i.e. the probe is directly above the nerve and the skin marked with an ink pen. The threshold will depend on the proximity of the probe to the nerve. Below 10mA suggests a close proximity or a shallow location of the nerve, whereas above 10mA suggests the nerve is deep to the skin. The probe is then moved sequentially along the path of the nerve, always probing back and forth at 90 degrees to the nerve to determine the exact location of the nerve beneath the probe and the skin marked with ink. When a nerve branches, each branch should be followed. Sometimes a nerve will run deep in the tissue or more specifically will run beneath a tumor and a CMAP cannot be elicited. In this event, the probe is moved distally to locate a distal component of the nerve by the back and forth method described and then the nerve mapped as far proximal as it can. The facial nerve and spinal accessory nerve and even the brachial plexus are most fruitfully mapped in this manner.

A map of the facial nerve is shown in **Figure 9**. The map shows the zygomatic branch traveling up towards the eye and forehead and the buccal branch projecting towards the nose and mouth. The mandibular branch begins along the jaw line but shows a break then reappears closer to the mouth. The break in the mandibular suggests its path is deep in the tissue and later during dissection was found to course under the tumor (indicated with an X). With the course of the facial nerve mapped in ink on the skin, the surgeons may plan their approach for most efficient dissection. Once the nerves are exposed, a quite close approximation of the actual anatomy to the maps has been observed. **(Fig. 9).**

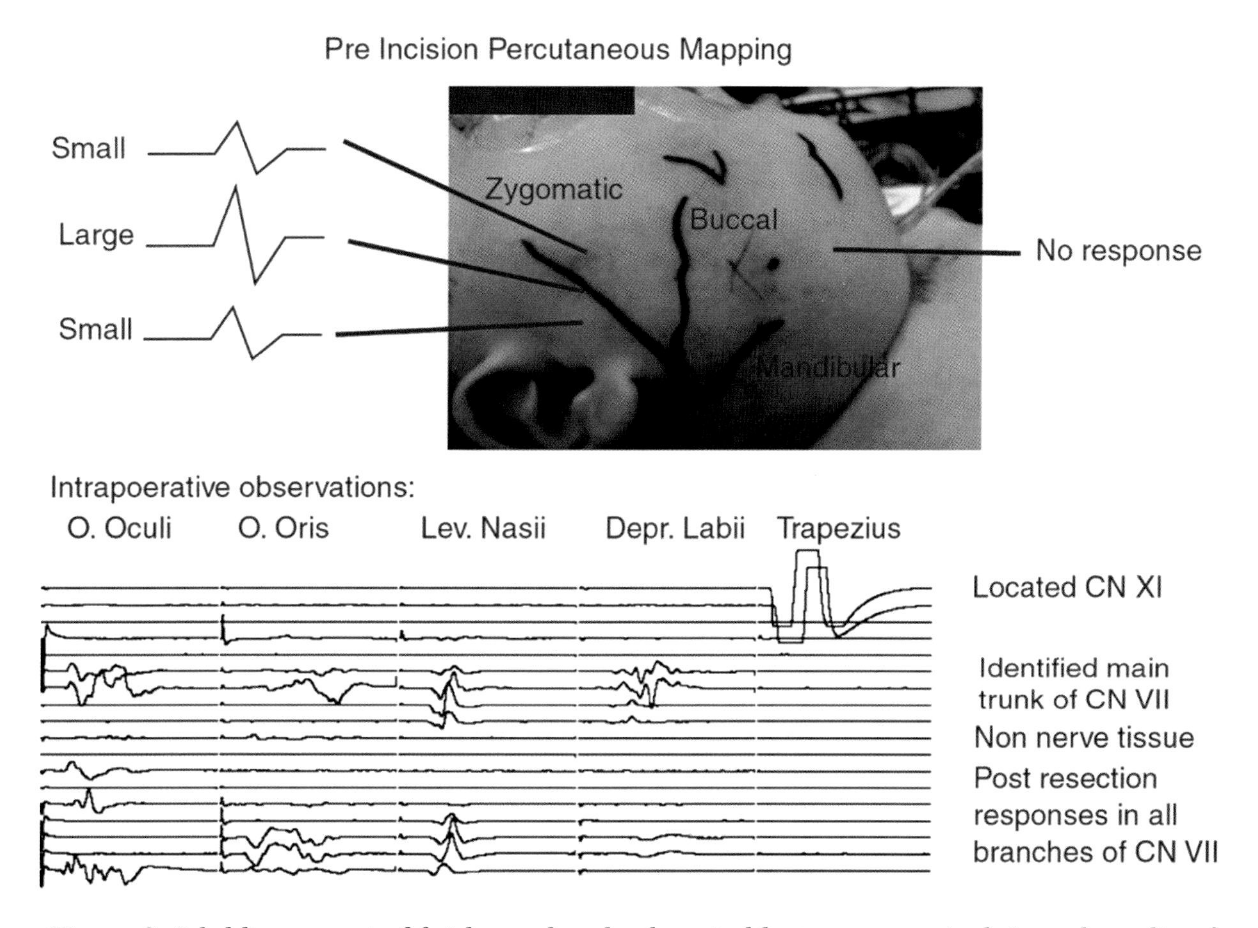

Figure 9. *Inked lines are map of facial nerve branches determined by percutaneous stimulation and recording of CMAPs in facial muscles.*

Once the skin is reflected, intrafield stimulation with a small ball tip or flush tip probe should be used to search and locate the nerves and their branches. When dissection is underway, frequent testing with an intrafield probe should be used to guide dissection and to test any tissue that is removed for motor fibers prior to its resection. Additionally, spontaneous EMG provides useful information to warn the surgeon of impending mechanical irritation of a nerve related to dissection. Often spontaneous EMG is heard, the tissue tested with intrafield stimulation and a fine nerve branch, heretofore not detected visually, is revealed.

With this technique, the facial nerve and its branches, the spinal accessory nerve, and the components of the brachial plexus can be successfully identified, the nerves isolated and separated from the tumor, and tumors resected with good preservation of nerve function.

Nerve Entrapments:

Nerve entrapments may occur on many nerves but the ones that may benefit from neurophysiological techniques are severe ulnar entrapment at the elbow, entrapment of the axillary nerve, and entrapments of the peroneal nerve. Monitoring can be most beneficial when the entrapment is due to prior injury or surgery. The resultant scarring can obscure the nerve and make its exposure a long and tedious dissection. The dissection process can be shortened significantly with less chance of nerve injury by electrophysiological localization of the nerve in the scarred tissue. Probing the tissue with intrafield stimulation and locating the largest amplitude CMAP in target muscles, as described in the previous section on nerve action potentials and shown in Figure 21-2 of Slimp and Kliot, 1995, will identify the closest point to the underlying nerve and reduces the time to locate and dissect the nerve. Monitoring with SSEPs or MEPs lends assurance that the manipulations to free the nerve are without consequence.

Final thoughts:

Intraoperative monitoring primarily is used to protect the nervous system from iatrogenic nerve injury and with the peripheral nervous system it is no different. SSEPs, MEPs, and spontaneous EMG have a role in protecting the peripheral nerves from injury due to compression, stretch, dissection, and Ischemia. However, in addition to keeping nerves safe, neurophysiologists may use the electrophysiological tools or SSEPs, MEPs, spontaneous EMG, triggered EMG (CMAPs), and nerve conduction studies, applied appropriately, to aid and assist the surgeon in the treatment of peripheral nerve disease. The coordinated efforts of both professionals will assure the best outcome for their patients.

Selected Bibliography:

Chiara J, Kinney G, Slimp J,, et. al. Facial nerve mapping and monitoring in lymphatic malformation surgery. Int. J. of Ped. Otorhinolaryngology. 73: 1348-1352

Crum BA, Strommen, JA. Peripheral nerve stimulation and monitoring during operative procedures. Muscle and Nerve. 35:159-170, 2007

Holland NR. Intraoperative electromyography. J. clin. Neurophysiology. 19: 444-453, 2002

Kliot M, Slimp JC. Techniques for assessment of peripheral nerve function at surgery. IN: Intraoperative Monitoring Techniques in Neurosurgery. C. Loftus, V.C. Traynelis (eds.) Mc Graw-Hill. 1993

Kwok K, Slimp J, Born D, et. al. Evaluation and management of benign peripheral nerve tumors and masses. In: Berger MS, Prado M (eds) Textbook of Neuro-oncology. Philadelphia: W.B. Saunders Co. 2003

Robert EG, Happel LT, Kline DG. Intraoperative nerve action potential recordings: Technical considerations, problems, and pitfalls. Neurosurgery. 65:A97-A104, 2009

Slimp JC, Kliot M. Electrophysiological monitoring: peripheral nerve surgery. In: R.J. Andrews Intraoperative neuroprotection. Williams and Wilkins, 1995

Slimp JC. Intraoperative neurophysiological monitoring of the spinal cord and nerve roots. Spine Line. 7: 6-15, 2006

Yuen E, Robinson LR, and Slimp JC. Electrodiagnostic evaluation of peripheral nerves: EMG, SSEP, NAP – Clinical and intraoperative application. H. Richard Winn (Ed.) Youmans Neurological Surgery. WB Saunders, Philadelphia, 2003

CHAPTER 8
Monitoring the Pediatric Population

AN UNDERSTANDING OF maturational issues of the pediatric population is not only helpful, but may be considered a prerequisite, for anyone monitoring children. The changes in electrophysiological signals observed from preterm through adulthood reflect the growth and development of a child's nervous system and body. From the earliest growth plates to final myelination, the nervous system is a changing landscape that influences recording technique and requires appropriate interpretation. Anesthetic management of younger patients has its own implications for monitoring that require certain adjustments that can impact electrophysiological recordings. The scope of pathology in children is different than in adults. Brain tumors, for example, are much more prevalent in the posterior fossa, neuromuscular disorders are common, spine curvature prevails over spine degeneration, and unique congenital defects may be encountered. Monitoring techniques, while essentially the same in children and adults, may require modification. This chapter is intended to identify these differences and provide a framework for approaching monitoring the pediatric patient.

Development of the nervous system:

Figure 1 is a timeline highlighting the essential aspects of growth of the brain, myelination of the nervous system, and growth of sensory and motor systems. Development of brain and its neural circuits begins at 2-3 weeks gestational age (1) with the formation of the neural tube by the folding and fusion of the ectoderm. At one end of the neural tube, the three vesicles that will become the forebrain, the midbrain, and the hindbrain are evident at GA week 4. By GA weeks 5-6, the neuroblasts, precursors of neurons and macroglia, have begun to proliferate and migrate from the ventricular lining of the cerebral ventricles outward to form the preplate, the first recognizable cortical layer. Axonal processes are evident in this early cortical layering as are synaptic connections. By week 8, another layer, the cortical plate, forms and splits the preplate into a subplate below and a marginal zone above. As in the preplate, synaptic connections in the subplate serve as placeholders for the eventual connections from thalamus and subcortical structures. The cortical plate continues to mature and acquires the appearance of mature cortex, while concomitantly the subplate dissolves. In weeks 10-11, the cortical plate thickens and compacts with further migration of cells into the external zones of cortex in weeks 13-15. The time from week 16 to the postnatal period is one of continued thickening of the cortical plate and differentiation of cortical layering. As early as GA week 25, the sensory and motor cortices show lamination, vertical organization, and a diversity of glia and neuronal types.

Following the general plan of a posterior to anterior scheme of development, other areas of cortex similarly develop but in later weeks.

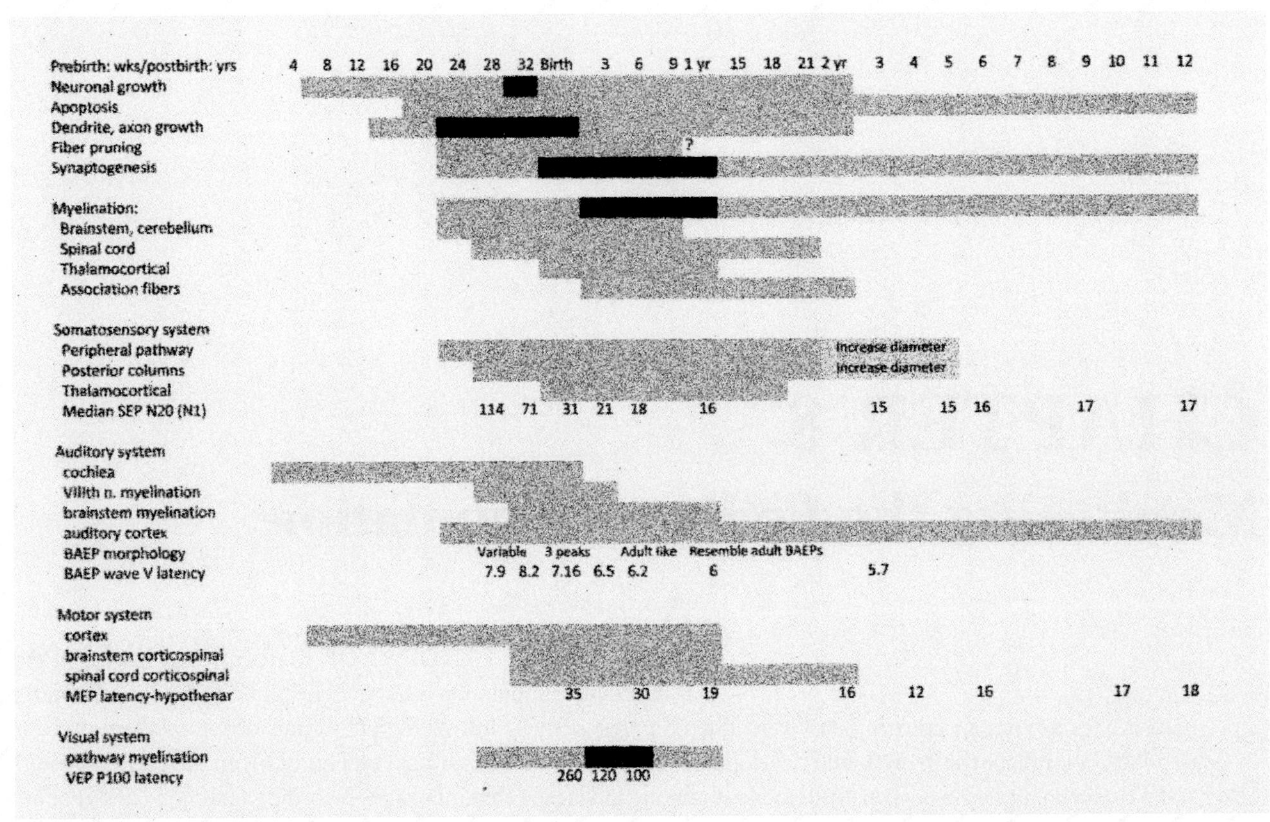

Figure 1. *Timeline of maturation of neuronal growth, myelination, and selected sensory and motor regions of the nervous system*

Differentiation into cell types, their migration and aggregation into cytoarchitectonic organization is governed by recognition molecules under genetic control. Once neuronal cell bodies have reached their destination they begin to form dendritic and axonal processes according the particular type of neuron of that region. Dendritic growth is through growth cones at the end of a dendrite that expands and grows to its destination. Once the dendritic process is at or near its completion, dendritic spines, if present, will form. The maximal dendritic growth occurs at about GA weeks 18 -24 and is mostly completed at birth, although some dendritic growth continues into adulthood. Axonal growth follows a similar pattern with growth cones defining the extent and destination of axons. Axons tend to reach their targets before full dendritic growth is achieved and before afferent processes have arrived. Thus, at birth, the infrastructure of neurons, axons, and dendrites is mostly complete.

Synaptogenesis begins in the early stages of cortical development and progresses evenly until about GA weeks 26-28 at which time dendritic arborization and growth of synapses increases several fold with peak synaptogenesis at GA week 34 when 40,000 synapses per second are produced. Synaptogenesis continues into the postnatal period. Similar to cortical lamination, synaptogenesis occurs in an inside-out manner with some suggestion that motor and sensory areas form the earliest synaptic connections.

Neuronal and synaptic structures are over produced. Following the peak growth periods of these structures, a process of cell death (apoptosis) and retraction of axons, and pruning of dendrites begins. The number of neurons is maximal at about GA week 28. This refinement of neural structures and pathways through growth, apoptosis, and pruning continues into the postnatal months. By adulthood about half of these neurons will have succumbed to apoptosis.

Following the overproduction and culling of neuronal structures, the process of myelination begins. The glial components of the nervous system, microglia, oligodendrocytes, and astrocytes provide essential support functions for neurons. Glial cells are present early on to guide the direction of growth of neuronal structures. They regulate the extracellular environment, modulate synaptic connections, and clear neurotransmitters from the extracellular space. Oligodendrocytes, which will produce myelin, are present in developing cortex. Mature myelin appears in subcortical regions between GA weeks 20 and 28 and in motor and sensory cortices at GA week 35. However, at a normal birth time of 36-40 weeks the vast majority of white matter is unmyelinated. In the first postnatal months myelination is intense, increasing myelinated white matter as a proportion of total brain volume by 1-5%. Myelination becomes less in toddlerhood and slower yet into young adulthood. In general, myelination follows a pattern of inside-out and posterior to anterior development, similar to maturation of neural circuits. For example, the medial longitudinal fasciculus, brainstem, medial lemniscus, lateral lemniscus, and inferior cerebellar peduncle show beginning myelination in in the second trimester followed by the optic tract and chiasm, posterior limb of the internal capsule, central coronal radiate and mesencephalon corticospinal tract in the third trimester. Sensory pathways become myelinated first then motor pathways and lastly association areas. Completion of myelination, particularly of the spinal cord and the corticospinal tracts, is not achieved until 2-3 years postnatal.

The early postnatal period shows tremendous change in brain structure and function. At birth, the brain is about one-third adult size and even a cursory look will reveal that the young brain is not simply a miniature adult brain but a brain in various stages of growth. Primary cortical areas, including the sensory and motor areas, show clear cytoarchitectonic formation, but association areas are still in the process of differentiation. In the first year, the brain grows to about 70% of adult size and in the second year reaches 80% of adult size. While growth of neuronal structures and white matter is occurring, it is myelination that is contributing most significantly to growth of the brain. Synaptogenesis and dendritic growth also proceeds at a rapid rate, reaching peaks in the first year of life that exceed adult levels, necessitating a concomitant process of pruning of dendritic structures and death of neurons.

In addition to the development of the nervous system, another aspect that impacts electrophysiological signals is body growth. Paralleling the rapid growth of the brain is the rapid increase in head circumference (Figure 2). Head circumference increases about 22% from birth to six months and by about 7% from 6 months to the first year of age. After year one growth in circumference becomes relatively steady until adulthood. Body length for these first years as can be seen in Figure 2 also undergoes a more rapid change early on.

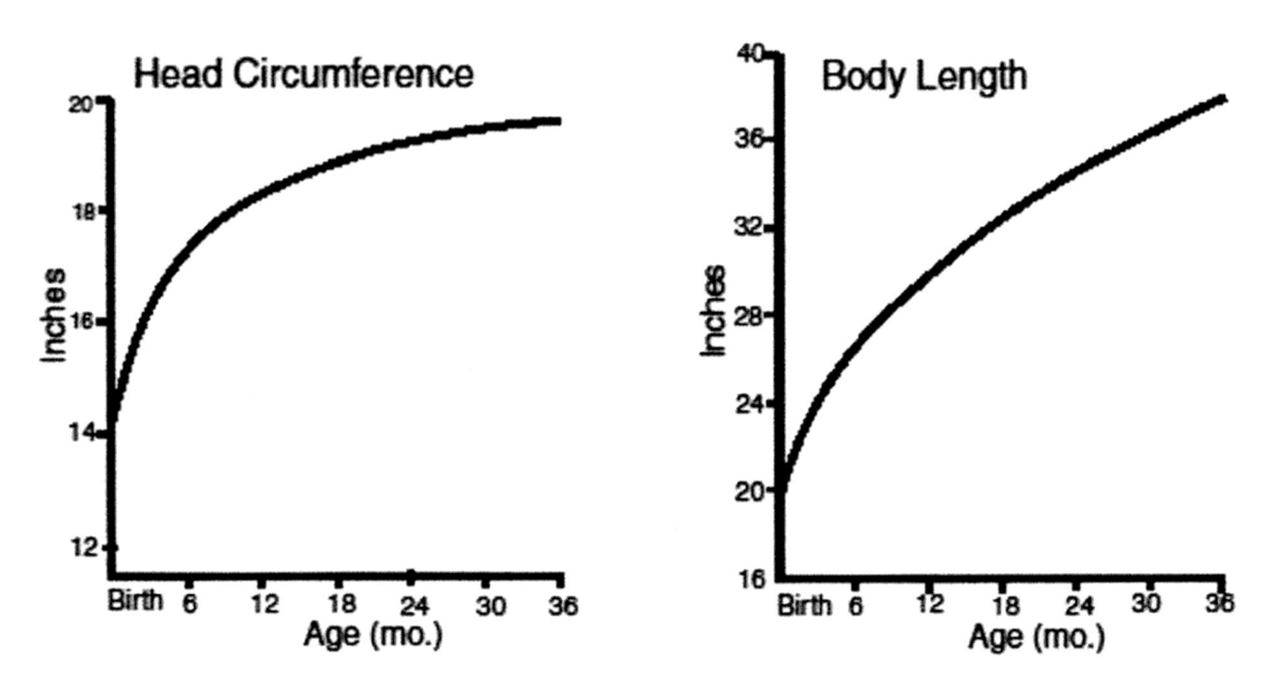

Figure 2. *Growth of head circumference and body length is most rapid during the first year of life. Data derived from standard World Health Organization charts.*

In fact, as can be seen on the left in figure 3, the overall body length, which increases by more than threefold from infant to adulthood, undergoes two rapid growth phases. The most rapid increase in length occurs during the first and second years, leveling off until another more rapid increase occurs during puberty. The average length of an infant is about 19-20 inches. Infants grow about 10 inches during the first year of life. Growth slows during toddlerdom to about 3 inches per year. At about 30 months a child is about one half adult height. Growth from four years to puberty is slower at about two inches per year before the pubertal growth spurt. Girls level off a near adult height before boys, about 14-15 years compared to 17-18 years.

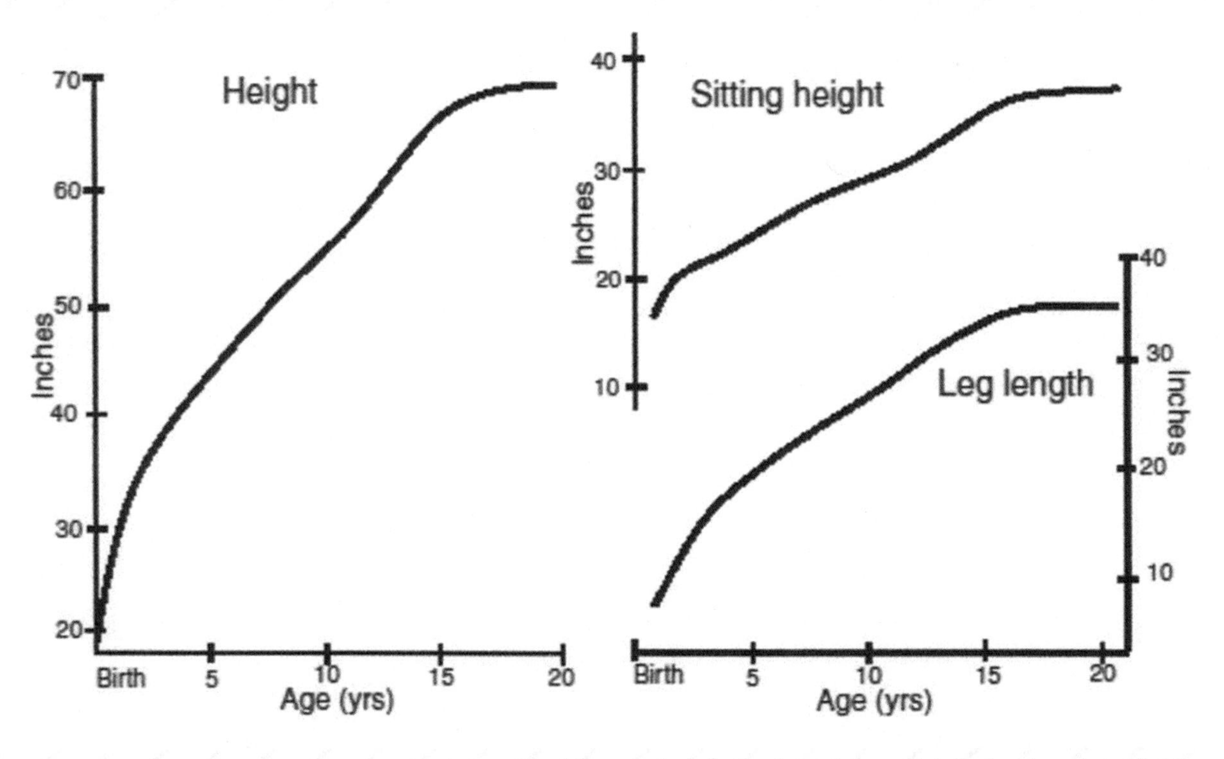

Figure 3. *Growth of height or overall body length is shown on the left. On the right growth is broken into sitting height (body plus head) and leg length. Data derived from Fredriks et al, 2005.*

Not only does the length of a person change with age but body proportion does as well. At birth, head length is about one quarter of total body length, giving the infant a "peculiar" look compared to adult proportions. It is not until young adulthood that individual's head length becomes the more ideal one eighth of body length that Da Vinci developed as the rule for human body proportions. Human body height should be eight head lengths with the neck one quarter of a head length and the legs as four head lengths. This means that leg length is about one half of body height. The right half of figure 3 show the growth curves for sitting height (body plus head) and for leg length. These data suggest that limb length increases proportionally more that body length. The curve for leg length is quite similar to overall height, namely showing a rapid change in first two years, levelling off with steady growth until the pubertal growth spurt where another rapid change takes place. The curve for sitting height, while similar, shows rapid growth for only the first year, then a steady but slower rate than leg length until puberty where another rapid growth is seen. Another way to look at these data is to consider the sitting height to height ratio (SHR). High SHR reflects short legs and low SHR reflects long legs. Infants have a SHR of about 0.68, which decreases rapidly until about 5-6 years becoming about 0.54 with a much slower change over year 6-12 when adult levels of 0.51 (boys) -0.52 (girls) are reached. Genetic, environmental, nutritional, and social factors contribute to variation in body growth and SHR seen in men, women, ethnic groups, and geographical groups.

Somatosensory System:

Maturation of somatosensory pathway:

The neuronal suprastructure of the somatosensory pathway is essentially in place by GA 20-24 weeks. Myelination of the cuneate fasciculus and dorsal roots begins at about 24 weeks and progresses rostrally to the medial lemniscus and internal capsule in the weeks following. At birth, myelination of the somatosensory is quite substantial but continues into the first 1-3 years postnatally. Growth and incorporation of myelin continues through infancy and becomes complete at about five to six years of age. The peripheral nerves and posterior columns complete their myelination at about 2-3 years, whereas central pathways in the brain are not fully complete until five to six years. Additionally, in the first postnatal years, there is a substantial increase in the length of the fibers as well as in increase in the diameter of fibers in the spinal cord.

The peripheral portion of the somatosensory pathway achieves full myelination before the central segment but this growth is accompanied by a steadily increasing length of the peripheral segment as limb length increases with age. The central portion (brain and spinal cord), while taking longer to fully myelinate, proportionally does not increase in length as much as the peripheral portion does. Thus, while the somatosensory pathways of younger children are shorter, they are incompletely myelinated and the latencies of SSEPs are longer than expected. As the pathway lengthens with growth in early years, it becomes more myelinated and conduction velocities increase. These observations have implication for electrophysiological studies of infants and children.

Evoked Potentials:

SSEPs recorded from the scalp reflect cortical activation. The latency of the first component (N1 equivalent to N20) represents the conduction time of the peripheral, spinal, and brain components. SSEPs may be recorded as early as GA week 28. The latencies are very long; amplitudes are small; and the duration of the waveform in spread out. Over the prenatal period, latencies shorten, reflecting the maturation of the somatosensory pathway (Figure 4). From term to several months, the latencies continue to decrease, amplitudes increase, and the waveform duration becomes less spread out. Latencies continue to decrease through age three to four as myelination proceeds and then show a steady increase commensurate with increased growth of limb length. These findings are essentially the same for upper (median n) and lower (tibial n) extremities.

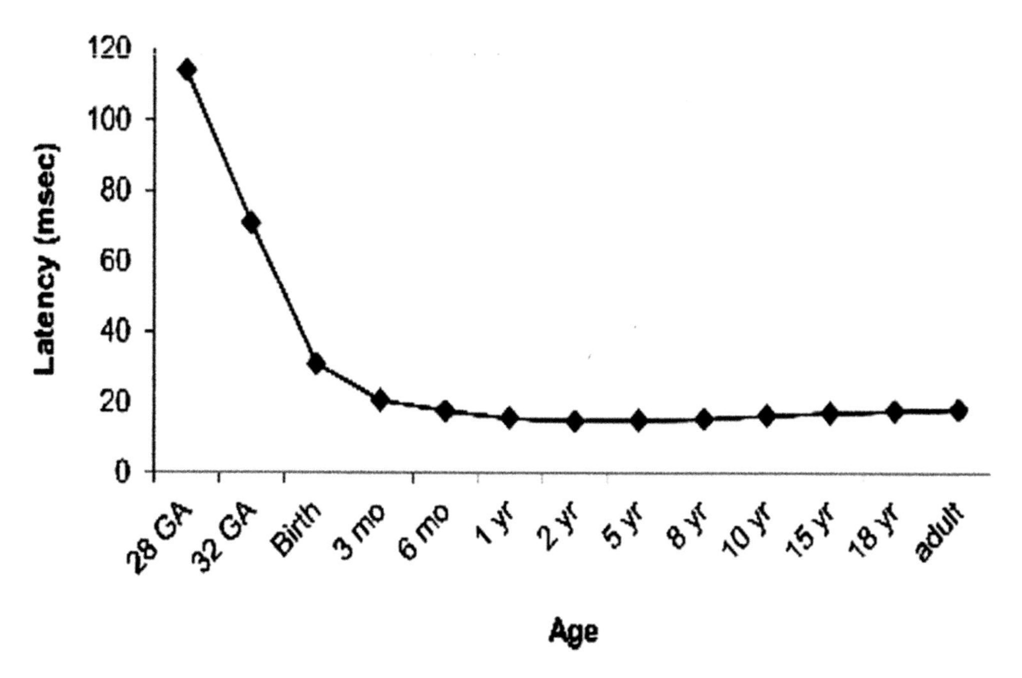

Figure 4. *Typical latencies of median nerve SSEP N20 (N1) from preterm though puberty. Data derived from multiple sources.*

The change in latency of the cortical level responses during preterm and the first two years postnatal reflects maturation of both the central and peripheral portions of the somatosensory pathway. Measurements can be taken over peripheral nerves and over the spinal cord that reflect responses in these two structures. Interpeak times may be calculated between the peripheral response and the spinal cord and between the spinal cord and the cortex.

During the first year of life, peripheral conduction time as reflected by the latencies of peripheral recordings decreases initially, as expected with increase myelination and increased conduction velocity. Then, after one year when myelination is more complete, the latencies of peripheral recordings increase linearly with age given that fiber diameter and myelination have reached full maturation.

In contrast to peripheral conduction time, central conduction time, measured as the time from the spinal cord response to the cortical response, shows a decrease from birth to two years of age. After two years the central conduction time remains constant at adult levels. The central pathway (length of spinal cord) is increasing during this time (see increase in sitting height in figure 3 but the expected increase in latency is offset by the myelination such that the conduction time remains essentially constant. If one divides central conduction time into spinal and brain components, one sees that spinal cord conduction time, measured by the interpeak latency between the spinal cord cervical or lumbar entry zone potentials and the brainstem potentials, is nearly constant from birth onward related to the increase in length of spinal cord pathway balanced by increasing myelination, whereas central conduction time, measured between brainstem and cortex tends to decrease with age as expected by develop of myelination and essentially constant pathway length and isn't complete until 6-8 years.

Technique Implications:

The developing physiology of the somatosensory pathway has implications for neurophysiological technique. For example, even in adults stimulation rate affects amplitude of responses. Up to 3-5 Hz there is no appreciable reduction of amplitude in cortical responses; however at stimulation rates above 5 Hz a rate-related attenuation in amplitude is observed. Because subcortical responses are not attenuated until higher stimulation rates (10-15 Hz), the thought is that the synaptic relays in thalamus and cortex cannot follow high rates of stimulation. In preterm and young infants, the effect is more profound such that rates above 1 Hz will cause attenuation. Again this is likely due to inability of the synaptic relays to "follow" the higher stimulation rates, but an effect of reduced myelination, slower conduction speeds and dispersion of the incoming afferent volley may contribute as well. The more prolonged or dispersed waveforms, seen in preterm and young children, also has implications for bandpass filtering. High pass filtering on the order of 20-30 Hz may reduce amplitudes and alter latencies because the power in the slower, dispersed responses may be falling below the high pass filter cutoff. Better results may be obtained with high pass filters of 10Hz or less. However, during intraoperative monitoring responses produced with lower high pass filter cutoffs may be noisier due to inclusion of slow wave activity in the underlying EEG. Under anesthesia, EEG power shifts to slower frequencies. Children tend to produce more power in the slower frequency range than do adults and thus produce a very low frequency background activity that generates a slow wave baseline in evoked potential recordings.

Auditory system:

Maturation of Auditory Pathway:

The otic placode, a thickening of the epidermis on the side of the head, appears early in the first trimester and soon forms a vesicle that divides into the vestibular and cochlear segments. By 13 weeks, the cochlea is formed into a coiled structure with a cochlear duct and rudimentary organ of Corti and hair cell formation is apparent. The nerve cells of the spiral ganglion have also migrated into position, and distal and central processes that will form the cochlear nerve are advancing. Paralleling the peripheral growth, the auditory nuclei and pathways in the brainstem are recognizable and increasing in size. The auditory cortical structures are present but undifferentiated.

During the second trimester, cochlear growth continues and is nearly complete by the beginning of the third trimester. Similarly the cochlear nerve structures are nearly fully formed and myelination with Schwann cells is beginning near the end of the second trimester. Brainstem neurons are increasing in size and growth of dendrites and axons. Oligodendrocytes are forming and aligning but central myelination has not begun. Auditory cortex shows cell formation with beginning dendritic and axonal formation.

Through the third trimester and first six months postnatal, final maturation of the cochlea occurs. Brainstem neurons reach 60% of size by birth. Myelination increases and reaches adult levels by 6-12 months postnatal. Auditory cortex is still developing and doesn't achieve fully mature appearing neurons until one year postnatal. Connections of deeper layers reach adult levels by age 5-6 and maturation of superficial layer is achieved from 6-12 years but adult density of mature axons awaits prepuberty at 11-12 years.

Evoked Potentials:

The brainstem auditory evoked potential (BAEP) may be recorded at or before GA 28 weeks but is a weak, unreliable response, requiring a stronger than normal stimulus delivered at a slow rate in order to evoke with any ease. At 32 weeks, BAEPs are still variable but slightly more consistent with weak amplitudes, prolonged and distorted peaks compared to older individuals. At term, three peaks are clearly and reliably defined. By 3 months postnatal, BAEPs take on adult characteristics and by 1-3 years closely resemble adult BAEPs. Latencies of the main components (wave I, III, V) progressively shorten from preterm to 4 year postnatal. Wave I reaches adult values shortly after birth, reflecting the early myelination of the cochlear nerve. Waves III reaches adult levels at about 6 months and wave V at about two years, reflecting the balance of increased myelination and faster conduction velocities and the increased length of the pathway as brain growth proceeds.

Technique Implications:

As mentioned, preterm BAEPs benefit from stronger stimulation intensities and slower rates in order to be more easily obtained. In general, increasing the stimulus rate from 10 to 40 Hz increases latencies and decreases amplitudes. Only in newborns is there an increased rate dependent change in latency. From three weeks of age and older the increase in latency due to increased stimulation rate is the same as that seen in adults. Thus, only in newborns should stimulation rates need be held to slower values of 10 Hz or less.

Motor system

Maturation of motor pathways:

The corticospinal tract is the primary pathway for voluntary movement, providing a direct pathway from the pyramidal neurons in layer V of cerebral cortex to the spinal cord. It primarily but not exclusively serves the distal musculature of the hands and feet. Compared to the sensory systems, the corticospinal pathway develops later and with a protracted time course that extends through the second year postnatal, such that it is the last, long tract pathway to complete maturation.

In humans, the cortical plate develops by GA 7-8 weeks, nominally about the same time frame as sensory cortices. Consistent with an inside-out pattern of growth of cortical layers, layer V, the site of corticospinal neurons, differentiates fairly early and begins forming dendritic and axonal processes. Differentiation and functional maturation of motor cortical neurons is substantially completed by GA week 24. The most often quoted studies of Humphrey say corticospinal axons reach the medulla by GA week 8, the pyramidal decussation by week 15, the lower thoracic cord by week 18 and the lumbar cord by week 29. However, studies that measured the growth associated protein 43 (GAP43), which is linked to nerve growth cones, suggest a possible later time course but that corticospinal axons reach the lower cervical cord by at least GA week 24. The earlier axons are followed by waves of additional axons. Growth into the spinal cord gray matter does not occur for several weeks until about the time of birth. Myelination starts at the beginning of the third trimester and is present in the brainstem by GA week 25. Further growth of axons, increases in diameter, and myelination in the spinal cord continues for two years postnatal. Guidance of the growth and direction of the corticospinal axons from cortex, through the internal capsule to brainstem, decussation, and spinal cord is by a host of growth guidance molecules, chemoattractants, and neurite growth inhibitors.

As with the general scheme of over production and scaling back, corticospinal projections in the early stages of development are found from wide areas of cortex. In the rat, they have been identified from all of cortex. Whether they are as wide spread in humans as they are in rat is not known. The projections to the spinal cord are similarly broad with clear bilateral projections. During late prenatal and postnatal periods, axon withdrawal and apoptosis leads to withdrawal of corticospinal projections from whole areas of cortex, reduction in numbers from somatosensory areas, and withdrawal of nearly all ipsilateral projections to the spinal cord.

Motor evoked potentials:

Functional connections from motor cortex to spinal cord as measured by muscle MEPs to transcranial magnetic stimulation have been demonstrated as early as GA 32 weeks, although responses are not consistently produced until after one year of age. During the first two years postnatal, MEP latencies decline rapidly then progressively increase from two years through puberty then levelling off at adult levels. Comparison of central conduction times to peripheral conduction times shows that the progressive increase is due to changes in peripheral conduction. Central conduction times show a decline over the first two years postnatal but remain steady at adult levels after two years. Peripheral conduction time, on the other hand, decreases over the first two years then steadily increase through puberty. Since peripheral nerves attain maximum value for fiber diameter and conduction velocity by five years, the progressive increase is related to lengthening of the limbs with age. The stable central conduction time after two years is likely related to an increase in central pathway fiber diameter proportional to height. These finding in motor pathways are similar to that described above for the somatosensory pathway.

Thresholds for MEP stimulation are initially high and decrease from early years into adolescence in proportion to height. It is known that threshold for nerve stimulation varies linearly with fiber diameter. One explanation for decreasing threshold for MEP stimulation may simply be due to increasing fiber diameter of the corticospinal axons. Production of greater synchrony of activation due to increased myelination or development of more spinal motoneuron synapses may also influence MEP threshold.

As mentioned earlier, morphologically bilateral projections from cortex to spinal cord are evident up to two years. Consistent with these morphological observations, Eyre and colleagues have reported bilateral upper extremity MEPs with focal TMS in preterm and infants, demonstrating bilateral physiological connections. Longitudinal and cross-sectional studies of young children showed a significant reduction of ipsilateral response amplitudes with a concomitant increase in contralateral responses, longer latencies of ipsilateral response, and increased threshold to activation. These findings suggest withdrawal of ipsilateral projections. The projections appear to be withdrawn rather than masked because ipsilateral responses will persist if unilateral preterm or perinatal motor cortical damage occurs, whereas similar insults in adulthood do not elicit bilateral responses.

Technique Implications:

The aforementioned findings come from clinical MEP studies done with magnetic stimulation (TMS). For intraoperative monitoring, it is convention and more successful to use transcranial electrical stimulation (TES). As with adults, MEPs may be recorded from selected muscles (mMEPs) or from the spinal cord (D waves and I waves).

Recording of mMEPs can be done in children with either subcutaneous, surface electrodes, or fine wire electrodes. Preferred muscles are the hand muscles (abductor pollicis brevis, first dorsal interosseus, abductor digiti minimi) and foot muscles (abductor hallucis, tibialis anterior, extensor hallucis longus) but more proximal limb muscles, both flexors and extensor, will produce MEPs, although not necessarily with the same robustness as more distal muscles. The selection of distal extremity muscles is appropriate because the corticospinal pathway preferentially innervates the spinal motoneurons for these muscles.

The site of activation of the corticospinal tract is thought to be deeper in motor cortex with TES than with TMS. TES current penetrates deep in the tissue with activation at the axon hillock of even deeper along the corticospinal pathway. Some express the opinion that activation with TES may be as deep as the brainstem. The site of activation with TMS, on the other hand, is thought to be confined to the neuronal layers of motor cortex.

Similar to adults, TES stimulation is usually an anodal current delivered at any of the International 10-20 sites: C3, C4, C1, C2 or Cz. The more lateral locations (C3, C4) may elicit stronger muscle twitches and may activate more deeply in the brain than the more medial sites (C1, C2, Cz). The latter observation is not an issue for monitoring of the spinal cord but may be a false monitor for brain surgery as the activation would be below the surgical site of interest.

In most situations, a train of stimuli would be used to activate mMEPs, whereas single pulse stimulation is used to elicit D wave and I waves. As mentioned previously, train stimulation overcomes the depressing effect of anesthesia on the generation of multiple descending volleys that are required to activate the spinal motoneurons. The train of stimuli provides a succession of impulses that temporally summate on the motoneuron bringing it to firing threshold and generate a mMEP. In a recent study by Azabou in a cohort of children ages 6-19, the optimal train length was found to be 6 pulses with reasonable success at 5 and 7. For children less than 6 years old, the author's experience as well as that of others (Lieberman) have been that longer trains of 6-9 are needed to achieve consistent mMEPs.

Another variable is the interstimulus interval (ISI), that is, the time between pulses in the train of stimuli. The relevance of this variable becomes clearer if one considers the temporal summation of descending impulses necessary to bring the motoneuron to firing threshold. The ISI determines the timing of the descending volleys, which should be adjusted so that the next stimulus or descending volley will arrive at the maximal peak of activation of the preceding volley. Done appropriately, the 4-7 impulses will summate across time and create a stair step of increasing activation of a motoneuron's membrane until it reaches threshold for action potential activation. Several studies have shown the optimal ISI to be 2 or 3 msec with shorter ISI of 1-2 more effective for upper extremities and longer trains of 2or 3 for lower extremities.

The threshold of activation of mMEPs is a function of age. At first blush, one would think the threshold should be lower than adults because of a child's thinner skull but that is not the case. Thresholds tend to be higher than adults because the thinner skull is offset by the immaturity of the corticospinal pathway that requires larger currents to activate smaller diameter, less myelinated nerves. Threshold with TES show a similar pattern to those mentioned above for TMS, namely that threshold are higher in infants and progressively falls until puberty. As with TMS, TES elicits mMEPs at all ages postnatal (see Fulkerson). Several studies (e.g. Lieberman) have confirmed that the inverse relationship between threshold and age. For children less than 3 years of age, mean thresholds are over 500 V. The thresholds fall about 50 V every 4 years until thresholds in late teens is about 200 V. Train length in these studies was 5-6 impulse and pulse duration was 50usec. Muscle selection was distal extremity but not precisely the same muscles. Anesthesia regimens, which will be discussed momentarily, were similar.

Direct cortical stimulation (DCS) may be used in children as in adults to map areas of motor cortex during epilepsy surgery and during tumor removals. DCS has historically been done using the Penfield technique which involves a sustained train of pulses, 0.5 to 1.0 msec duration delivered at a 20 and 60Hz rate at up to 20mA intensity. The techniques are essentially the same for children as adults. The main drawbacks to this technique is the occurrence of seizures, which can occur up to 20% of the time and necessitates concurrent electrocorticography to identify seizures and delineate misleading after discharges that lead to errors in interpretation of the actual site of activation. This technique may also fail to elicit motor responses in a goodly number of children and especially in anesthetized patients. For similar stimulating parameters, it has been found that the threshold for motor activation is higher in younger children. Identification of language cortex occurs with a higher percentage, when patients are over ten years of age. The immaturity of the nervous system and susceptibility to anesthetics are likely explanations for the failure to elicit responses.

Short train stimulation is rapidly becoming the method of choice for DCS, wherein a short train of 4-5 pulses of short duration (0.5 msec or less) and an ISI of 3-4 msec. is delivered usually with a monopolar stimulating electrode. Stimulation is focal, causes fewer seizures, and may be more successful in eliciting motor responses than the Penfield technique, although there is little information involving direct comparisons. Confounding such a comparison is the tremendous variability observed in thresholds across motor cortex and the functional variability or human motor cortex that does not follow the classic homunculus model, for example a single area that may elicit responses in multiple body regions or variability in the topographical map.

Subcortical mapping using the short train technique may be done with parameters similar to those used for cortical stimulation. Bipolar stimulation may be considered in subcortical applications for more precise localization. Localization of corticospinal fibers may be done in the internal capsule and in the brainstem. Tumor removal is more precise and more complete if corticospinal tracts are localized, identified, and monitored during resection. Train stimulation can also be used in the spinal cord but interpretation must be cautious because motor responses may result from stimulation of posterior columns or other motor pathways and not solely from corticospinal tract stimulation.

Single pulse TES, which is much less effective and not typically used to elicit mMEPs, elicits D waves and I waves recorded with epidural electrodes on the spinal cord. D waves and I waves were first described over 60 years by Patton and Amassian at the University of Washington. D waves (direct waves) and I waves (indirect waves) reflected different mechanisms of activation. D waves reflect the direct, nonsynaptic activation of rapidly conducting corticospinal fibers near the axon hillock. I waves, on the other hand, represent the elicitation of corticospinal activity through trans-synaptic activation of neurons within motor cortex, likely involving the connections of neurons from layers II and III to corticospinal neurons in layer V. These synaptic connections are quite susceptible to anesthetic and other variables. They are not as reliable an electrophysiological tool as D waves. Recording D waves is limited to cases whose focus is above the T11-12 level before significant numbers of corticospinal fibers have left the tract and synapsed in the spinal cord and above the cauda equina as the D wave does not occur below it. Furthermore, if spinal cord ischemia is the likely mechanism of injury, D waves are probably not going to be more sensitive that either mMEPs or even SSEPs since the white matter has a higher resistance to ischemic insult than gray matter. For this reason, D wave is not a technique of choice, and could even be considered inappropriate, for spinal cord monitoring of spine surgeries. However, D waves are most appropriate and quite useful for spinal cord tumor resections.

In children, the success rate for recording D waves is a notable limitation. In contrast to adults, where D waves can be successfully recorded in 75% or better of cases, in children D wave can be recorded in about 50% of cases over the age of 21 months and much less so under 21 months. Over seven years old it seems the success rate approaches adults, although data are limited. The immature, still growing and myelinating nervous system is offered as the reason for the lack of success.

Visual system:

Visual evoked potentials are generally not part of intraoperative monitor armamentarium. VEPs may be recorded to either a reversing checkerboard pattern or flash stimulation. The classic pattern reversal stimulation used in clinic is not feasible in the operating room due to inability to keep the pattern focused on the macular area. Flash stimulation may be relatively easily delivered even through closed eyes but the illumination of the retina is variable from trial to trial in part due to the changing luminosity in a flash stimulus dark to bright back to dark, reflection and scatter of the light by the environment around the eye, and the accompanying unpredictable activation of retinal elements. Thus, flash VEPs are difficult to obtain, variable, susceptible to anesthetic and other variables, and may not accurately reflect visual function. To this writer's knowledge no single study has focused on intraoperative VEPs in children, although reports have included older children with adults, in part, possibly because of the immaturity of the visual system. To be sure, the visual system, both at the eye and in the brain, are not completely developed at birth. Basic visual capacity begins shortly after birth and develops over the first year. The peripheral portion of the retina is nearly completely developed at birth but the foveal and macular areas, while starting in the third trimester, don't experience growth until a few weeks postnatal, and don't achieve full growth until about eight months. Visual cortex begins to develop at least as early as somatosensory cortex following the same general principles of development, namely inside out growth of cortical layer, migration and growth of neuronal structures prior to myelination. The optic tract and optic chiasm begin myelination at the end of the second trimester and is not complete until 2 years postnatal. Thalamocortical axons begin myelination before birth and continues until at least four months post natal. Synaptogensesis starts in the later part of the third trimester and reaches its peak at about eight - nine months. Flash VEPs can be recorded as early as GA 24 weeks with a rudimentary one long peak. A positive peak doesn't appear until GA 34 weeks with a long latency that shortens in the first months postnatal. Pattern reversal VEPs can be recorded in children at or just prior to birth but the latencies of the classic P100 wave are prolonged occurring at over 250 msec. The P100 latency shortens rapidly during the first months and reaches 100 msec at about 6-8 months post natal.

Intraoperative electroencephalography:

Intraoperative electrocorticography has been used in children as in adults with medically refractory epilepsy to assist in the surgical management of this disease by localizing the areas of focal seizure activity, assisting the locus and extent of surgical resection, and provide functional mapping of cortical areas. While the express utility of intraoperative electrocorticography may be questioned, it seems to be clearly beneficial in resection of epileptic foci in the hippocampus and in resection of tumors and other cortical malformations and for monitoring afterdischarges during functional mapping. Since the early reports by Penfield and Jasper in the 1950's, electrocorticography has been applied in the operating room to assist in the removal of epileptic foci with minimal postoperative deficits. Intraoperative electrocorticography is not typically used to identify the epileptogenic areas, a task better accomplished by preoperative methods, because of the limited time intraoperatively to observe an ictal event and the relatively limited success of eliciting epileptic activity with stimulating agents or testing for afterdischarge threshold. Rather, intraoperative electrocorticography's value is in guiding resection and preventing inadvertent postoperative deficits.

In the adult awake state, EEG is characterized by beta rhythm (fast, >13 Hz frequency and low amplitude,) activity. With eyes closed, alpha activity (slightly slower, 8-13 Hz frequency and higher amplitude) predominates. During sleep the rhythm progresses from theta (slow, 4-7 Hz frequency and larger amplitudes) to delta activity (very slow <3 Hz, large amplitude) interspersed at periodic intervals with paradoxical or rapid eye movement sleep with rhythms similar to awake state.

Knowing the maturing condition of the child brain, it is no surprise that pediatric EEG takes on different characteristics. In the awake state, newborns show a predominately slow wave, large amplitude EEG. With age, the primary frequency increases and amplitudes decrease becoming adult like after puberty. Early months are characterized by delta-like slow activity with sleep spindles appearing by 3 months. Up to 3 years, theta activity is primarily present with the beginning of alpha rhythm in posterior areas. From 3 to 10 years, alpha rhythm gradually becomes the predominate rhythm with much less theta and no slow activity. From 10 years through puberty, alpha rhythm predominates but with amplitude gradually reducing to adult levels and theta only occasionally present.

Anesthesia:

Predictably, anesthetic agents alter EEG. GABAergic anesthetics (inhalational gases such as isoflurane and sevoflurane, intravenous agents such as propofol, barbiturates, etomidate, and benzodiazepines) slow the rhythm and increase amplitude in manner reminiscent of sleep. The degree to which these drugs accomplish this effect is directly related to their potency of GABA stimulation. As their dose is increased the slowing of activity and increase in amplitude proceeds until at large doses activity is burst suppressed (isoelectric with periods of slow activity) or completely eliminated. Non GABAergic drugs, typically NMDA blockers such as ketamine and nitrous oxide, have much less effect on EEG and, in fact, activate EEG as shown by increased beta activity. Narcotics in normal dosage have no appreciable effect on EEG. At very high doses, slowing of EEG may be seen.

The changes induced in EEG by anesthetic agents such as propofol and sevoflurane, the two main agents used for anesthesia, are considered to be the similar in 1 year and older children as in adults. One notable difference is the concentration needed to achieve a certain level of anesthesia is correlated with age, higher in younger children decreasing with increasing age. To be sure, each agent has its own effect on EEG. With propofol, EEG progressively slows during intubation and into a steady state until at high dosages burst suppression occurs. Sevoflurane, on the other hand, may show a transient increase in beta rhythm during the initiation of induction followed by slow activity later in induction and into the steady state.

Anesthesia in children is presents challenges on several levels. Of immediate concerns during surgery are sufficient amnesia, sedation, and analgesia to provide no pain, little risk of recall, minimal movement, and appropriate respiratory and cardiovascular regulation. Also of concern are long term effects of anesthesia on physiological function and nervous system development. Recent studies in animals have shown impaired synaptic development, increased apoptosis, and impaired cytoskeleton formation following exposure to anesthetics such as isoflurane, sevoflurane, ketamine, propofol, and barbiturates. While studies are limited in humans, there is some suggestion of deleterious behavioral changes following general anesthesia in children, especially younger than four years. Consequently, the choice of the "best" anesthetic regimen for intraoperative evoked potentials must reflect a balance of these considerations. What may be best for evoked potentials may or may not be best for the child.

The effect of anesthetic agents on electrophysiological measures is well appreciated. In general, most anesthetic agents, whether inhalational or intravenous, reduce amplitude and increase latency of evoked potentials compared to the unanesthetized condition. For sensory evoked potentials, cortical level signals are affected more by anesthetic agents than are subcortical signals. Anesthetic agents have little effect on SSEPs generated from the spinal cord or near the cervicomedullary junction but significantly depress signals originating in somatosensory cortex. Similarly, BAEPs are resistant to anesthetic effects because of the brainstem origins. The explanation these differential effects is the disruptive effect of anesthetics on synaptic transmission. Thus, the more synapse intervening between the stimulation site and the neural generator of a signal the greater the disruptive effect of anesthesia.

For the same reason, MEPs are disrupted by anesthetic agents. The pathway for MEPs terminates and synapses at the spinal motoneuron, a synapse that seems particularly susceptible to the disrupting influence of anesthetic agents, as shown by the fact that the H reflex (Hoffman reflex) is also quite sensitive to anesthetic effects. The I waves recorded from the spinal cord to transcranial stimulation are also suppressed by anesthetic agents, which is due to disruption of synaptic transmission of cortical interneurons responsible for activation corticospinal neurons. The D waves, on the other hand, are not affected by anesthetic agents because there are no synapses between stimulation site and the recording site. The corticospinal neurons are activated at the axon hillock and the D wave recording reflects activity in the cortical spinal tract.

In addition to anesthetic effects on synaptic activity, there can be influences on other pathways as well. For example, the effects of anesthetic agents on motor potentials may also be due to the influence on other neural pathways, such as the propriospinal pathways in the cord that serve to support and facilitate the excitability of motoneurons. Additionally, anesthetic agents may have a general effect on the brain that has ramifications for sensory responses along the lines of their suppressive effects of reflex pathways. Finally, there can be a significant effect on electrophysiological signals related to the secondary effects of anesthetic agents on blood pressure, oxygenation, and temperature.

The consensus of the monitoring field is that total intravenous anesthesia with a propofol/narcotic mixture is the regimen of choice for anesthesia with intraoperative monitoring. The recommendation is the same with children as with adults; however, as with adults children present circumstances that temper this approach. Anesthesiologists each have their own approach to achieve good amnesia, sufficient analgesia, and proper sedation. Induction of anesthesia in most cases is with an inhalational agent, e.g. sevoflurane, although some will do pure intravenous. Most pediatric centers use a restricted dose of inhalational agents or pure intravenous anesthesia for maintenance of anesthesia agent along with ketamine and/or dexmedetomidine as adjunctive agents. For children, propofol is the most commonly used intravenous anesthetic agent. All regimens use an opioid for analgesia, usually remifentanyl. When motor responses (EMG, mMEPs, or H reflex) are recorded, it is advisable to abstain from using neuromuscular blocking agents, beyond the use of a short acting agent for induction. Some people use a partial neuromuscular blockade but this requires constantly monitoring the level of blockade and, if not done well, can lead to false interpretations.

Since the nervous system is in the process of maturation even through puberty, the age of the child may determine what anesthetic protocol to choose. For example, some have advocated using ketamine routinely for children under 6 and to replace propofol with ketamine when MEPs cannot be obtained in older children. Ketamine acts by decreasing NMDA receptor activity, decreases presynaptic glutamate release, inhibits nicotinic receptors and is hallucinatory in adults but much less so in children.

Pediatric procedures:

The procedures presented here are typical for pediatrics. These procedures are by no means unique to pediatrics, although some present relatively more often in pediatrics, and the list is by no means a complete list. The choice of modalities used for these procedures is as with any monitored case dependent on the neural elements at risk and the goals for neuroprotection. The anesthesia regimens used are also determined by the goals intended. For example, if a procedure presents risk during the positioning phase of the procedure and MEPs are intended then the regimen should insure a level of anesthesia appropriate for MEPs such as induction with total intravenous anesthesia or with a low dose, short duration of inhalational agents and no neuromuscular blockade.

Perhaps even more so than in adults, communication and cooperation between neurophysiologists, anesthesiologist, surgeons, and the operating room staff is paramount to achieve common goals.

Spinal Deformity Surgery:

Spine fusions are the mostly commonly monitored pediatric surgery. The usual deformity is scoliosis but kyphosis and kyphoscoliosis form a small percentage of the patient population as well. Surgeries for disk disease are also not frequently. Spinal deformities are most often idiopathic but may result from congenital abnormalities, neuromuscular disorders, or neurodegenerative disorders (e.g. cerebral palsy, muscular dystrophy, Prader-Willi syndrome Friedrich's ataxia, Down syndrome,). The most common risks during spine procedures are screw placements, laminotomies, placement of sublaminar wires, and correction of the deformity with distraction, rotation, or compression. Injury to the spinal cord may come from compression of the canal contents, stretching the spinal cord, or concussion of the cord or nerve roots, but many investigators feel ischemia p for the simple reason that insults seem to take several minutes to evolve.

Increasingly, these procedures are monitoring with multiple modalities: SSEP, MEPs, and EMG. Initially in the late 1970's, monitoring of the spinal cord was done solely with SSEPs (see Slimp and Holdefer). A plethora of studies such as those by Nuwer et al, Forbes et al, and Thuet et al have been published supporting the utility of SSEPs as a means to identify potential neurological complications. It was not long before SSEPs were proposed as a viable alternative to the wake up test and as a standard of care for spine surgeries by individual publications and by the Scoliosis Research Society. While notable in its success as a means of avoiding complications, SSEPs were not without some limitations. Several studies reported false negative cases in which SSEPs were normal throughout a case that resulted in a neurological complication (see Wiedemayer).

In part motivated by the limitation of SSEPs, MEPs were developed as a means to test directly the motor pathways. MEPs have been shown to be an effective monitor of spinal cord motor function in pediatrics. Notable studies, such as those by Schwartz et al, Thuet et al, and MacDonald et al, have consistently demonstrated the ability of MEPs to detect impending neurological injury during scoliosis surgery. The sensitivity and specificity of MEPs with regard to motor function has, not surprisingly, been higher on the order of >95% than that of SSEPs, confirming that MEPs are a direct assessment of motor pathways while SSEPs directly assess the posterior columns. While there are several instances of changes in MEPs without concomitant changes in SSEPs, the reverse has also been reported, lending support for the use of both modalities as a comprehensive monitor of spinal cord function. Further discussion of the relative effectiveness of SSEPs and MEPs may be found in the chapter on spine monitoring.

Growing rods is one method of spinal instrumentation that is unique to pediatrics. Management of early onset scoliosis can be complicated. Casting techniques are not always successfully and early posterior fusion with result in a short trunk and disproportionate body appearance. Growing rods or expandable implants have been used for many years with recent versions of instrumentation gaining more usage. The concept is to use hooks or screws at each end of an adjustable rod that can be lengthened as the child grows. Different systems each has advantages and disadvantages as to orthopedic intent, but all of them present some risk to the nervous particularly the upper extremities where hooks inserted near the first rib and threaten the brachial plexus. Similarly, hook placements in the spinal canal or lengthening of the rods could threaten the spinal cord. For these reasons, monitoring of both upper (ulnar nerve) and lower extremities is advisable.

While SSEPs and MEPs may quite successfully be recorded in normal children, neuromuscular, congenital, and neurodegenerative disorders present some limitations to success. Generally speaking according to DiCindio, SSEPs may be recorded in a higher percentage of cases than MEPs (80% versus 40-60%). The more dense the neurological involvement the less likely signals will be recorded. Occasionally, some have shown that a more proximal muscle than the standard abductor hallucis for MEPs may be more successful. At our institution, routine use an anal sphincter lead often yields a workable and relevant endpoint on neuromuscular cases. Down syndrome patients, who tend to have cervical instability issues, show reasonably normal SSEPs and MEPs but the data must be cautiously interpreted as Patel has shown MEP changes but no neurological complications. Neurodegenerative disorders, such as Friedrich's ataxia, involve both sensory and motor pathways, such that over 90% of cases show no SSEPs and reduced or no MEPs. Whether it is fruitful to monitor these patients is debatable from a cost/benefit perspective, but it must be remembered that some do have signals and likely attempts should be made to obtain signals even though the yield may be low.

Spine at risk:

Our institution has evolved a "spine at risk" classification for those children with unstable spines that may place the child at risk for spinal cord injury due to positioning during any type of surgical procedure even non-spine related procedures. The spines of these patients may have congenital spinal abnormalities or fail to maintain spinal alignment when supporting muscles relax under anesthesia and so risk compression or ischemia of the spinal cord. Spine at risk patients are monitored with upper and lower SSEPs and MEPs. Evaluations are made prior to position, and in some cases prior to induction of anesthesia, then immediately after positioning and throughout the remainder of the procedure. For these patients it is paramount to use an anesthetic regiment that is most compatible with MEPs and SSEPs and at the same time serve the patient's needs. Any alert to change in signals necessitates either repositioning the spine or return of the patient to a supine position.

Arnold Chiari Malformation:

Children with Arnold Chiari malformation are fairly common and in a way are similar to spine at risk children in that the initial concern is positioning of Arnold Chiari patients in the prone position. The suboccipital decompression phase and, if done, dural opening and decompression of soft tissue are also points of risk but not as concerning as positioning. Reports in the literature by Anderson et al and Danto et al focus on changes in SSEPs related to positioning. At our institution, Arnold Chiari decompressions are monitored with upper and lower extremity SSEPs upper and lower MEPs, BAEPs, and spontaneous EMG of lower cranial nerves. While a few cases have shown changes in the EMG activity, the more concerning have been the few cases SSEP changes with patient positioning.

Tethered cord:

A normal spinal cord is free to move up and down within the spinal canal. In tethered cord syndrome, the spinal cord is abnormally attached to the spinal canal. In children, this anchoring prevents the spinal cord from migrating cephalad as the child grows, stretching the spinal cord and nerve roots leading to neurological impairments. Most typically one sees, leg weakness, sensory loss in the lower extremities, disturbed gait, and bladder and bowel dysfunction. Tethered cord syndrome is often but not exclusively associated with spina bifida. It can be associated with a tight filum terminale or with injury. Most typically the spinal cord is tethered at the caudal end of the spinal cord as is associated with spina bifida but on rare occasions the cord can be tethered at the cervical and thoracic regions.

Monitoring of tethered cord release depends on the level of involvement. The vast majority of cases involves the conus medullaris and the cauda equine and requires an EMG recording array of muscles innervated by the L2 through sacral nerve roots. Typical muscles sampled are Iliopsoas (L1/2), hip adductors (L2), vastus medialis (L3/4), extensor hallucis longus (L5), biceps femoris (S1), abductor hallucis (S1), gluteus maximus (S2), and anal sphincter (S2-4). Spontaneous EMG monitoring of these muscles alerts the surgeons to potential mechanical irritation or injury to nerve roots of the cauda equina. Triggered EMG to direct nerve stimulation is essential for the surgeon to differentiate nerve roots in the abnormal and often scarred tissue prior to dissection and certainly prior to any resection. SSEPs to tibial nerve or peroneal nerve stimulation can be helpful but is not a timely monitor for potential mechanical injury. Also, SSEPs are not particularly suited for single nerve roots and only address L5 and S1 levels. MEPs can be reassuring but again may not be a timely monitor of impending mechanical injury but rather will give information about functionality after the fact.

A few studies (e.g van Koch et al, Pouratian et al, and Beyazova et al) have reported on the pediatric population. Only one of the 111 total cases showed a postoperative complication; the rest were either improved or stable. In total, the studies were supportive of the use of neurophysiological tools to identify the location of nerve roots so as to avoid their injury and to monitor overall motor and sensory function.

Spinal Cord Tumors:

Spinal cord tumors are classified according to their location within the spine. Intramedullary spinal cord tumors are rare neoplasms, occurring in less than 5% of central nervous system tumors. They are located within the spinal cord and are usually gliomas such as astrocytomas, emendymomas, and dermoid cysts. Intradural, extramedullary tumors are found within the dura but outside of the spinal cord itself and are usually benign such as meningiomas. These tumors occur infrequently and most often in older children. Epidural tumors occur outside the dural membrane and may be bone tumors or metastatic tumors. Their occurrence is also fairly infrequent.

All spinal cord tumors carry a degree of risk for spinal cord injury. There is little good evidence outlining the relative risk of epidural, intradural, extramedullary, and intramedullary tumors, but logic would dictate that intramedullary tumor surgery can be quite concerning. For tumors outside the spinal cord, the techniques for monitoring are similar to those for spine surgeries in general, i.e. upper and/or lower extremity SSEPs and MEPs. EMG may be added for additional coverage of specific nerve roots. Intrafield stimulation and recording of target specific CMAPs can benefit a surgeon's identification and location of a nerve roots, particularly when that nerve root is engage around or within a tumor.

Monitoring of intramedullary spinal cord tumors (ISCT) by contrast presents different challenges and has led to refinements in monitoring protocols. The overall goal of monitoring is protection of the spinal cord from iatrogenic damage. Not only should the tools used accurately predict a neurological deficit but they should identify impending injury in time for corrective actions to be taken. The need for good electrophysiological tools is motivated by the surgeon's goal of maximizing tumor resection in order to insure maximal long term survival.

SSEPs were the first neurophysiological tool applied to intramedullary spinal cord tumor surgeries but were quickly found to have limited value. SSEPs monitoring of sensory pathways appropriately identified impending impairment of the posterior columns but provided little accurate information about motor function. With the advent of MEPs and additionally EMG, relevant information about motor function was provided and now it is common to see multiple modality monitoring applied to ISCT surgeries. Recent studies have reported high sensitivity and specificity values in predicting neurologic outcome when using multimodality monitoring.

During ISCT surgeries there are two times that IOM techniques can play a significant role in the course of the surgery. The first is with identification of the midline of the spinal, which is important to minimize damage to the posterior column during the midline myelotomy for the approach to a tumor. The midline sulcus that divides the posterior columns is nearly virtually impossible to determine by visual inspection. It can be inferred by examining dorsal root anatomy, vascular anatomy, and overall shape of the spinal cord but an accurate identification can only be made electrophysiologically. Two methods have been described. One is to utilize stimulation of the posterior tibial nerves and record SSEP responses directly from the spinal cord with an array of tiny electrodes embedded in a silastic strip that is placed across the posterior spinal cord. Left side stimulation will elicit responses in the left posterior column and right side stimulation in the right posterior column. An analysis of the sequence of responses record across the strip electrode should show responses to left posterior tibial nerve stimulation on the left contacts with response gradually fading to little or no responses as one proceeds to the right side contacts. Right tibial nerve stimulation will do the obverse. The contact that yields small responses to both would indicate the midline.

A second approach is to use a small stimulating probe such as a flush tip probe to stimulate in a sequence of sites across the posterior columns and record the cortical SSEPs from the scalp at lateral locations (C3' referred to C4' and C4' referred to C3'). Recordings in this manner augment the paradoxical lateralization seen with cortical recordings to posterior tibial nerve stimulation, namely that the appropriate initially positive (P37) wave is recording over the ipsilateral cortex. So, when stimulating the left posterior columns one would expect the initially positive wave to appear on left side leads (e.g. C3' referred to C4'). As one proceeds to stimulate across the posterior column, when stimulation reaches the right posterior columns one would expect the presentation of cortical responses to reverse and be initially positive over the right hemisphere (or reverse polarity on the left hemisphere leads). An example is shown in Figure 5. Left tibial nerve stimulation was done with a scalp montage of C3'-C4'. Stimulation of the left posterior columns should and did elicit an initially positive component with a latency of about 10 msec. Stimulation of the right posterior columns elicited as expected a reversal of response polarity such that an initially waveform was generated. Stimulation at the suspected midline (labeled Lt of Midline in figure 5) elicited a positive waveform suggesting that location was left of midline. A more right side stimulation (labeled Rt of Midline) was suggestive of right side stimulation at that location. An intervening vessel was moved and stimulation below the vessel elicited an essentially flat waveform suggesting that location was the midline. Subsequent midline myelotomy at the electrophysiologically determined midline was without consequence to standard left and right tibial nerve SEPs indicating that a midline location was accurately determined.

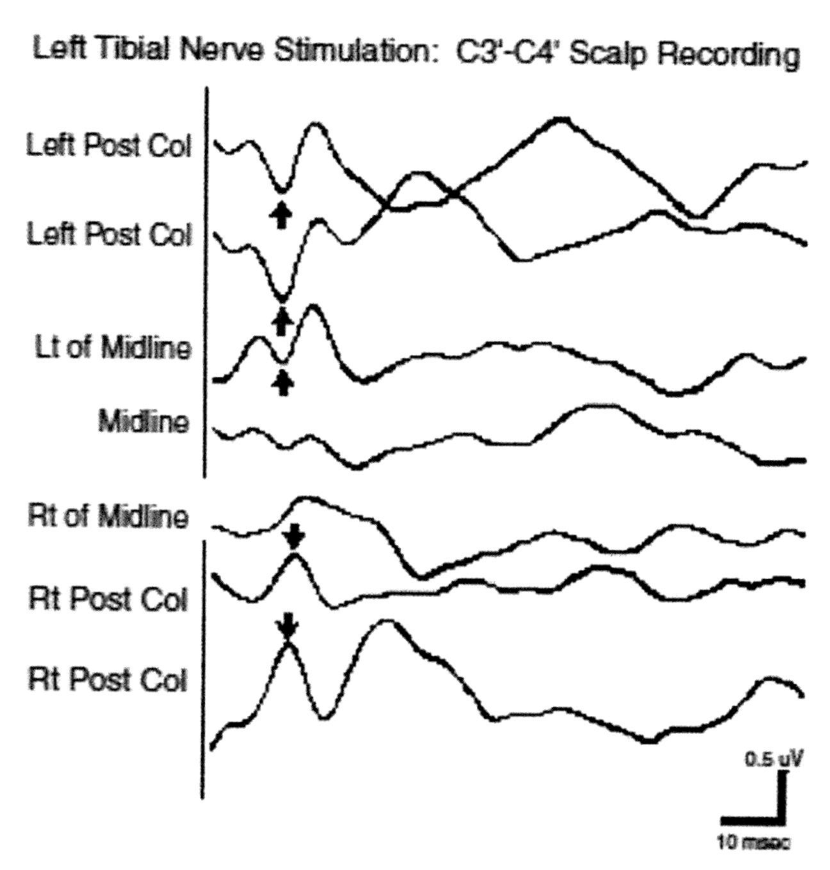

Figure 5. *The midline of the thoracic spinal cord was electrophysiologically determined by stimulation of the spinal cord with a flush tip probe at a series of locations across the posterior columns orthogonal to the midline and recording at the scalp with a C3'-C4' montage. The midline was located by reversal of the response polarity as the stimulation was moved from left to right.*

A second way IOM techniques may impact ISCT resections is providing information about the extent of the surgical resection. As the resection nears its completion, contact with the spinal cord is imminent. Too much or inappropriate resection may damage the spinal cord. Too little action and tumor will be left to enjoy another day. Again SSEPs are less suited for this task but may not be totally irrelevant. As indicated above, it is common to lose SSEPs if a midline myelotomy is not on the midline. Second, SSEPs are often lost when the incised spinal cord is retracted, especially if held retracted by sutures for a long period of time. However, if the SSEPs are maintained then they can be used to monitor posterior column and their loss during resection would imply injury to the posterior columns with possible implications for motor pathways. With the advent, though, of MEPs to transcranial electrical stimulation, a more direct measure of motor function was developed. MEPs recorded from muscle are strong indicators of postoperative outcome. Intact MEPs are associated with good outcome, whereas loss of MEPs may indicate possible loss of function. D waves recorded with epidural electrodes provide a further refinement in predicting postoperative outcome. Deletis and colleagues assert that the presence of D waves in the presence of the loss of MEPs means a transient paralysis that should recovery within weeks. Loss of both D waves and MEPs, on the other hand, implies permanent paralysis. Observation of spontaneous EMG of target muscles related to the surgical site may also provide feedback to the surgeon of irritation or potential injury to either the motor nerve roots or to the anterior horn cells for the target muscle.

An example of transient functional loss associated with loss of MEPs in the presence of stable D waves is shown in Figure 6. Stable D waves were recorded above a thoracic intramedullary tumor as a control and below the tumor as a monitor. At baseline, the D waves recorded below the lesion showed smaller responses compared to those above the lesion, which may have reflected the patient's preoperative condition of hyperreflexia and clonus but with essentially normal strength. Both of the D waves were stable throughout giving no indication of impairment on the corticospinal pathway. The MEPs recorded from muscles in the lower extremities, on the other hand, showed a loss of signal in the left abductor hallucis during the tumor resection that remained absent through closing with no changes in the right abductor hallucis MEP. Postoperatively, the patient had no movement in the left leg and some weakness in the right leg, all of which recovered completely after two days.

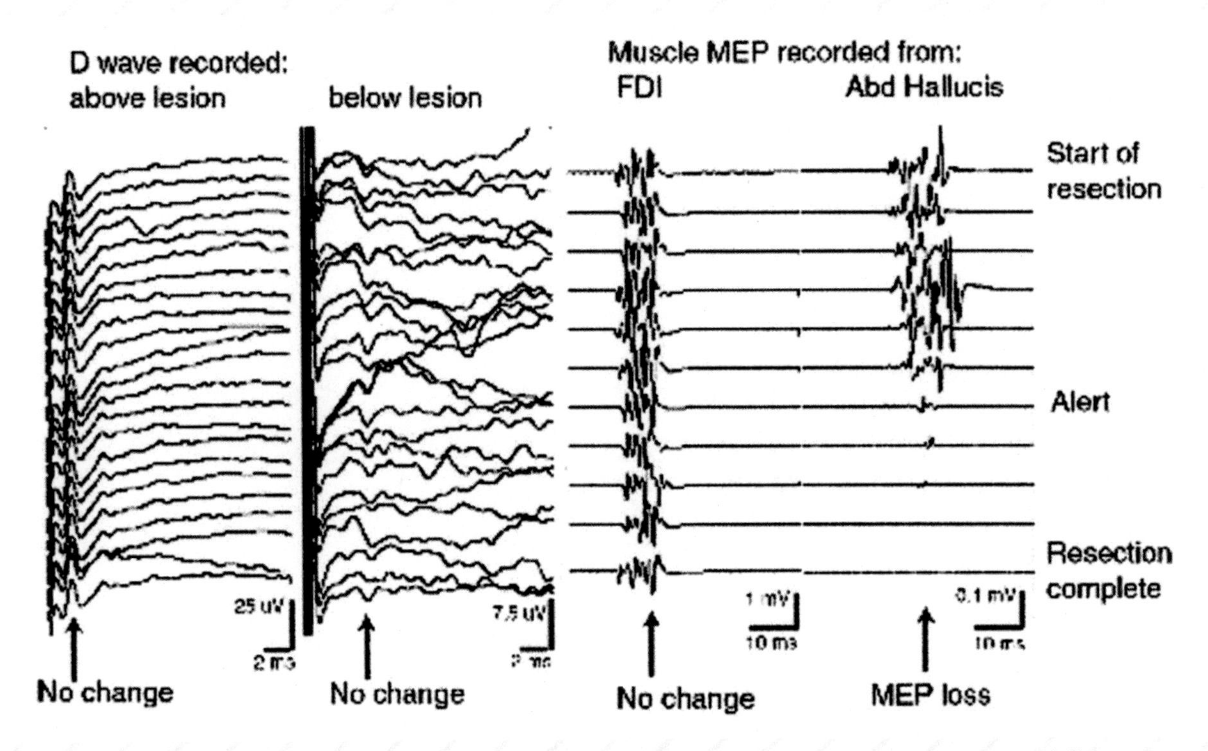

Figure 6. *Stability of D waves recorded with epidural electrodes above and below a mid-thoracic Intramedullary tumor and loss of muscle left lower extremity MEPs correlated with transient paralysis of left leg.*

In addition,times associated with midline myelotomy and tumor resection, IOM can be important during the initial approaches to and resection of a tumor. Often pulling, lifting, or pressing on a tumor can affect nerve roots or the spinal cord and induce changes in signals. Lastly, monitoring during closure should be conducted to protect against bleeding or compression effects on the spinal cord.

Brain tumors:

Brain tumors are the second most prevalent cancer in children, accounting for about 21% of childhood cancers. Central nervous system tumors in children are associated chiefly with the posterior fossa, occurring in about 90% of the cases in contrast to adults where posterior fossa tumors occur much less frequently. Multiple structures are at risk during posterior fossa surgery, including the dorsal column nuclei, cranial nerve nuclei located on the floor of the 4th ventricle (IX, X, XI, XII), the facial nerve at its genu below the floor of the 4th ventricle, the brainstem auditory pathways, and less likely but still at some risk the corticospinal pathway. Consequently, monitoring of posterior fossa tumors utilizes multimodality monitoring with upper and lower SSEPs, upper and lower mMEPs, BAEPs, and spontaneous EMG and triggered EMG of facial and lower cranial nerves (X, XI, XII). Triggered EMG may identify and monitor cranial nerves and be used to map the facial nerve and cranial nerve nuclei on the floor of the 4th ventricle. More recently procedures have described monitoring MEPS from cranial nerves VII, X, XI, and XII.

Face and Neck Tumors:

Tumors of the face and neck in children, such as hemangiomas, lymphatic malformations, and parotid tumors, are often intimately associated with the facial nerve. These tumors infiltrate the surrounding tissue including the facial nerve distorting and displacing the structures. The facial may be displaced deep to the tumor, stretched and elongated around the tumor, and sometimes surrounded by tumorous tissue. Moreover, these types of tumors often require repeated surgeries, which mean scar tissue in addition to the tumor itself. Facial nerve paralysis often occurs with resection of these tumors and has become an acceptable consequence of these surgeries because of the difficulty to visualize and separate the facial nerve and its branches from surrounding tissue.

At our Children's hospital, we routinely use continuous EMG monitoring, both with spontaneous and triggered EMG, of target muscles for the branches of the facial nerve as described in the chapter on peripheral nerve monitoring. Comparison of the path of the facial nerve branches identified by preoperative percutaneous stimulation to that derived by intrasurgical stimulation shows a remarkable similarity. In a recent evaluation of a small sample of our patients, it appears that facial nerve monitoring reduces the incidence of facial nerve injury. However, the most distinct advantage is the reduction of time in the surgery. The facial nerve preoperative mapping allows the surgeon to plan a quicker approach to the tumor through areas devoid of facial nerve branches and with intrasurgical mapping to define the location of the facial nerve more quickly and reliably than by simple visual inspection. It is estimated that surgical time is reduced by at least two hours.

Brachial plexus tumors in children are typically associated with neurofibromatosis, due to sarcoma, and occasionally are nerve sheath tumors. Brachial plexus are most often associated with brachial plexus birth palsy. For these disorders, a similar monitoring approach is taken. The brachial plexus may be mapped with percutaneous stimulation, particularly in large, widespread tumors. Additionally, the spinal accessory nerve may be mapped so its location can be defined and avoided during approach to a brachial plexus tumor. Intrafield stimulation is also used to define and locate nerve elements during dissection and to test tissue for presence of absence of motor fibers prior to transection. Although samples are small, our experience has resulted in little or no postoperative deficits when these cases are monitored.

Selective Dorsal Rhizotomy (SDR):

The intent of the surgical procedure, selective dorsal rhizotomy, is to reduce spasticity in the lower extremities of children with cerebral palsy, spinal cord injury, or brain injury. The underlying aberrations in neurological function in spasticity is still unfolding but is thought to be due to abnormal regulation of spinal motoneurons through loss of descending spinal inhibition that results in increased responsiveness of the segmental reflex to sensory stimulation primarily of large diameter sensory fibers.

Some have considered it mandatory to do electrophysiological monitoring during SDR, but the effectiveness of electrophysiology in SDR is not universally agreed upon. Nevertheless, a carefully prepared protocol and clear criteria for nerve root evaluation will ensure better results. Whether electrophysiologically determined procedures actually provide superior outcomes over simple random transection of nerve roots is still an open question. However, since most institutions employ electrophysiology in some form it is unlikely that question will be specifically addressed. The best SDR teams utilize electrophysiology and its continued use suggests there is benefit in the process.

The SDR procedure involves a laminectomy for exposure of the lumbosacral nerve roots. The neurosurgeon isolates the nerve roots of the cauda equine. Electrophysiological testing requires sampling EMG activity from a comprehensive array of muscles that reflect innervation of the L1 to S4 nerve roots. Hook electrodes are used to deliver electrical stimulation (0.1 msec duration pulses at varying intensities or a one sec train of pulses delivered at a rate of 50Hz) to selected nerve roots. Larger nerve roots are divided into smaller rootlets. Each nerve root or nerve rootlet is evaluated and either transected, partially or totally, or left alone.

The first evaluation involves whether a nerve root is a ventral or dorsal root. This decision is based on threshold of CMAP elicited in a target muscle. Typically, ventral root thresholds are 0.1 mA or less but our criterion for a ventral root is less than 0.5mA. These roots are set aside with no other treatment. Thresholds above 0.5 mA are considered to be dorsal roots. Following identification of a dorsal root, a one second train of 0.1 msec duration pulses at a rate of 50Hz and intensity just above threshold for single pulse activation of a CMAP is delivered. An EMG response to train stimulation is considered normal if it shows a steady response in the possible target muscle for that root, but is considered abnormal if sustained responses shows waxing and waning amplitudes or steadily increasing amplitudes and/or if sustained responses spread to non-target muscles and/or if the EMG responses persist significantly after the stimulation train ends. Nerve roots may be subdivided and retested for threshold and for train responses. Roots or rootlets are partially or totally transected depending on the degree of abnormal responses and depending on whether the affected nerve root/muscle has significant spasticity. If there are no abnormal responses, roots are typically minimally transected. Any dorsal roots associated with S2-4 nerve roots (anal sphincter EMG) are spared entirely.

The effectiveness of using electrophysiology in SDR has been questioned over the years (see McLaughlin et al, and Steinbok). Patients with SDR plus physical therapy show similarly effectiveness to patients with physical therapy alone or patients with electrophysiologically determined SDR show similar outcomes to patients with no electrophysiologically determined SDR. While these findings are not encouraging, recent anatomical evidence by Fukuhara indicated a significantly higher occurrence of anatomical abnormalities is electrophysiologically identified nerve roots compared to normal nerve roots. The latter finding has encouraged our institution to continue using electrophysiology with early results suggesting its effectiveness.

Selected Bibliography:

Anderson RC, Emerson RG, Dowling KC, et al:. Attenuation of somatosensory evoked potentials during positioning in a patient undergoing suboccipital craniectomy for Chiari I malformation with syringomyelia. J Child Neurol. 2001; 1,12,936- 939

Azabou E, Manel V, Andre-obadia N, et al. Optimal parameters of transcranial electrical stimulation for intraoperative monitoring of motor evoked potentials of the tibialis anterior muscle during pediatric scoliosis surgery. Neurophysiol Clin. 2013; 43,4, 243-250

Beyazova M, Zinnuroglu M, Emmez H., et al. Intraoperative neurophysiological monitoring during surgery for tethered cord syndrome. Turk Neurosurg. 2010; 20,4,480-489

Boor R, Goebel B, Doepp M., et al. Somatosensory evoked potentials after posterior tibial nerve stimulation--normative data in children. Eur J Paediatr Neuro.,, 1998; 2, 3, 145-152

Boor R, Goebel B, Taylor MJ. Subcortical somatosensory evoked potentials after median nerve stimulation in children. Eur J Paediatr Neurol. 1998; 2, 3, 137-143

. Brody BA, Kinney HC, Kloman AS., et al. Sequence of central nervous system myelination in human infancy. I. An autopsy study of myelination. J Neuropathol Exp Neurol. 1987; 4,3, 283- 301

Constant I, Sabourdin N. The EEG signal: a window on the cortical brain activity. Paediatr Anaesth., 2012; 22, 6, 539-552

Deletis V, editor. Corticospinal tract monitoring with D- and I-waves from the spinal cord and muscle MEPs from limb muscles. 2008; New York, Elsevier

Deletis V, Sala F . Intraoperative neurophysiological monitoring of the spinal cord during spinal cord and spine surgery: a review focus on the corticospinal tracts. Clin Neurophysiol. 2008; 119, 2, 246-264

DiCindio S, Theroux M, Shah S., et al. Multimodality monitoring of transcranial electric motor and somatosensory-evoked potentials during surgical correction of spinal deformity in patients with cerebral palsy and other neuromuscular disorders. Spine (Phila Pa 1976). 2003; 28, 16, 1851-1855

Dubois J, Dehaene-Lambertz G, Kulikova S., et al. The early development of brain white matter: A review of imaging studies in fetuses, newborns and infants. Neuroscience, 2013

Eyre JA, Miller S, Ramesh V. Constancy of central conduction delays during development in man: investigation of motor and somatosensory pathways. J Physiol., 1991; 434, 441-452

Fagan ER, Taylor MJ, Logan WJ. Somatosensory evoked potentials: Part I. A review of neural generators and special considerations in pediatrics. Pediatr Neurol. 1987; 3, 4,189-196

Ferguson J, Hwang SW, Tataryn Z., et al. Neuromonitoring changes in pediatric spinal deformity surgery: a single-institution experience. J Neurosurg Pediatr. 2014; 1,(3, 247 -254

Forbes HJ, Allen PW, Waller CS., et al. Spinal cord monitoring in scoliosis surgery. Experience with 1168 cases. J. BoneJoint Surg. Br. 1991; 73, 3, 487-91

Fredriks AM, van Buuren S, van Heel WJ., et al. Nationwide age references for sitting height, leg length, and sitting height/height ratio, and their diagnostic value for disproportionate growth disorders. Arch Dis Child. 2005; 90, 8, 807-812

Fukuhara T, Nakatsu D, Namba Y., et al. Histological evidence of intraoperative monitoring efficacy in selective dorsal rhizotomy. Childs Nerv Syst. 2011; 27, 9, 1453-1458

Langeloo DD, Lelivelt A, Louis Journee H., et al. Transcranial electrical motor-evoked potential monitoring during surgery for spinal deformity: a study of 145 patients. Spine (Phila Pa 1976). 2003; 28, 10, 1043- 1050

Lieberman JA, Lyon R, Feiner J., et al:. The effect of age on motor evoked potentials in children under propofol/isoflurane anesthesia. Anesth Analg., 2006; 103, 2, 316-321

MacDonald DB, Al Zayed Z, Khoudeir I., et al. Monitoring scoliosis surgery with combined multiple pulse transcranial electric motor and cortical somatosensory-evoked potentials from the lower and upper extremities. Spine (Phila Pa 1976). 2003; 28,2, 194-203

McLaughlin J, Bjornson K, Temkin N., et al. Selective dorsal rhizotomy: meta-analysis of three randomized controlled trials. Dev Med Child Neurol. 2002; 44, 1, 17-25

Moore JK, Linthicum FH, Jr. The human auditory system: a timeline of development. Int J Audiol. 2007; 46, 9, 460-478

Nuwer MR, Dawson EG, Carlson LG., et al. Somatosensory evoked potential spinal cord monitoring reduces neurologic deficits after scoliosis surgery: results of a large multicenter survey. Electroencephalogr Clin Neurophysiol. 1995;96, 1, 6-11

Padberg AM, Wilson-Holden TJ, Lenke LG, et al. Somatosensory- and motor-evoked potential monitoring without a wake-up test during idiopathic scoliosis surgery. An accepted standard of care. Spine (Phila Pa 1976). 1998; 23, 12, 1392-1400

Patel AJ, Agadi S, Thomas JG, et al. Neurophysiologic Intraoperative monitoring in children with Down syndrome. Childs Nerv Sys. 2013; 29, 2, 281-287

Pelosi L, Lamb J, Grevitt M, et al. Combined monitoring of motor and somatosensory evoked potentials in orthopaedic spinal surgery. Clin Neurophysiol. 2002; 113,7,1082- 1091

Pihko E, Lauronen L Somatosensory processing in healthy newborns. Exp Neurol., 2004; 190 Suppl 1, 2-7

Pouratian N, Elias WJ, Jane JA, Jr,. et al. Electrophysiologically guided untethering of secondary tethered spinal cord syndrome. Neurosurg Focu. 2010; 29,1, E3-

Sala F, Krzan MJ, Deletis V. Intraoperative neurophysiological monitoring in pediatric neurosurgery: why, when, how? Childs Nerv Sys., 2002; 18, 6-7, 264-287

Sala F, Manganotti P, Grossauer S, et al. Intraoperative neurophysiology of the motor system in children: a tailored approach. Childs Nerv Syst., 2010; 26, 4,:473-490

Salamy A. Maturation of the auditory brainstem response from birth through early childhood. J Clin Neurophysiol. 1984;1, 3, 293-329

Schwartz DM, Auerbach JD, Dormans JP, et al. Neurophysiological detection of impending spinal cord injury during scoliosis surgery. J Bone Joint Surg Am. 2007; 89, 11, :2440-2449

Schwartz DM, Sestokas AK, Dormans JP, et al. Transcranial electric motor evoked potential monitoring during spine surgery: is it safe? Spine (Phila Pa 1976) 2011;36,13, 1046-1049

Simon MV, Chiappa KH, Borges LF. Phase reversal of somatosensory evoked potentials triggered by gracilis tract stimulation: case report of a new technique for neurophysiologic dorsal column mapping. Neurosurgery 2102; 70, 3, 783-788

Slimp JC, Holdefer RN:. Somatosensory Evoked Potentials: An Electrophysiological Tool for Intraoperative Monitoring, in Intraoperative Monitoring, edited by BJ Loftus CM, Baron EM. New York, McGraw Hill, 2014:405-412

Sloan T. Anesthesia and intraoperative neurophysiological monitoring in children. Childs Nerv Syst., 2010; 26, 2, :227-235

Steinbok P, Tidemann AJ, Miller S, et al. Electrophysiologically guided versus non-electrophysiologically guided selective dorsal rhizotomy for spastic cerebral palsy: a comparison of outcomes. Childs Nerv Syst. 2009; 25, 1091- 1096

Tau GZ, Peterson BS. Normal development of brain circuits. Neuropsychopharmacolog. 2010; 35, 1, 147-168

Taylor MJ, Boor R, Ekert PG. Preterm maturation of the somatosensory evoked potential. Electroencephalogr Clin Neurophysiol. 1996; 10, 5, 448-452

Taylor MJ, Fagan ER. SEPs to median nerve stimulation: normative data for paediatrics. Electroencephalogr Clin Neurophysiol, 1988; 71,5, 323-330

ten Donkelaar HJ, Lammens M, Wesseling P. et al. Development and malformations of the human pyramidal tract. J Neurol 2004; 251,12,,1429-1442

Thuet ED, Winscher JC, Padberg AM., et al. Validity and reliability of intraoperative monitoring in pediatric spinal deformity surgery: a 23-year experience of 3436 surgical cases. Spine (Phila Pa 1976), 2010; 35,20,1880-1886

Turner RP. Neurophysiologic intraoperative monitoring during selective dorsal rhizotomy. J Clin Neurophysiol;, 2009; 26,2,82-84

von Koch CS, Quinones-Hinojosa A, Gulati M., et al. Clinical outcome in children undergoing tethered cord release utilizing intraoperative neurophysiological monitoring. Pediatr Neurosurg;, 2002 37,:81-86

Wiedemayer H, Sandalcioglu IE, Armbruster W., et al. False negative findings in intraoperative SEP monitoring: analysis of 658 consecutive neurosurgical cases and review of published results. J. Neurol., Neurosurg., and Psychiatry. 2004; 75, 280-286

CHAPTER 9

Electrocorticography and Brain Mapping. Epilepsy and Brain Tumor Surgery

Introduction

ELECTROCORTICOGRAPHY IS BASED upon electroencephalographic techniques rather than evoked potential techniques. The basic instrument used in electroencephalography is the differential amplifier. The differential amplifier has qualities which only recognize differences between Input 1 and Input 2. If an event such as a large electrical artifact is seen equally in Input 1 and Input 2, then no output occurs. The differential amplifier utilizes this feature to for the development of a high common mode rejection ratio. This type of amplifier increases the size of the measured wave forms through its "Gain" settings. This type of amplifier requires a high impedence input. Almost all modern electroencephalographs are digital in nature and no longer give the curvilinear and blocking artifacts generated by pen systems in the older analog systems. Electroencephalography polarity has been standardized for many years. Electroencephalography convention is that a Negative response in Input 1 gives an upward deflection. Figure 1 shows the results of this convention.

EEG Polarity Convention

1. **Negative in Input 1 = upward deflection**
2. **Positive in Input 1 = downward deflection**
3. **Negative in Input 2 = downward deflection**
4. **Positive in Input 2 = upward deflection**

Figure 1. *Polarity Convention with an upward deflection occurring when a negative event is seen at Input 1.*

Utilization of this convention allows the establishment of montages when scalp electrodes are placed on the head in a standardized method. The 10-20 electrode system has been used for many years, over the past 10 years or so a slightly modified system has been used which renames certain electrodes and provides a shorthand naming system for electrodes that are placed ½ way in between the 10-20 system and is known as the 10-10 system. Even numbered electrodes are over the right hemisphere, mid-line electrodes have a Z designation, for example Fz, Cz, Pz, and Oz) and over the left hemisphere the electrodes are odd numbered. Montages consist of straight lines of electrodes in both an anterior-posterior and lateral-medial-lateral configurations. How the electrodes are linked determine how the recorded information is used for localization**. (Fig. 2)** demonstrates a typical bipolar montage with each electrode in the chain linked to its adjacent neighbor.

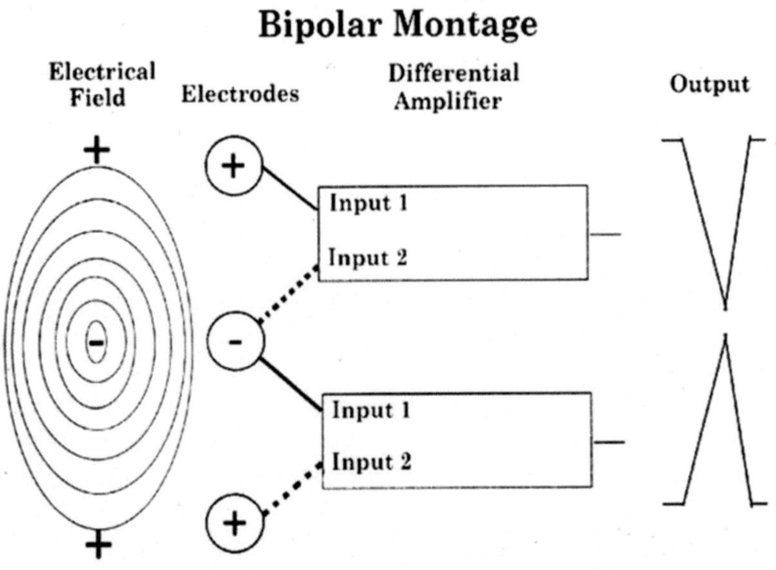

Figure 2. *Electrical spike with major negativity at the second electrode in the chain. Output shows a phase reversal which is used to localize the source of the epileptic spike.*

(Figure 3) demonstrates a slight variation in the location of the spike and how the flat line surrounded by a phase reversal is used to localize the source of the epileptic spike.

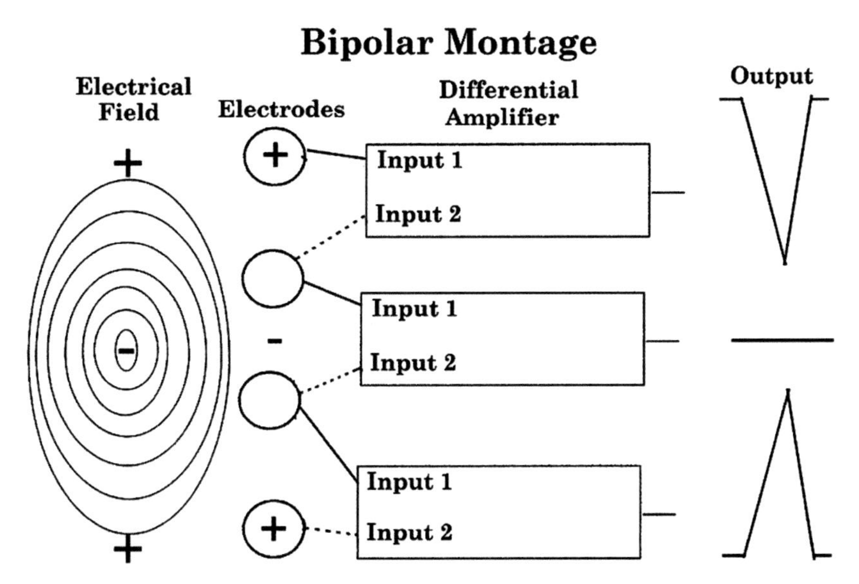

Figure 3. *In this example, the source is equidistant between Input 1 and Imput 2 of the second amplifier. This means that an equal charge will be seen by the two imputs and an isoelectric signal will be the output. The isoelectric signal is surrounded by the typical phase reversal. In this instance the source of the epileptic spike is the isoelectric activity.*

The second major method used to localize an electrical source is by using a referential montage. A number of different systems have been used in referential recordings. None of them are perfect and each has certain weaknesses. For example, reference to an ear or mastoid process is an acceptable montage in a child with seizures, but not in an adult. The difference being that in children the temporal lobe is an uncommon source of seizures, but in the adult the temporal lobe is the most common source. In an adult, the ear and/or mastoid electrodes would be active and utilization of these electrodes could electrically cancel the response. An average electrode reference ties all the electrodes on the scalp together, then any active electrode can be removed from the average to eliminate the effects of the active electrode. With this reference, if the active electrodes are not removed, reference contamination makes it look like the source of the discharge is more widespread than it actually is. **(Figure 4)** is a demonstration of the use of the average reference electrode. In this type of recording the electrode closest to the source will be the largest response.

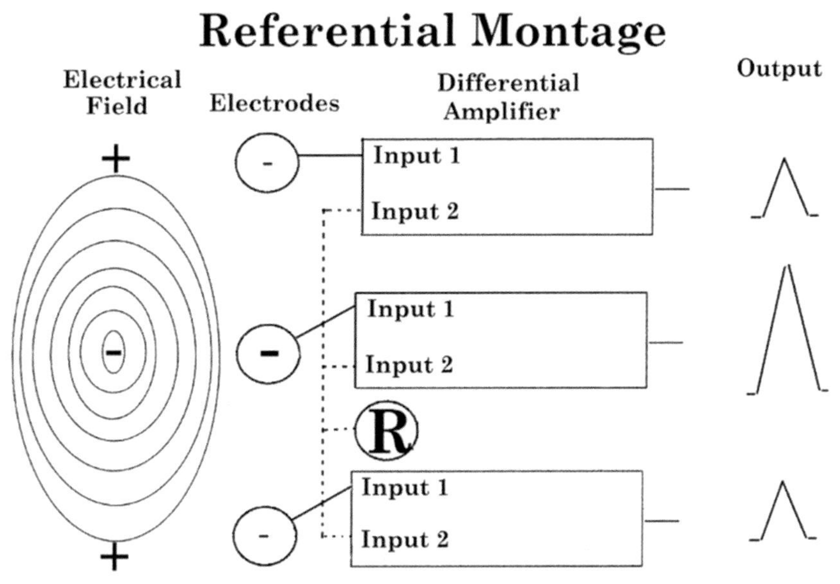

Figure 4. *Average reference electrode used for a referential recording.Note the source of the epileptic spike is localized by its amplitude.*

Electrocorticography almost always uses referential montages. The setup includes a "halo" to hold the cortical electrodes that are placed directly on the cortex. Two types of electrodes are available. 1) Cotton wick electrodes in which the wick is saturated with physiologic saline. 2) Carbon tipped electrodes, soaked in physiologic saline. I prefer the later electrode as they provide a stable recording, covers a much smaller area of cortex and the surgeon can work around the electrode without making contact with the electrode. Impedence is never checked in electrocorticography, it is too dangerous. (An impedance check sends current through the electrodes to obtain the impedance values of the electrodes.) The reference electrode can either be applied to the opposite scalp or use a tied pair of electrodes on each side of the neck. Sometimes the surgeon will place a sterile needle electrode in the tissue at the edge of the incision to be used as the reference. Electrode strips or various sized grids may substitute for the cortical electrodes or may be used in locations where the cortical electrode cannot be used. The limitation of the electrode strips is that the spacing of the inter-electrode distance is fixed and cannot be customized for each recording.

Epilepsy Surgery

Epilepsy is not a specific disease, but it is a symptom. It is seizure disorder that recurs. A single seizure is not epilepsy, but when a second event occurs it is then diagnosed as epilepsy. Epilepsy has a strong tendency to recur without warning. The classification of Epilepsy can be found in one of the standard textbooks on epilepsy, but for our purposes here will divide this condition into two types. First, the most common form of epilepsy in the adult that is frequently treated with surgery is mesial temporal epilepsy followed by neocortical epilepsy. Both of these types begin in a localized area of cortex and are classified as local or partial seizures. The generalized epilepsies are not generally considered for epilepsy surgery.

Mesial temporal epilepsy has had a number of names over the years and including: psycho-motor epilepsy and temporal lobe epilepsy. Mesial temporal epilepsy is preferred as this name gives the area where the recurrent seizures originate. The diagnosis of epilepsy is established on a clinical description of the clinical seizure and additional tests are used to establish the area of the brain where the event originates. The scalp EEG is of great importance in establishing the specific type of seizure and a single wake-sleep EEG will pick up about 50 percent of the cases. Two to three additional EEGs over a couple of years will get the diagnostic confirmation rate up to about 80 percent. Treatment is started as soon as a diagnosis is established and it is those patients who fail therapy with multiple drugs at an optimal therapeutic range who become candidates for surgical intervention. The typical inter-ictal EEG change seen in this condition is shown in **(Fig. 5).**

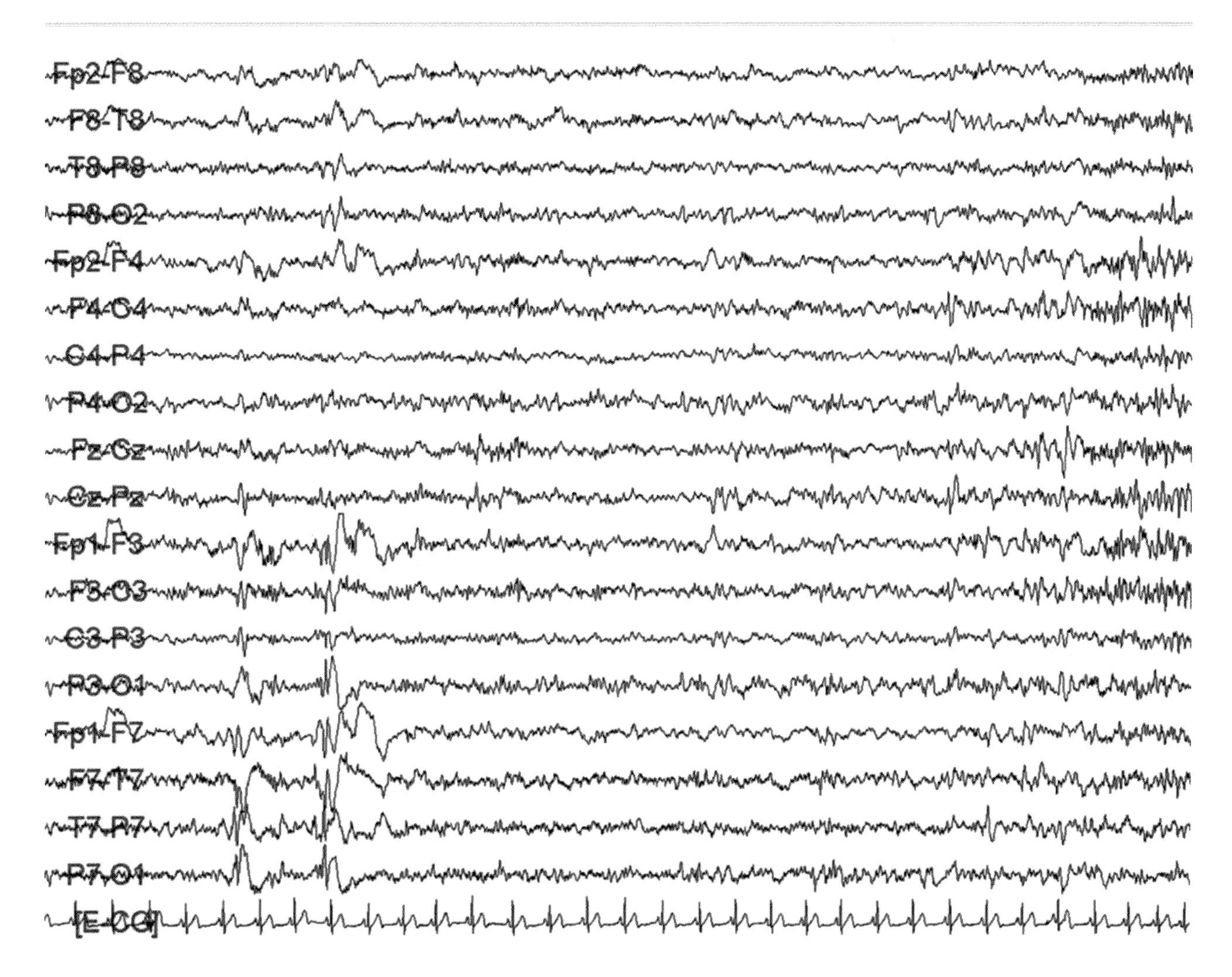

Figure 5. *Scalp EEG in drowsiness showing left temporal spikes followed by a trailing slow wave. In this bipolar montage the spikes phase-reverse at electrode T7.*

In about 80 percent of individuals with this condition, the MRI scan. will show atrophy of the hippocampus on the same side as the epileptic discharge. Figure 6 shows a typical MRI scan.

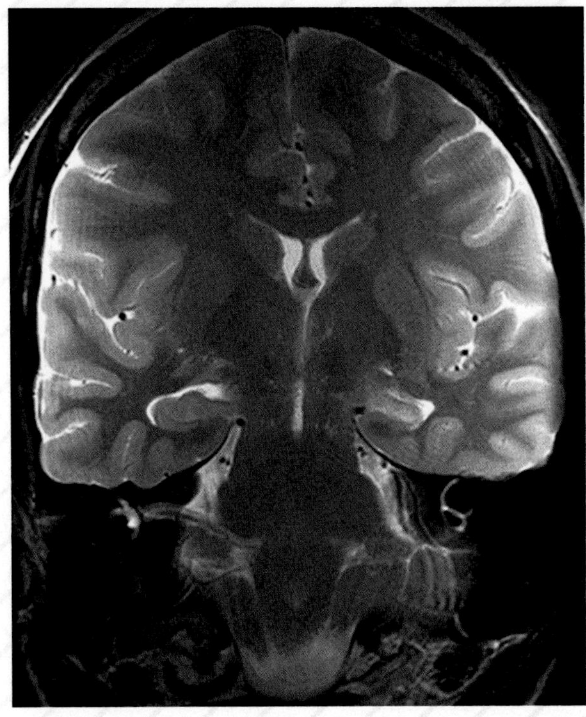

Figure 6. *Special MRI scans showing atrophy and sclerosis of the hippocampus this finding is found in about 80 percent of patient's being evaluated for mesial temporal epilepsy surgery.*

Rarely in the EEG laboratory, the patient will have an ictal event and the EEG shows the typical rhythmic theta (5-7 Hz discharge) over the temporal region, see **(Fig. 7).**

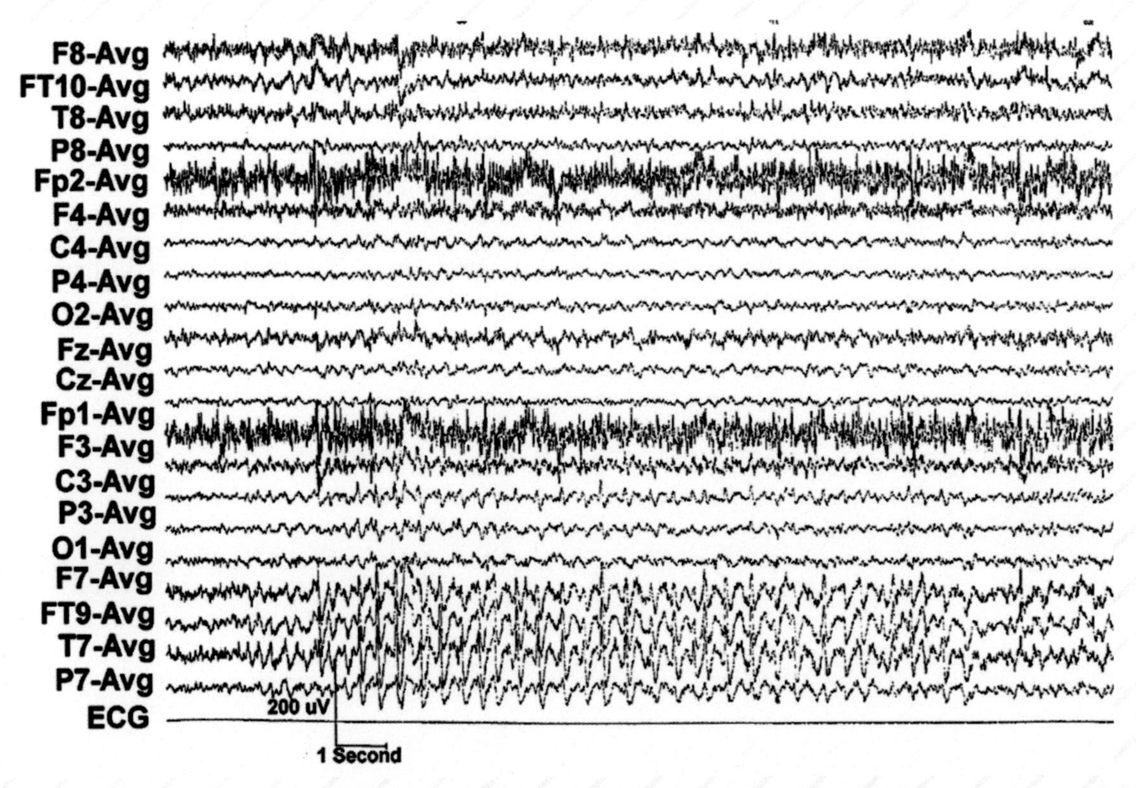

Figure 7. *Rhythmic theta discharge starting at electrodes FT9 & T7, but the epileptic field is seen over most of the left temporal lobe. This is not actually the start of the ictal event, but the earlier low voltage fast activity is filtered by the skull and scalp and not seen. Only those frequencies that are slow enough not to be filtered to any great extent are seen in scalp recordings.*

In a patient with medically intractable epilepsy whose clinical seizures are very stereotyped, all being identical, who has had a number of EEGs which all show unilateral temporal lobe interictal spikes and who's MRI reveals atrophy and gliosis of the appropriate hippocampus can proceed directly to surgery without the need for long term epilepsy monitoring. Other individuals who have a normal MRI scan or whose EEG shows bilateral spikes need to have long-term monitoring to establish the onset of their seizures. During long term monitoring, first with scalp electrodes including FT9 (≈T1) and FT10 (≈T2) ictal events must be recorded. The use of FT9 and FT10 eliminates the need to use invasive sphenoidal electrodes as the information obtained is the same. Long-term scalp recording includes the reduction of anticonvulsants and other stressors, such as sleep deprivation and exercise in an attempt to activate and record a series of seizures. Generally, one likes to obtain the seizures spread out over a number of days. When seizures come in clusters, there is a tendency for the onset of seizures to spread to additional areas of the brain and give a false localization. Ideally, one would like to record 4-6 seizures with an identical onset. During the hospital stay each patient undergoes neuropsychology testing, WADA language lateralization testing and on the day before surgery patients undergo language mapping using 40-80 line drawing slides. Any slide the patient cannot identify during this screening test is eliminated from the slide pool to be used for language mapping during the actual surgery. The slides include common animals, tools, transportation vehicles, clothing, household furniture, etc.

When scalp recordings and the MRI fail to provide a consistent lateralized epileptiform focus invasive long-term monitoring is then carried out. In general, these implanted electrodes should not remain in place for more than a couple of weeks as the incidence of infections increase dramatically. Epilepsy surgery should be done when the implanted electrodes are removed. About 20 percent of cases of mesial temporal epilepsy don't qualify for surgery following scalp long-term monitoring with an appropriate abnormal MRI, This group may undergo mre-monitoring with 4 to 8 contact strip electrodes placed through burr holes. In general 4 strips are located on each side of the brain. Four contact strips are placed in the following locations; 1 under the temporal lobes in a medial to lateral orientation; 2 along the lateral temporal lobe in an anterior to posterior orientation. From burr holes over each frontal region, an 8 contact electrode strip is placed over the anterior frontal extending to the orbital surface; a second 6 contact strip is placed in a anterior to posterior orientation over the mid frontal region. See **(Fig. 8)** for a schematic of these electrode locations.

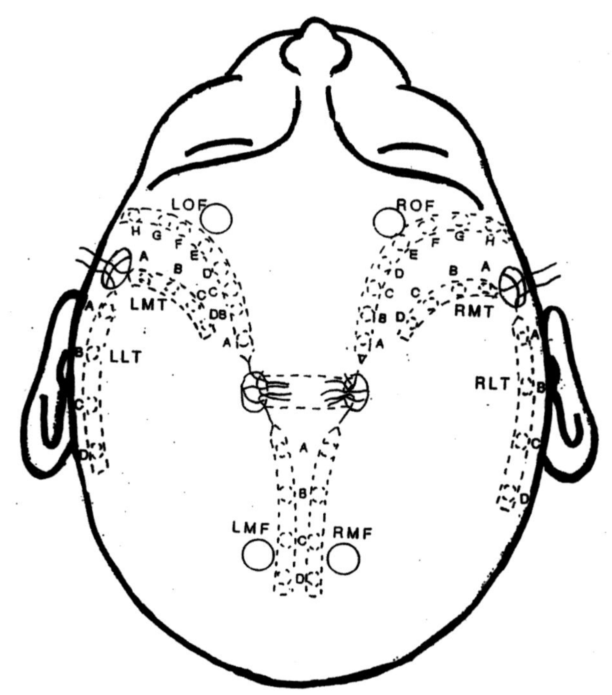

Figure 8. *Schematic map of the various strip electrodes for monitoring a patient who did not have consistent lateralizing MRI and ictal seizure activity. These cases represent about 20 percent of the cases studied for mesial temporal epilepsy. LMT and RMT = sub-temporal strips, LLT and RLT = lateral temporal strips, LOF and ROF = orbital frontal strips, and LMF and RMF = medial frontal strips. Each strip has a colored wire and in the OR the surgeon maps the colored wire to the individual strips so that the END technician can correctly identify the strips for long term monitoring.*

Anticonvulsant drugs are then reduced and then terminated while the patient undergoes long-term epilepsy monitoring which includes 24 hours of continuous Video-Telemetry-Electrocorticography for up to 2 weeks. Only those cases with a consistent unilateral onset are considered candidates for Epilepsy surgery. This later group of patients does not have quite as good an outcome as the 80 percent with unilateral scalp and MRI findings, but approximately 60 plus percent benefit measurable by the surgery.

Seizures beginning in the neo-cortex require a higher percentage of invasive monitoring. In these cases, scalp monitoring helps to localize the region of brain to be studied by invasive techniques. In these cases there are a high percentage of lesions that are recognized on MRI scans. One must remember that the epileptic area may or may not be associated with the lesion. Numerous examples of the epileptic area being found at a distance from the lesion are seen. The epileptic focus may originate at one margin of the lesion, in the ipsilateral mesial temporal lobe or even in the opposite mesial temporal complex. Long-term scalp monitoring is utilized to plan the placement of 1 or 2- 64 contact grids or parts of grids and how many strips will be necessary to cover the area of interest. A craniotomy is carried out and the dura reflected. The appropriate grid or grids and strips are placed directly in contact with the cerebral cortex. Photographs of the grid and strips placement is then carried out and the surgeon generates a detailed map of the electrode placement and this map is provided the END technologist to set up the patient for long-term epilepsy monitoring. The bone is replaced and the scalp wound closed. Color coded electrode wires are utilized to identify the grids and strips for recordings. The patient is then monitored for up to 2 weeks to collect a series of ictal events. At the end of this time, the patient is returned to the operating room where the craniotomy is reopened and the grid and strips removed followed by epilepsy surgery. Examples of the different types of epilepsy surgery are presented in the following sections.

Staged Hippocampal Resection for Mesial Temporal Epilepsy

This form of surgery with electrocorticography has been used by Dr, George Ojemann to treat literally hundreds of patients with medically-intractable epilepsy. This surgery tailors the amount of hippocampus to remove for each individual and allows uninvolved hippocampus to remain intact. The results of this type of surgery are virtually identical to those patients operated on by epilepsy surgeons who remove the entire hippocampus. The description of the 3 stages of electrocorticography will be provided for a case involving left mesial temporal epilepsy so as to include language mapping. Propofol is utilized as the anesthetic of choice as the patient can be awaken fairly rapidly and at certain levels, especially changing levels tend to activate epileptic discharges. Propofol generally causes a burst-suppression pattern in the electrocorticogram when the patient is fully anesthesized. After the Propofol is stopped to awaken the patient, the burst-suppression pattern reverses to a continuous pattern first involving the lateral hemisphere, then the lateral temporal lobe, then finally on the undersurface of the posterior temporal lobe. No language testing should be accomplished until an awake EEG pattern is obtained.

Stage 1. After exposing the left temporal lobe and injecting local anesthesia into the dura and scalp tissues of the surgical site, the patient is awakened. Continuous electrocorticography is obtained during this stage. In general the electrodes are placed as shown in **(Fig. 9)**.

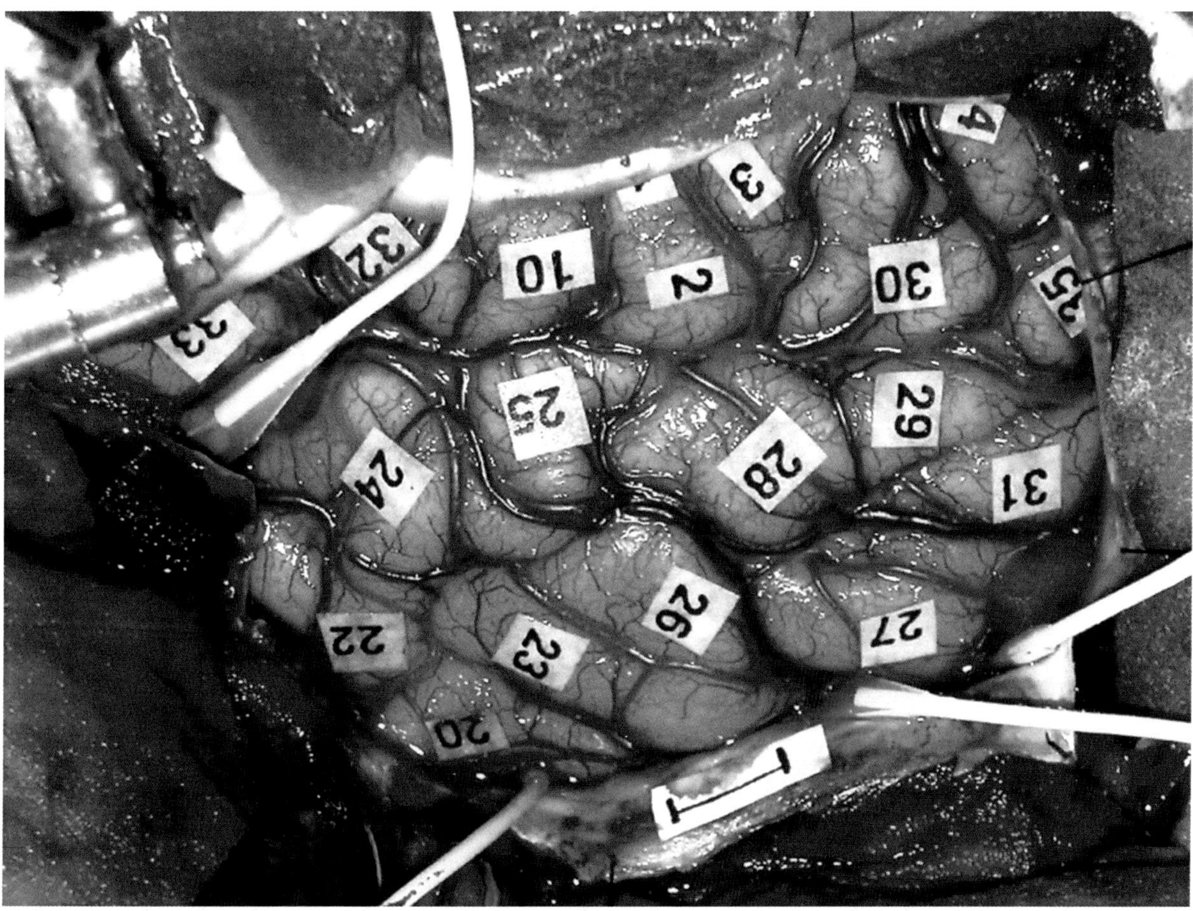

Figure 9. *This brain map is used during Stage 1 of the epilepsy surgery. 4 strip electrodes are inserted along the undersurface of the brain. Three strips (1, 2, 3,) are placed under the anterior, middle, and posterior temporal lobe and Strip 4 on the undersurface of the frontal lobe. Carbon-tipped cortical electrodes are placed on the superior temporal gyrus, the middle temporal gyrus and on the Suprasylvian fromtal lobe.*

Recording occurs as the patient awakens and in general, epileptiform discharges are typically in the sub temporal strips, contacts C and D. It is not uncommon to also see epileptiform spikes over the mid- to post-temporal electrodes on the middle gyrus. After recording an adequate sample of interictal spikes, after- discharge threshold levels are determined. A DC biphasic Ojemann stimulator is utilized to determine after-discharge thresholds. Starting with very low currents, approximately 2 milliamps, the areas around each electrode are stimulated and the electrocorticography monitored for after discharge. Care must be taken in using the Ojemann stimulator as this instrument delivers 2X the amperage indicated on the dial. It only shows the amperage of a one-half stimulation wave. The surgeon is notified after each stimulation, whether there are after-discharges or not. The current is gradually increased until good levels for language mapping (ideally 5-6 milliamps minimum) are obtained or after discharges develop and limit the amount of current that can be used.

LANGUAGE MAPPING:

Language mapping is carried out when operating on the dominant hemisphere for language. The language dominant hemisphere is on the left side in over 99 percent of right handed individuals and about 50 percent of left handed individuals. The preparation for language mapping includes the use of a frame attached to the operative bed to allow the surgical drapes to be elevated off the patient so that a "tunnel" can be developed for the patient to clearly see the lap top computer screen and for the END technologist to see the patient's face and arm. A microphone is attached to the patient's gown and his or her speech is amplified so that both the surgeon and the END technologist can clearly hear the responses.

This surgical setup is shown in **(Fig. 10)** where one END technologist presents the slides to the patient and scores the responses.

The surgeon or his assistant also records the sites of stimulation and the responses. The second END technologist runs the EEG equipment and observes the EEG screen for after discharges or epileptic discharges. Line drawings are used in mapping language of the exposed cerebral cortex during surgery for epilepsy and for cortical tumor removal and for cortical mapping of a 64 contact grid during long term monitoring. (See APPENDIX) The day before surgery, the patient is tested with the pictures and any that the patient cannot name, eliminated for surgery. During surgery language areas are located by direct stimulation of the cerebral cortex using a bipolar Ojemann Cortical Stimulator. First, the after-discharge threshold is determined by inducing after discharges. A slightly sub-threshold stimuli is used for mapping. The current is frequently over 4 milli-amps at a duration of 4 seconds. If a non-language area is stimulated the patient will correctly identify the line figure. If a language area is stimulated the patient develops either a complete speech arrest or produces paraphasias. Small numbered tags are used to identify the the language cortical areas. The END technologist keeps track of the errors made by the patient. Following language mapping, motor-sensory mapping of Rolandic cortex is accomplished. Generally a lower stimulus intensity about 2-3 milli –amps for just a short period of time is necessary for motor-sensory mapping. Again, a series of numbered tags (a different series than the language tags) are utilized to identify eloquent motor-sensory cortex.

The following intra-operative photos were taken at the time of epilepsy surgery and demonstrate the placement of the "numeric tags" for locating both the language and motor-sensory localization. **(Fig. 10).**

Figure 10. . *Tags to show localization of Language and Motor functions from Brain mapping during Surgery for Mesial Temporal Epilepsy.*

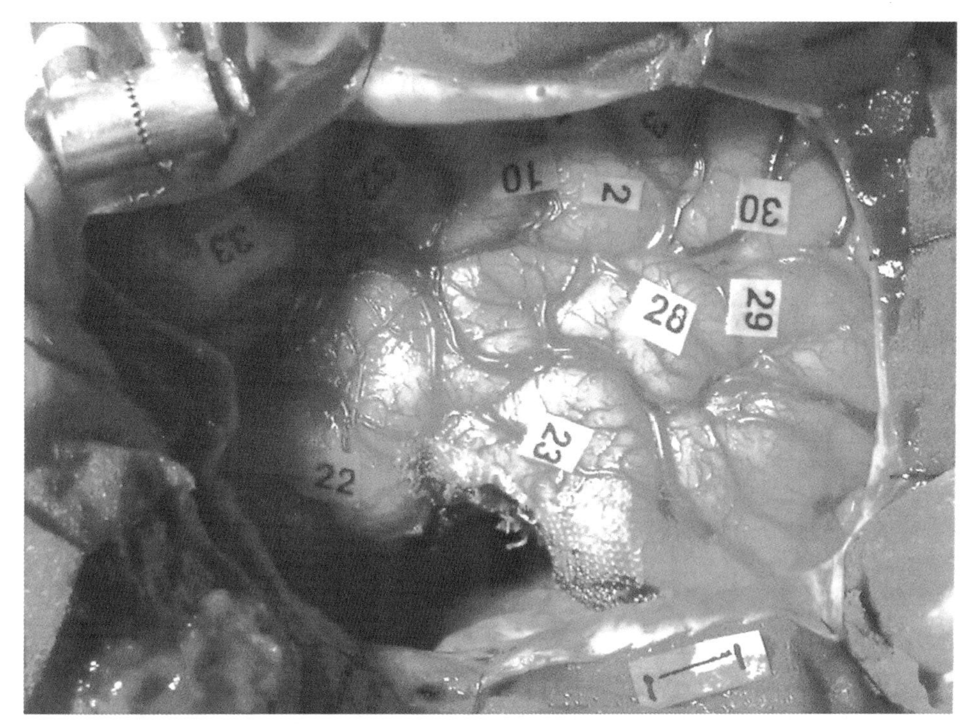

Figure 11.. *Temporal lobe after resection of inferior temporal area to expose the Temporal Ventricle. A 4 contact electrode strip is then placed on the lateral surface of the Hippocampus for determiningwhich portions of Hippocampus to remove.*

The APPENDIX not only contains copies of the 80 slides, but a Microsoft Power Point presentation of the above line drawings for use on a laptop computer. This program includes the sounds of a slide projector so the neurologic surgeon can tell when a new slide is presented to the awake patient.

The line drawings include a number of examples by object class. For example, there are a number of different slides of animals, articles of clothing, body parts, transportation items, food items, tools and musical instruments. Other objects are also included. Rarely, one finds a patient to have a deficit isolated to one of the major categories, for example an inability to name any animal, yet be able to name the other categories with accuracy. **(Fig. 12).**

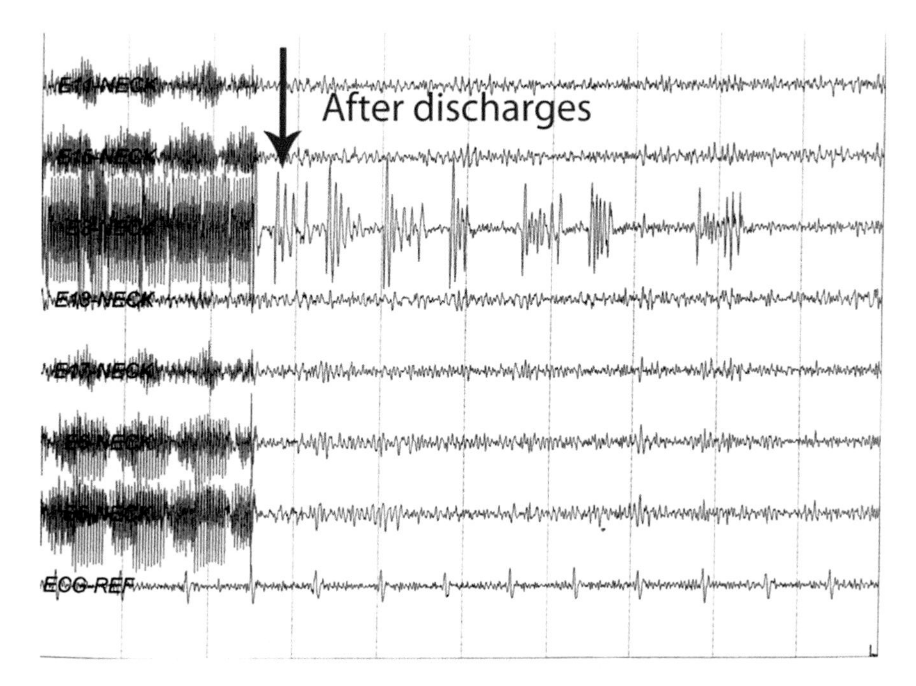

Figure 12. . *Example showing stimulation artifact followed by a short train of afterdischarges.*

Each area is stimulated for 4 seconds (or the duration between slide transitions for language mapping). Current values that cause afterdischarges are reduced until no after-discharges occur. At times, different currents must be used for different portions of the brain. For example, after-discharge threshold is generally higher over the frontal region than over the temporal lobe. It is important not to induce a seizure. Following the establishment of the after-discharge threshold and the amount of current to be used for language mapping, motor-sensory mapping is accomplished over the exposed fronto-parietal cortex. In the awake adult patient low currents generally in the range of 2-4 milliamperes is all that is necessary to induce movement or have the patient recognize sensory phenomenon. In the anesthesized patient and children the current levels needed are frequently much higher and at times motor responses cannot be obtained and evoked potentials are necessary to localize motor/sensory cortex.

Language mapping is then accomplished. **(Figure 13)** shows one END technologist scoring the language mapping while the other runs the electrocorticography equipment, looking for after-discharges.

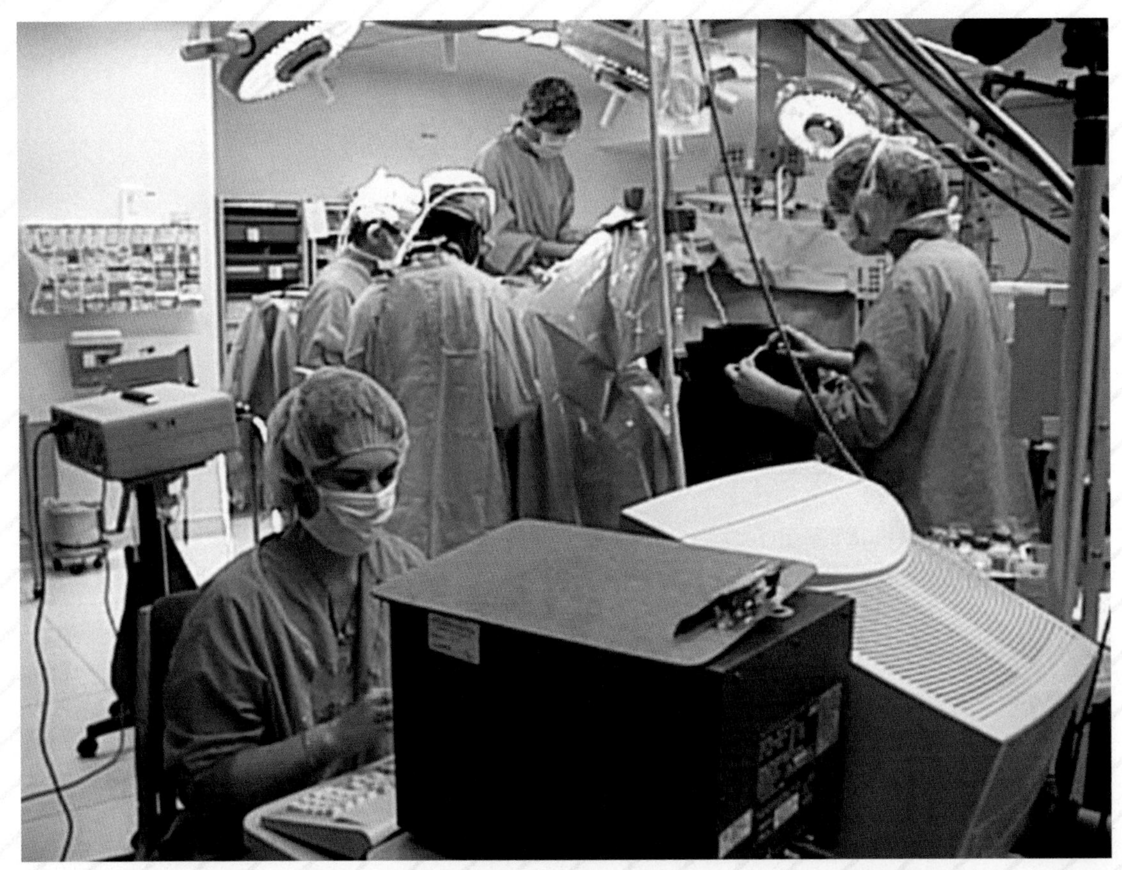

Figure 13. *Photo showing 1 END technologist standing next to the OR Table providing scoring for the language mapping session. A second technologist runs the electrocorticography equipment. The surgeon stimulates the various areas of cortex for 4 seconds and the technologist checks for language arrest or for aphasic errors. A total of 40 slides are projected by slide tray or laptop computer.*

After completion of the language mapping the patient is then re-anesthesized and the remainder of the surgery is done under anesthesia.

The surgeon then removes the anterior inferior portion of the temporal lobe gaining access to the ventricle in the temporal lobe.

Stage 2. The next stage of electrocorticography occurs after 4 contact electrode strips are placed in the ventricle on the hippocampus (Strip 1). Strip 2 is placed in an anterior-posterior orientation under the parahippocampal gyrus. Electrode strip 3 is oriented in a lateral to medial orientation at the posterior portion of the resection. Electrode strip 4 is again placed on the under surface of the frontal lobe as shown in **(Fig. 14).**

1ST POST-RESECTION MONTAGE DIRECT HIPPOCAMPAL RECORDING

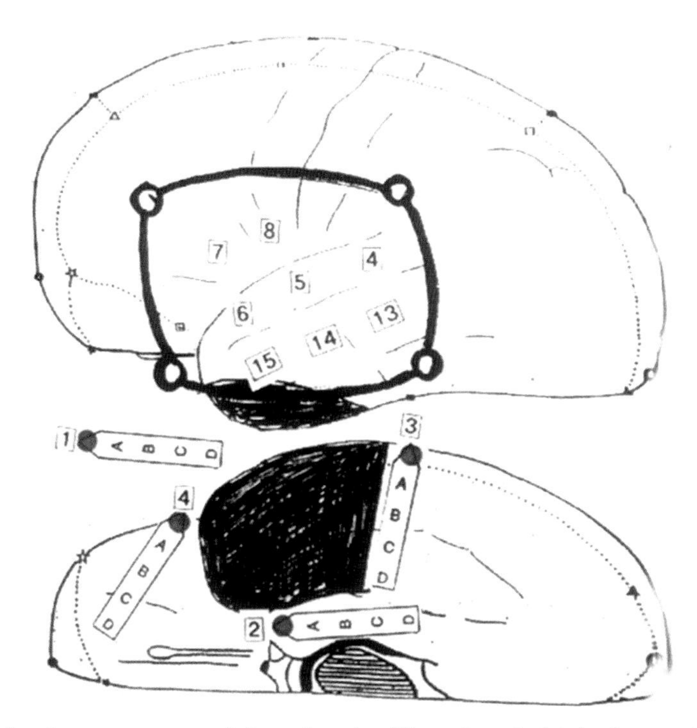

Figure 14. *Brain map showing arrangement of electrode strips. Electrode strip 1 is in the ventricle on the hippocampus while strip 2 is under the parahippocampal gyrus. Strip 3 in an lateral – medial orientation at the posterior margin of the resection and strip 4 on the under surface of the frontal lobe. Cortical electrodes are placed in the pre-resection montage.*

The Electrocorticography recording is then carried out. Figure 15 shows a typical recording showing interictal epileptiform spikes originating in the hippocampus.

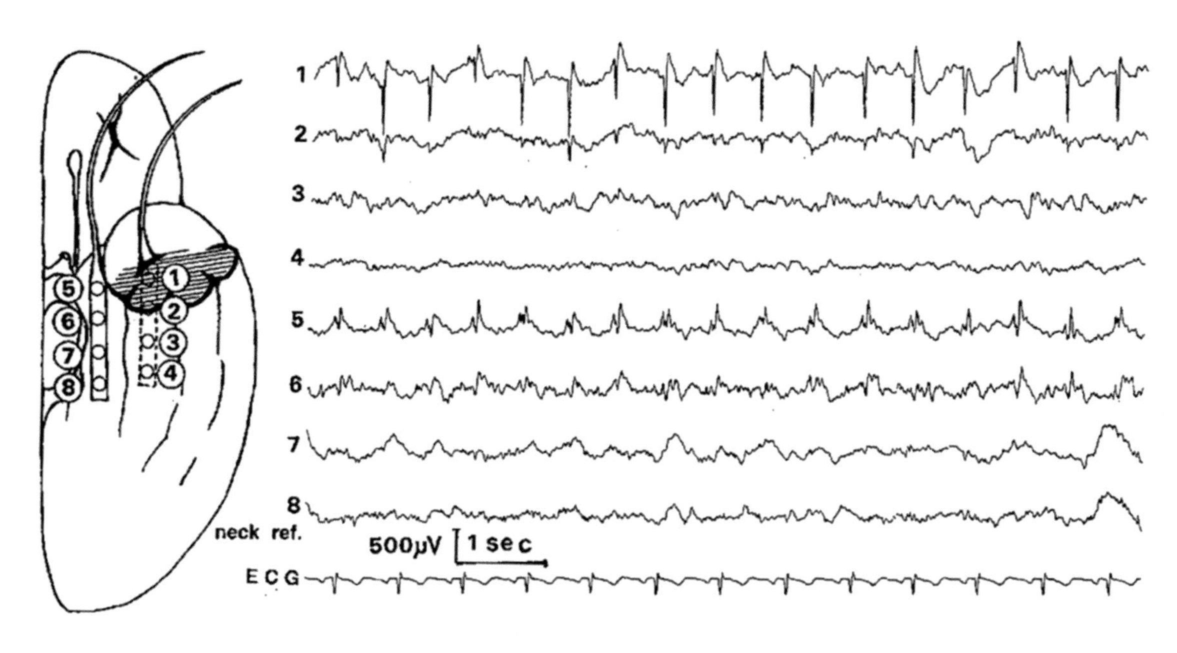

Figure 15. *Artist (E. Lettich) rendition of the Hippocampal (1-4) and parahippocampal Electrodes (5-8). The epileptic spikes are limited to the anterior electrode contacts In the hippocampus (1) and parahippocampus (5-6). Note the phase reversal (dipole) between the hippocampus and parahippocampus. On the hippocampus surface the Epileptiform spike is surface positive while the same spike is surface negative on the Parahippocampus..*

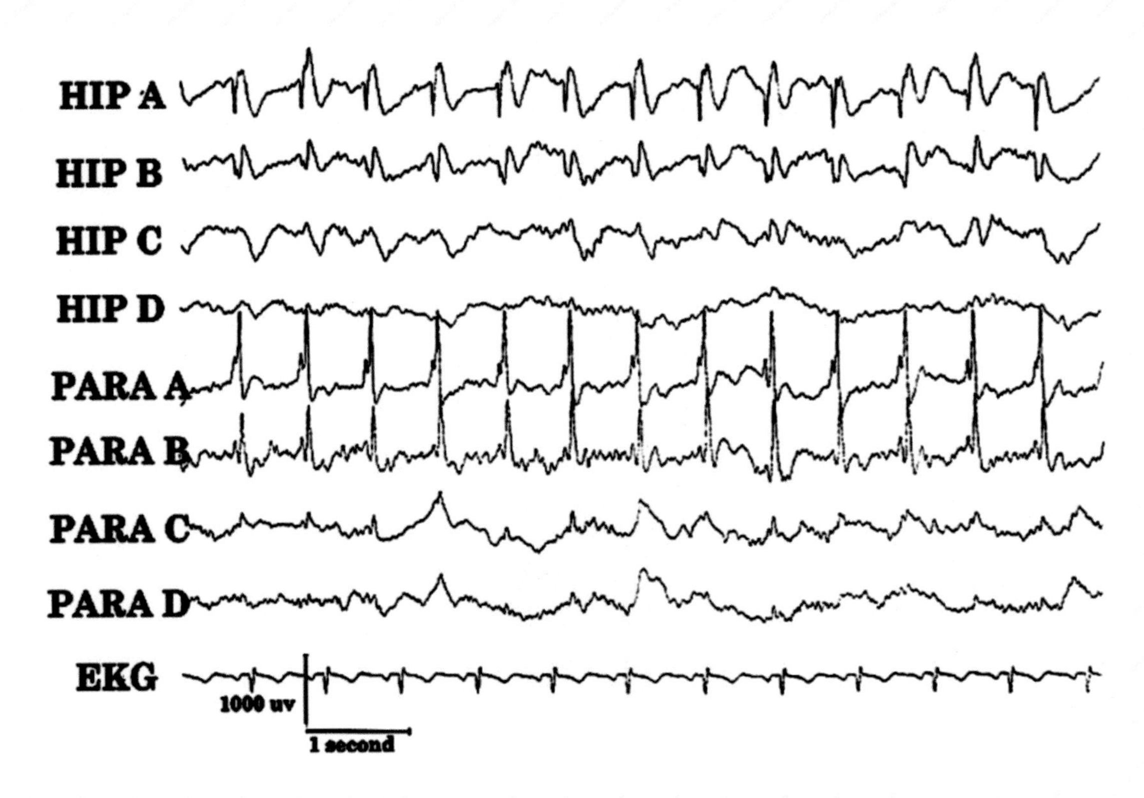

Figure 16. *A second example of frequent Hippocampal and parahippocampal discharges. This example shows the epileptiform spikes to extend further Posteriorly in the hippocampus and parahippocampus. Again note the phase reversal between the hippocampus and parahippocampus.*

After obtaining an adequate sample of epileptiform discharges the affected hippocampus and parahippocampus are removed. A tissue sample is taken for histology. The neuropathologic changes consisting of neuron loss between CA1 and the subiculum are shown in Figure 17.

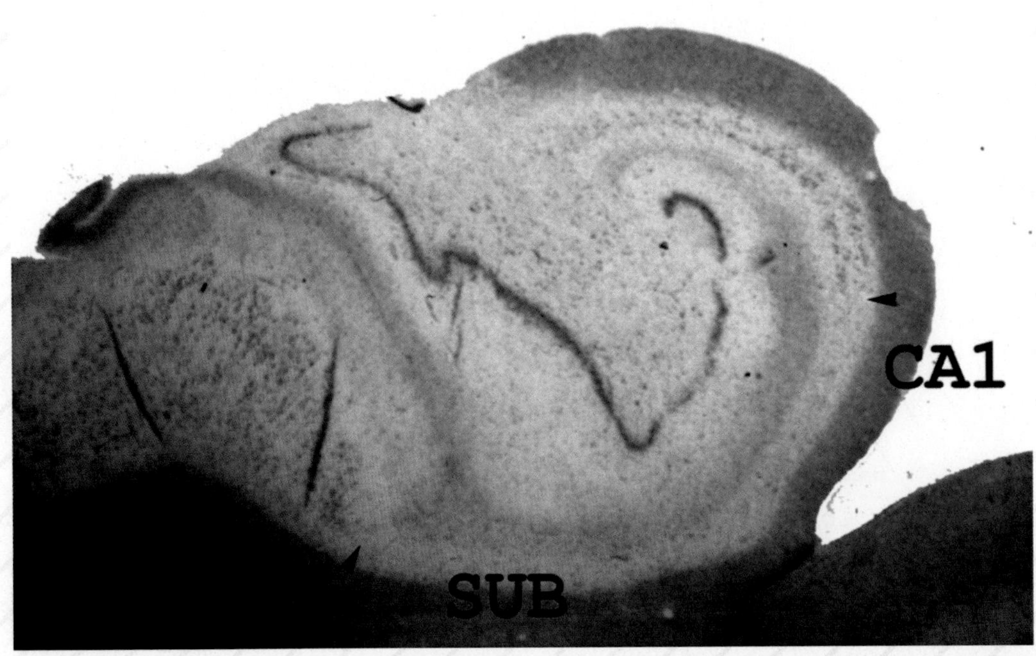

Figure 17. *Microscopic slide showing abnormal hippocampus. There is a severe loss of neurons in CA1 to the subiculum. The region of neuron loss is marked with small arrows. This is the typical pathology of mesial temporal sclerosis.*

SECOND POST-RESECTION MONTAGE

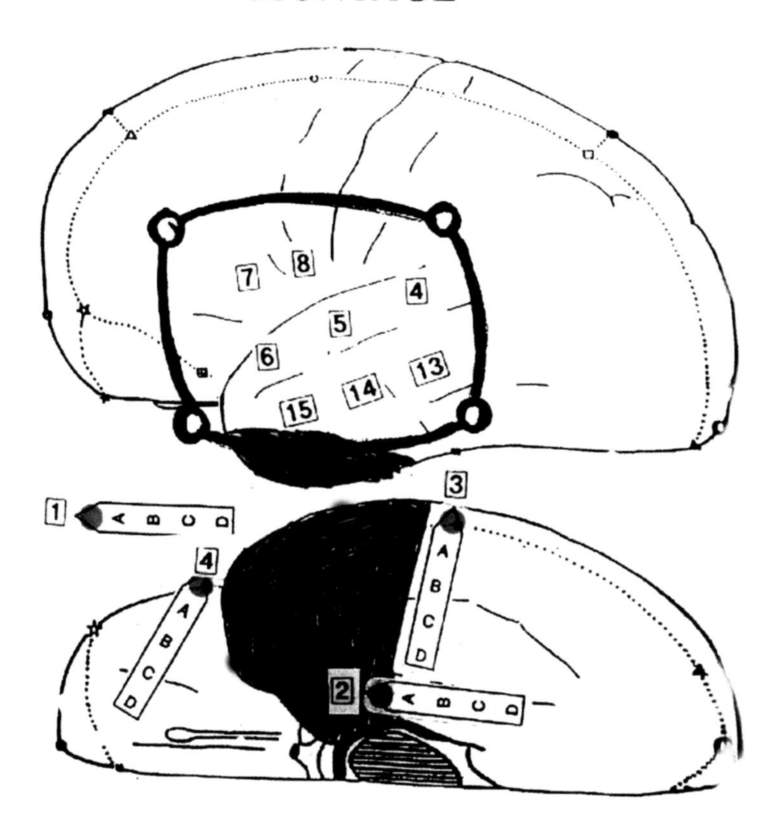

Figure 18. *Map of Stage 3 electrode placement. Strip 1 is in the Ventrical on the unresected portion of the hippocampus. Strip 2 On the undersurface of the unresected portion parahippocampal gyrus. Strips 3 and 4 and the cortical electrodes remain as before.*

The stage 3 Electrocorticography. Following removal of the hippocampus and parahippocampus one additional electrocorticography session is accomplished to make sure that no residual Hippocampal spikes remain. If no spikes remain, the closing begins. At times there are new spikes that appear at the margin of the resection on the under surface of the temporal lobe. These are post-resection spikes and should not be removed as they will spontaneously disappear with time and do not influence the outcome of the surgery. The most recent results from tailored epilepsy surgery as outlined above and include 141 individuals show that with a Hippocampal resection greater than 19 millimeters (N=125) 83 patients had a good outcome (66 percent). Those who had a Hippocampal resection less that 20 millimeters (N16) 11 had a good outcome (69 percent). There were no statistically significant differences between groups. However, when residual Hippocampal spikes were left major differences in outcome occurred. No residual Hippocampal spikes (N = 120) 72 percent had a good outcome. If residual spikes were left (21) only 33 percent had a good outcome. These results are statistically significant at the P 0.0001 level.

Non-tailored Epilepsy Surgery for Mesial Temporal Epilepsy

With this type of epilepsy surgery, electrocorticography is used to determine after-discharge thresholds just as in the tailored form of surgery. Once the after-discharge threshold is determined language and motor mapping is carried out in the awake patient. Following the identification of eloquent cortex, no more recordings are used. The hippocampus is then removed in its entirety. The improvement rate for epileptic events is virtually identical to the tailored form of surgery except that more hippocampus is removed. One of the advantages of this form of surgery is the amount of time required to do the resection is much shorter. The disadvantage is that by the removal of more hippocampus there is a tendency for more memory loss than with the tailored surgery.

Neocortical Epilepsy Surgery

The causes of neocortical epilepsy are many and include low grade astrocytomas, oligodendrogliomas, and mixtures of the two; vascular malformations, trauma, cortical dysplasia and no lesions identified (idiopathic). New onset seizures are evaluated by MRI scan as well as scalp electroencephalography. It is at this time that the lesions causing the seizure disorder are identified. The vast majority of these lesions and non-lesion cases will have to undergo long-term epilepsy monitoring to identify the location of the "epileptic focus". In these patients, as mentioned earlier, a 64 contact grid plus a variable number of 4-6 contact strips are surgically placed in the sub-dural space according to the results obtained by scalp electroencephalography and/or MRI scans. The patient then undergoes long-term epilepsy monitoring with electrocorticography of the grid and strips and video-taped recording of the seizures. In general about 1 week of monitoring is required to record an adequate number of seizures and to map the grid for motor, sensory and language function. At times, the grid must be left in place longer **than a week to collect enough seizures, but the longer the grid and strips remain in place the** more likelihood for the development of an infection. One should also remember that the lesion may not be the source of the epileptic event and there are many examples of a patient having a neocortical lesion and the seizure originates in the hippocampus on the same side or even on rare occasions from the opposite sided hippocampus.

At the time of surgical removal of the grid and strips the epilepsy surgery is carried out. In the next few figures some examples are used to demonstrate the principles of this type of surgery. **(Fig. 19., Fig. 20, and Fig. 21.).**

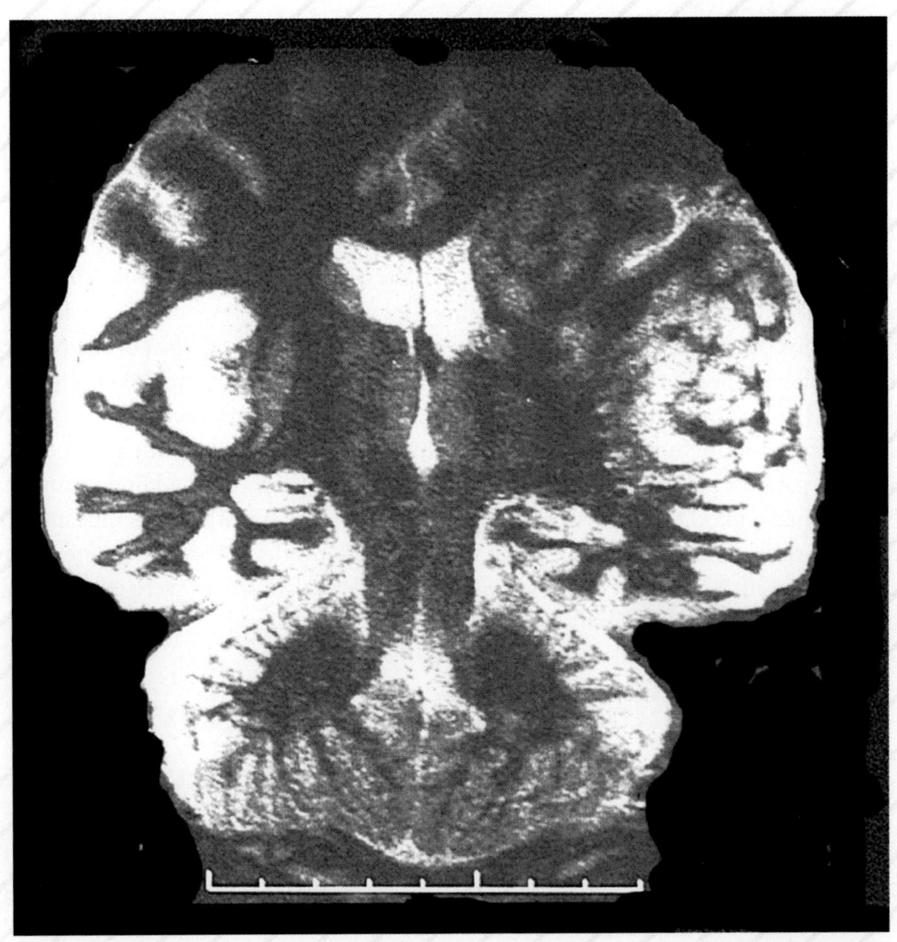

Figure 19. *MRI scan shows a very large cortical dysplasia. The dysplasia has multiple layers of cortex in it and is almost a large as a lobe of the brain.*

Cortical dysplasias are being recognized as being the cause of neo-cortical epilepsy in increasing numbers. The MRI scan identifies many of these, but the dysplasias may also be microscopic in nature and not seen with MRI.

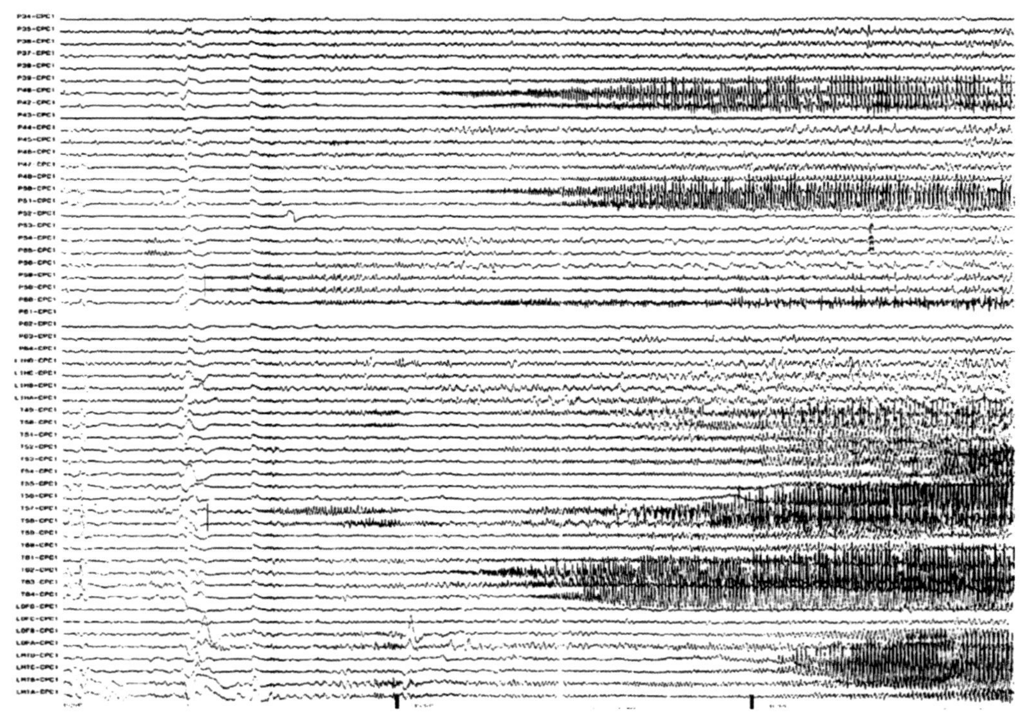

Figure 20. *Electrocorticogram of a 64 contact grid showing the onset of the seizure in this large cortical dysplasia. The onset of electrical activity is complex and widely separated over the grid.*

At the time of surgery, the origin of the epileptic activity appeared to be from 2 different locations. One area included the margin of the dysplasia and apparently normal brain (superior resection). The second area was from the dysplasia itself (inferior resection).

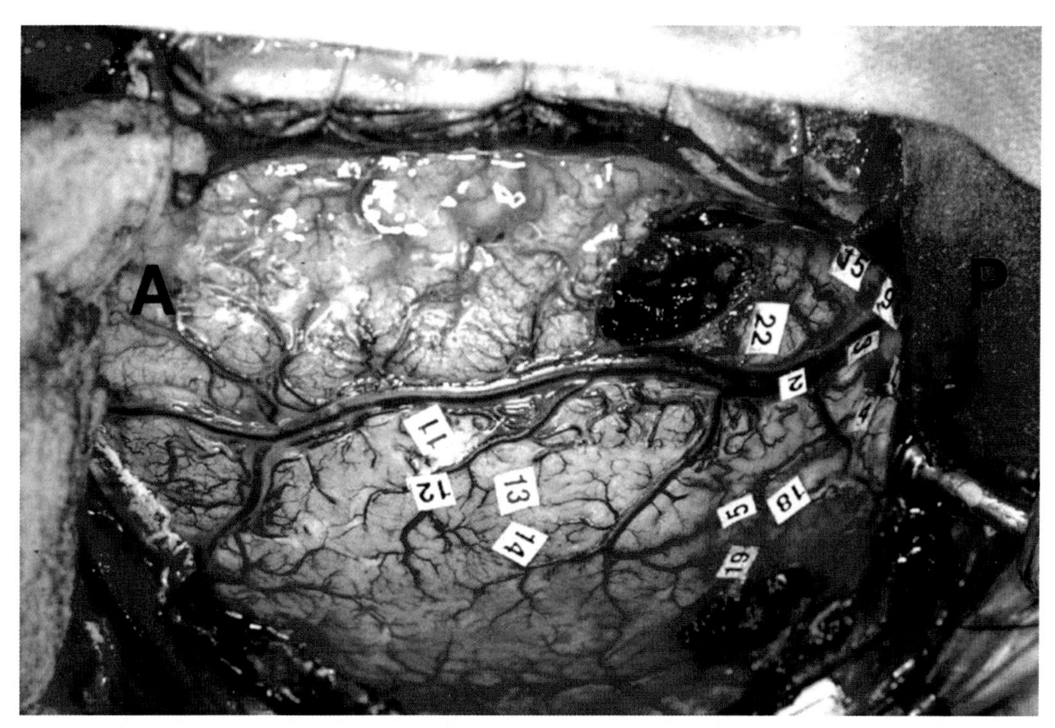

Figure 21. *Intraoperative photo of very large cortical dysplasia, which is discolored, and shows polymicrogyria. The identified epileptic origins have been removed. One area is in the dysplasia itself while the other is at the superior photo*

of very large cortical dysplasia, which is discolored, and shows polymicrogyria. The identified epileptic origins have been removed. One area is in the dysplasia itself while the other is at the superior margin of the dysplasia.

The next example is that of a cortical dysplasia that turned out to be microscopic and not seen with the MRI scan. **(Fig. 22 and Fig. 23).**

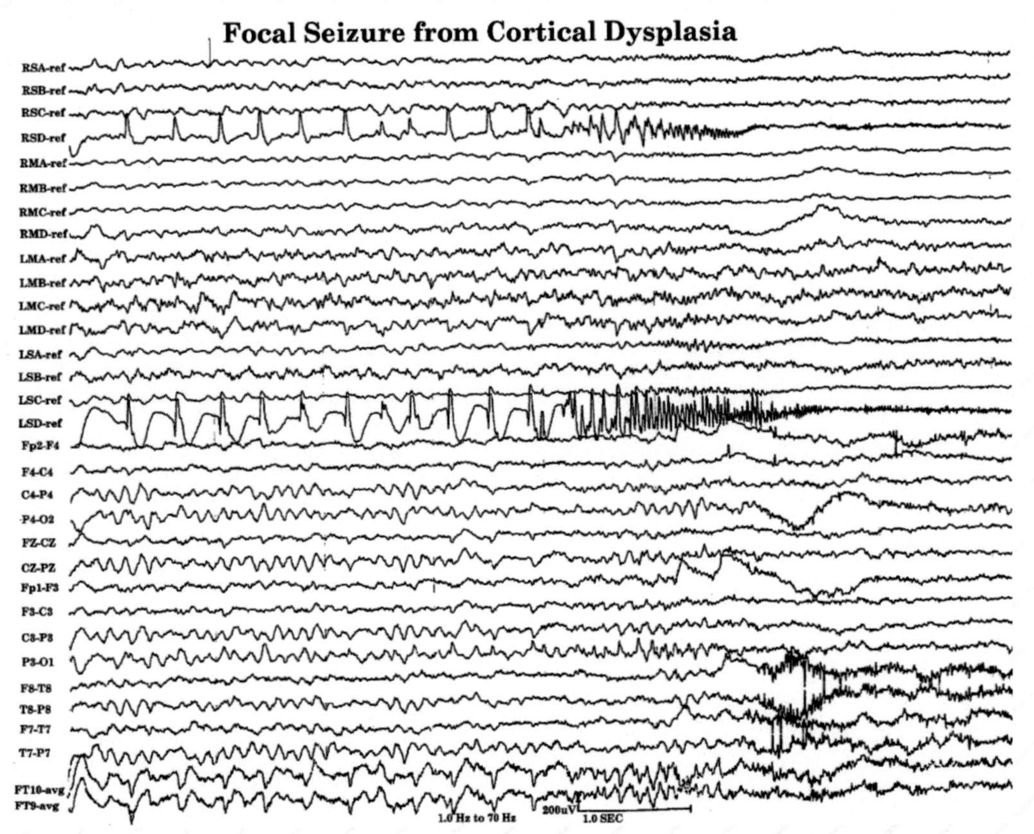

Figure 22. *Combination of sub-dural strip electrodes and scalp electrodes. The subdural strip electrode LSD shows a continuous spike and slow wave discharge which is interictal. This continuous interictal activity is very common in cortical dysplasias and when seen should suggest the presence of cortical dysplasia. The Ictal event which was very stereotyped with a subjective complaint of unsteadiness consisted of the short burst of fast spikes followed by a very low amplitude fast discharge. This discharge originated over the left parietal region. The scalp recording is not of much value asonly the interictal spikes are seen over both temporal electrodes at FT10 and FT9. The ictal event is not seen in the scalp electrodes.*

Tuberous sclerosis is one of the more common genetic disorders that cause a number of different types of seizures. Early in life this disorder may cause infantile spasms, but later in life tends to cause focal neocortical seizures. The clinical manifestations vary greatly from individual to individual, both in physical appearance and intellectual capacity. One individual may have such a mild case that the diagnosis is not expected until an infant is born with more severe features. **(Figure 23 and Fig. 24)** shows an example of a severe form of the disorder.

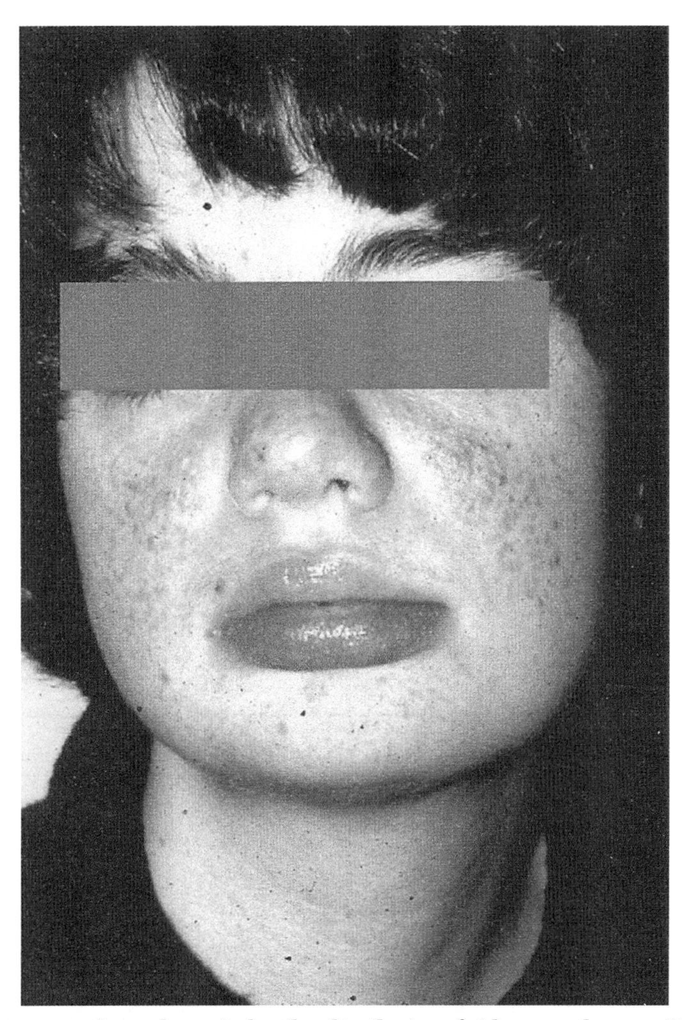

Figure 23. *This young woman shows the typical malar distribution of Adenoma sebaceum. These are small red tumors, not acne. These patients have tubers (cortical dysplasia) of the brain and may have hamartomas of the heart, kidney or other organs. Subungual fibromas are found in the majority of cases.*

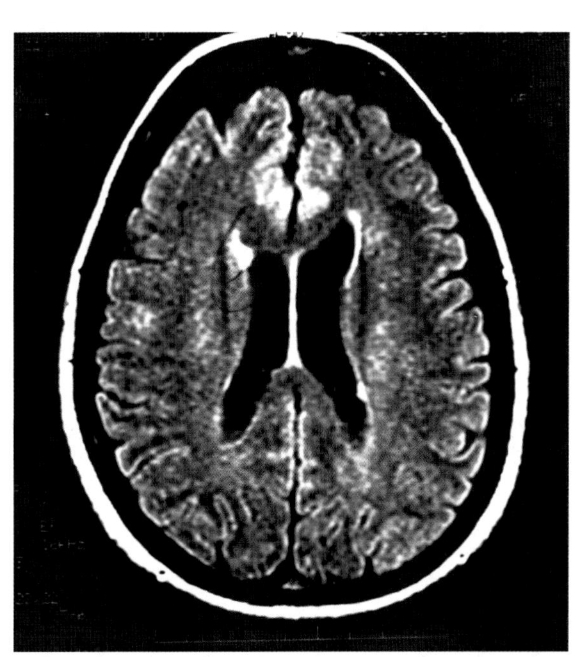

Figure 24. *MRI scan demonstrating changes typical of Tuberous Sclerosis. The cortical tubors are areas of cortical dysplasia.*

Gelastic Epilepsy is a rare seizure disorder that results from a hypothalamic hamartomas. These lesions are benign, but are located in a very critical area of the nervous system and may be difficult to completely remove. Epileptic activity frequently originates within the hamartoma.

The limits of resection are related to the closeness of the hypothalamus.

The hypothalamus must not be damaged as it is the part of the nervous system that controls appetite, temperature, and water metabolism. **(Fig. 25 and Fig. 26).**

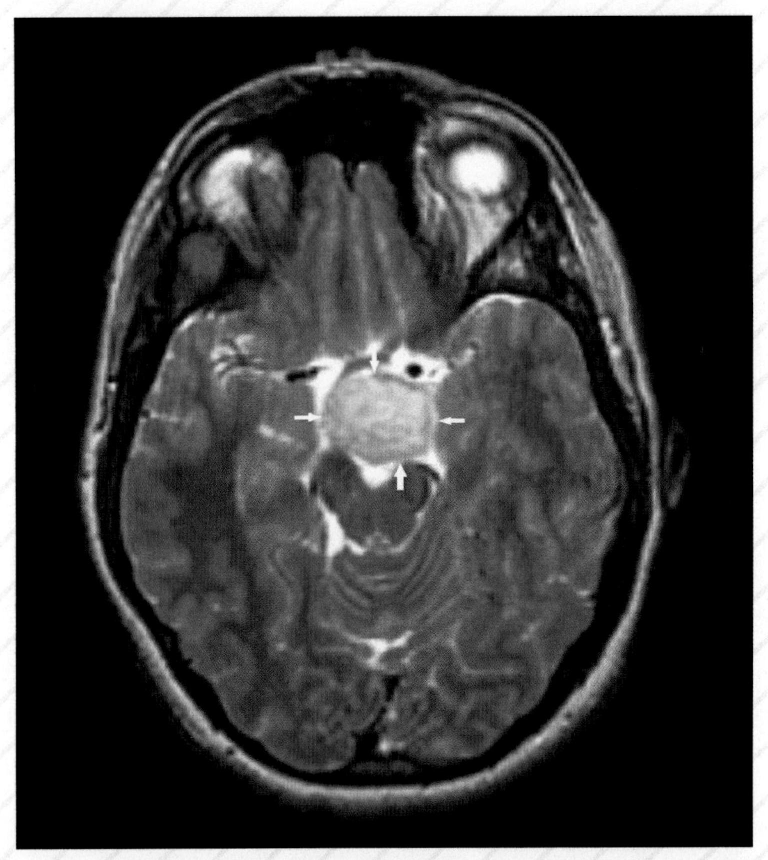

Figure 25. *MRI scan showing a large hypothalamic hamartoma in a patient who had Gelastic epilepsy. Arrows indicate limits of hamartoma.*

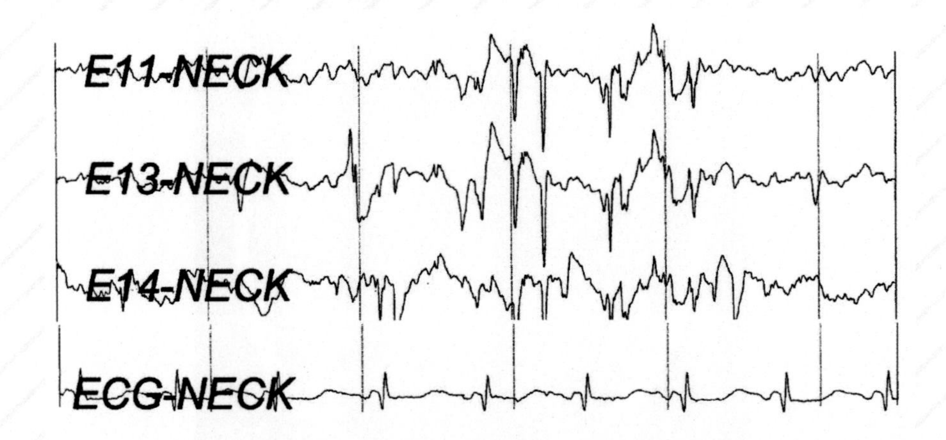

Figure 26 . *Direct recording from the hypothalamic hamartoma. Note the multi-focal epileptic spikes originating in the hamartoma. Following removal of a large part of the hamartoma, seizures were diminished in frequency although the residual seizures retained their typical characteristics.*

At times the source of the epileptic discharge originates in the motor-sensory cortex. There are surgical limitations in these regions, but it is possible to remove the non-dominant sensory cortex for the face. **(Figure 27)** is such an example.

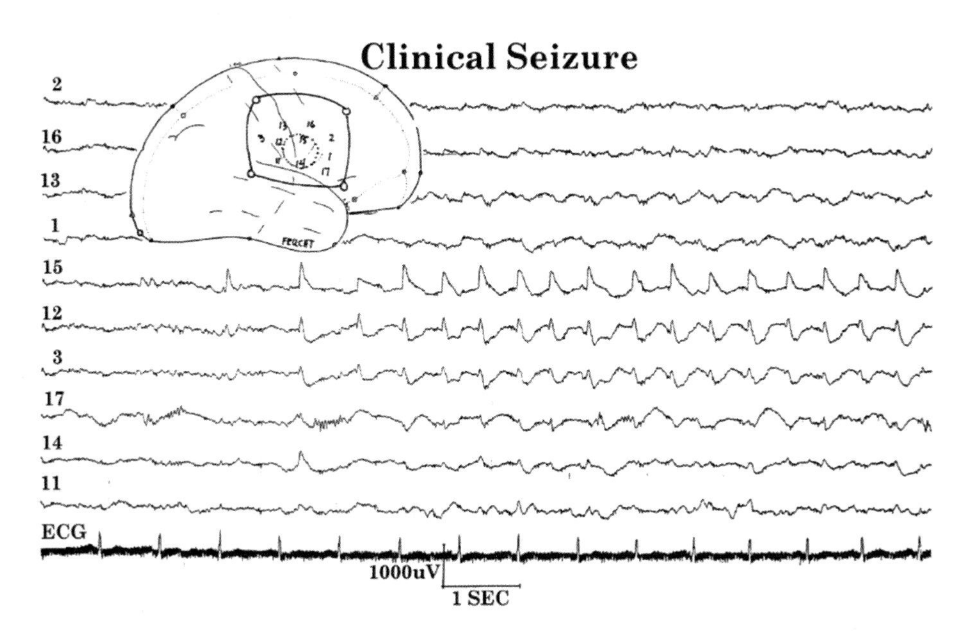

Figure 27. *Electrocorticography demonstrating a seizure that starts at the margin of a lesion represented by electrode 15 and to a lesser extent electrode 12. These electrodes are in the non-dominant face region and can be surgically removed without significant morbidity.*

Sub-pial transection is one surgical technique that can be carried out in eloquent cortex (motor, sensory and language regions). This type of surgery leaves the vertical connectivity intact while interrupting in horizontal connectivity. This surgical technique is based on the observation that horizontal connectivity is important in sustaining an ictal event. It is the vertical connectivity that is responsible for normal function. Frequently there are transient neurologic abnormalities following this procedure, but complete recovery generally occurs in these patients. While this procedure has been available for about 40 years, its role in epilepsy surgery remains some what controversial.

Electrocorticography in Brain Tumor Surgery

Neocortical brain tumors are relatively common in the adult and increase in frequency with age. The vast majority of these tumors are astrocytomas of various grades of malignancy. Seizures are more likely to occur in the lower grade tumors (grades 1 and 2) while grades 3 and 4 are much less likely to present with a seizure. The low grade tumor may be present for many years before being discovered as the cause of a seizure disorder. MRI scans in new onset seizure disorders should eliminate the long delay in establishing a correct diagnosis. The treatment of the malignant astrocytomas is still unsatisfactory. Ideally, current treatment includes maximum resection of the tumor without causing new symptoms, followed by irradiation and/or chemotherapy. Even under the best circumstances increased life expectancy following treatment for the Grade 3 and 4 astrocytomas (glioblastoma multiforme) are generally measured in months. Electrocorticography and the use of median nerve Somatosensory evoked potentials as well a language and motor mapping allow the surgeon to identify eloquent cortex, remove tumors that are adjacent to these vital areas, maximizing the amount of tumor removed without causing injury to the patient. Electrocorticography is important in establishing the current threshold necessary for language mapping.

One of the major differences from standard threshold testing as described for epilepsy surgery is the fact that neo-cortex which has been undercut by a tumor tends to be very epileptic and long runs of epileptic discharge can be triggered with very low currents. **See (Figure 28)** for an example.

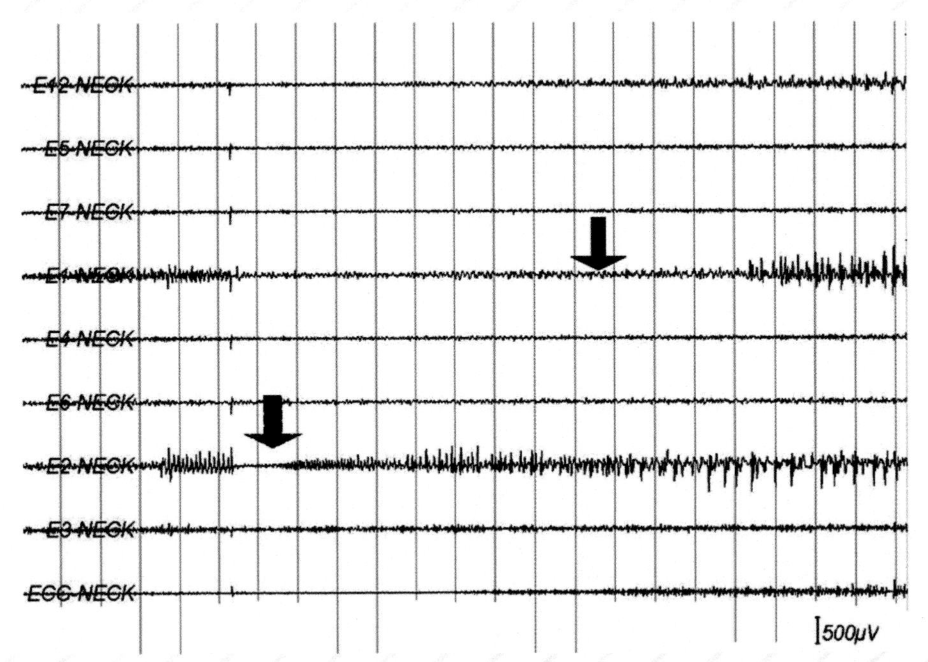

Figure 28. *Electrocorticography in a patient with a brain tumor. This very low Voltage fast discharge was triggered by stimulation with low currents. After the initial discharge begins at the lower arrow spread to adjacent cortex develops. These discharges may go on for many minutes and can generally be stopped with cold irrigation fluid.*

These discharges rarely progress to a clinical seizure, but the anesthesiologist should be prepared for such an event should it occur. The development of a clinical seizure or its treatment would interfere with critical brain mapping.

In addition to electrocorticography, brain mapping using evoked potentials has turned out to be a very valuable tool. Median nerve SSEPs are by far the common used. This allows one to identify both the sensory and motor hand region. A 4 contact electrode strip is used to record the responses as shown in **(Fig. 29)**; however, an 8 contact strip is more frequently used for this purpose.

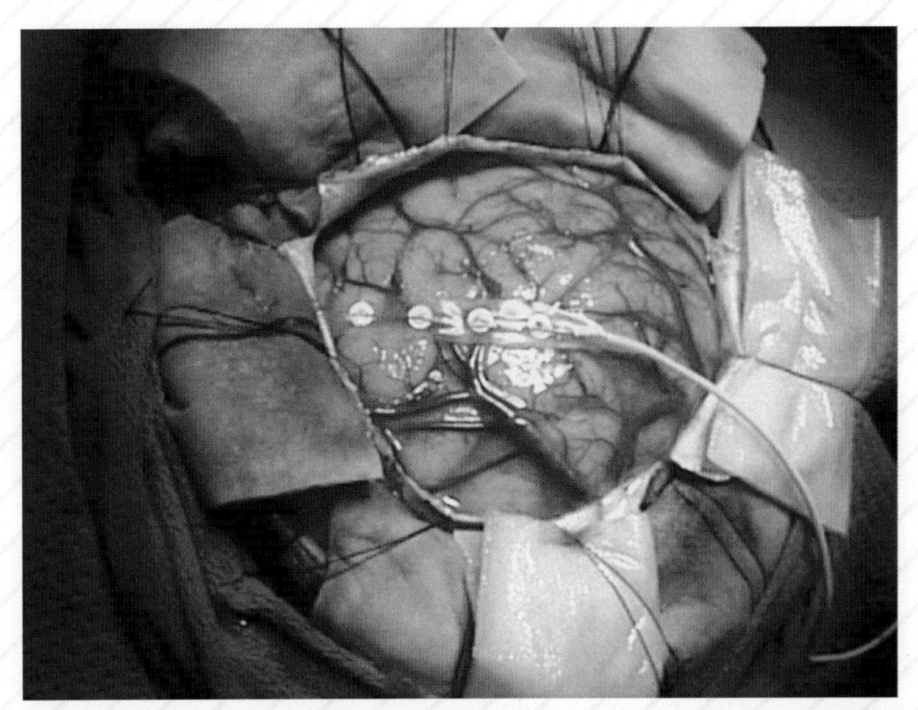

Figure 29. *Intraoperative photograph showing a 4 contact strip In the approximate location to identify the hand region of the cerebral cortex.*

The recorded response can be divided into two separate entities. Over the sensory cortex is found the N20 (a negative response with a latency of about 20 milliseconds). Over the motor cortex just medial to and anterior to the sensory N20 is the P22 (a positive response with a latency of about 22 milliseconds). See **(Figure 30).**

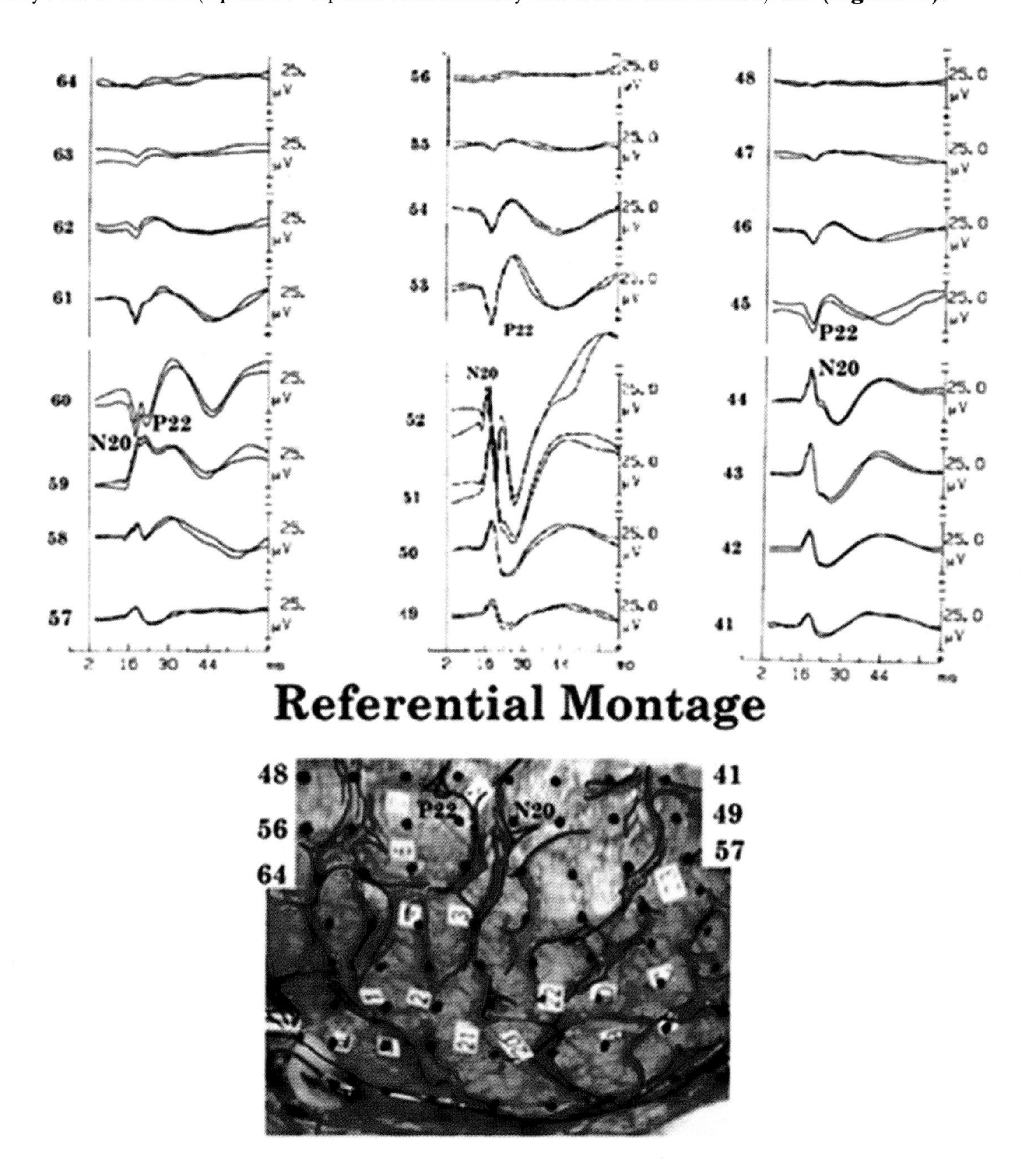

Figure 30. *Photograph of cortex and the median nerve responses over the exposed cortex. This example was actually recorded by a 64 contact grid, but the responses are the same. The shortest latency and smaller response is seen in the first panel. The N20 is seen in contacts 59-57 and the P22 in 60-62. The second panel shows the largest response with a slightly longer latency and the N20 is seen in 52-49 and the P22 53-54. The third panel shows the N20 in 44-40 and The P22 in 45-46. The Central Sulcus lies between the phase reversal created by the N20 and P22.*

These two responses are generated independently as one can be absent and the other remains. The hand area varies from individual to individual, both in the amount of brain involved with the response and the location.

In a large series of cases we found that the hand area can be located anywhere from 2 centimeters to 8 centimeters from the interhemispheric fissure. The location of the central sulcus is shown in **(Fig. 31 and Fig. 32).**

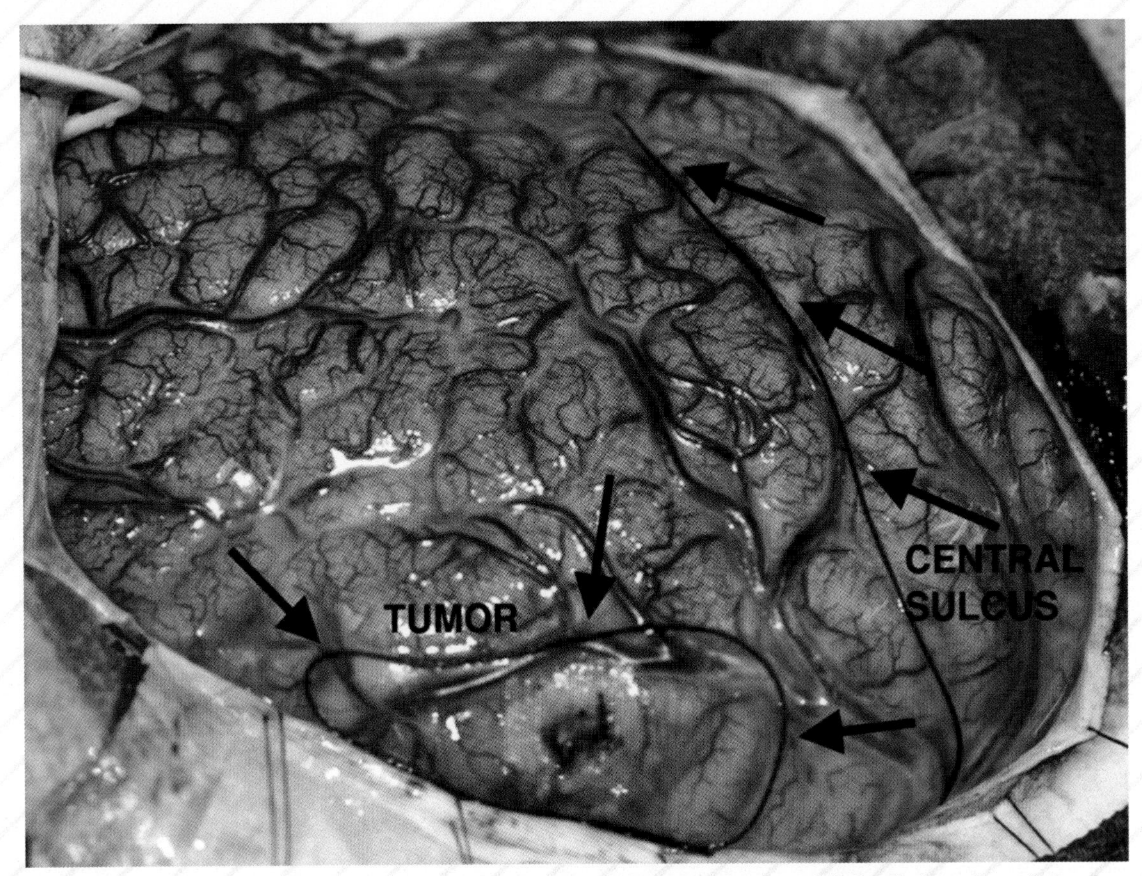

Figure 31. *Intraoperative photograph showing the location of the central sulcus (black silk thread) and the closeness of the tumor to the sensory gyrus.*

When the tumor is this close to the sensory gyrus, continuous median nerve SSEPs are carried out during the tumor removal. This is especially important once the cortex has been removed and the surgeon is working in the sub-cortical white matter. If any of the thalamo-cortical fibers are resected, the amplitude of the N20 will be decreased and the surgeon notified.

Direct cortical stimulation with a biphasic DC stimulator (Ojemann stimulator) can also be used to map the Motor/Sensory cortex. This in general requires an awake patient who is able to report sensory as well as motor phenomenon elicited by the stimulus. If the patient is anesthesized only the motor components can be identified. When stimulating the motor-sensory cortex in the anesthesized patient false identification is possible as motor responses are at times elicited from sensory cortex and sensory responses may be described by the patient when the motor strip is stimulated. The evoked potential responses do not suffer from this short-coming.

Rarely PTNSSEPs are used for brain mapping, but this requires the surgeon to insert a 4 contact strip in the interhemispheric fissure to obtain the recording. The response is a positive response at about 37 milliseconds latency. There is no response over the motor cortex as found in the median nerve response so the central sulcus cannot be recognized by this test.

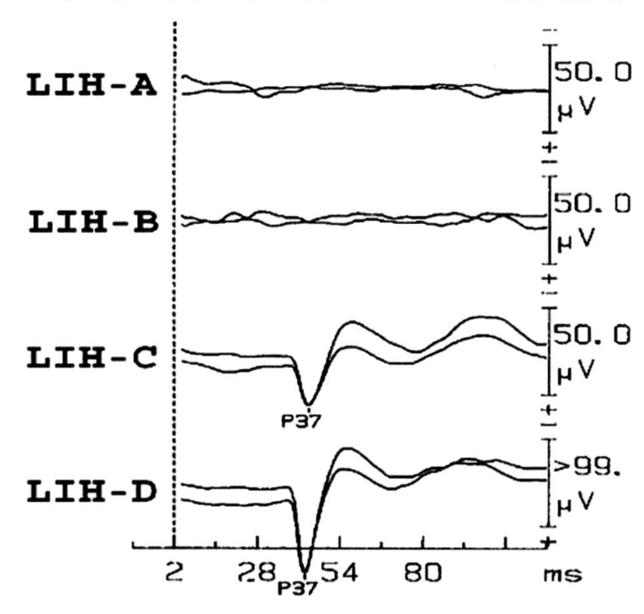

Figure 32. *Direct cortical recording on a 4 contact strip placed in the interhemispheric fissure over the foot area. Only a sensory P37 is recorded, the A and B contacts are presumably over motor strip and there is no response in that location. Contacts C and D are over sensory cortex.*

A combination of all of these tests are of value to the surgeon and allows maximum tumor removal while keeping patient morbidity to a minimum.

Selected Bibliography

Penfield, W. and Jasper, H.H. Epilepsy and the Functional Anatomy of the Human Brain. (1954) Little, Brown, Boston.

Ludwig, B.I., Ajmone-Marsan, C. and Van Buren, J. Depth and direct cortical recording in seizure disorders of extratemporal origin. (1976) Neurology 26: 1085-1099.

Lesser, R.P., Lüders, H., Dinner, D.S., et. al. The location of speech and writing functions in the frontal language area: Results of extra-operative cortical Stimulation. (1984) Brain, 107:275-291.

Wyler, A.R., Ojemann, G.A., Ettore, L. and Ward, A.A. Subdural strip electrodes for localizing epileptogenic foci. (1984) J. Neurosurg., 60: 1195-1200.

Ojemann, G.A. and Whitaker, H.A. Language localization and variability. (1978) Brain Lang. 6: 239-260.

Ojemann, G.A. Individual variability in cortical localization of language. (1979) J. Neurosurg. 50: 164-169.

Grant, G.A., Farrell, D.F., Silbergeld, D.I. Continuous SSEP monitoring during brain tumor resection. Case series and review of the literature. J Neurosurg 2002; 97: 709-713.

Farrell, D.F. Description of Electroencephalographic Technique for Tailored Mesial Temporal Epilepsy Surgery. In "Epilepsy Surgery: Principles and Controversies." J. W. Miller and D.L. Silbergeld (eds.) pp. 430-435. Taylor and Francis, New York. 2006.

Kutsy, R.L., Farrell, D.F. and Ojemann, G.E. Ictal Patterns of Neocortical Seizures. Correlation with Surgical Outcome. Epilepsia, 1999, 40:257-266.

Doherty, M.J., Jayadev, S. Miller, J.M., Farrell, D.F., Holmes, M.D. and Dodrill, C.B. Age at focal epilepsy onset varies by sex and hemispheric lateralization. Neurology 2003, 60:1473-1477.

Farrell, D.F., Leeman, S., and Ojemann, G.A. Study of the Human Visual Cortex: Direct Cortical Evoked Potentials and Stimulation. (2007) J. Clin. Neurophysiol. 24: 1-10.

Farrell, D.F., Burbank, N., Lettich, E., and Ojemann, G.A. Individual Variations in the Motor-Sensory (Rolandic) Cortex of Man. (2007) J.Clin. Neurophysiol. 24: 286-293.

Doherty, M.J., Simon, E., DeMenzes, M.S., Kuratani, J.D., Saneto, R.P. Holmes, M.D., Farrell, D.F., Watson, N.F., Dodrill, C.B., and Miller, J.M. When might hemispheric favoring of epileptiform discharges begin? Seizure 2003, 12:595-598.

McKhann, G.M. 2nd, Schoenfeld-McNeill, J., Born, D.E., Hagland, M.M., and Ojemann, G.A. Intraoperative hippocampal electrocorticography to predict the extent of hippocampal resection in temporal lobe epilepsy. J. Neurosurg. (2000) 93:44-52.

CHAPTER 10

Intraoperative Monitoring for Carotid Endarterectomy

ENDARTERECTOMY HAS BEEN used to treat internal carotid stenosis for a number of decades. Symptomatic carotid stenosis has utilized electroencephalography as a method to monitor blood flow to the homolateral cortex. The stenosis almost always affects the internal carotid artery at or just above the bifurcation of the common carotid into its internal and external branches. The narrowing of the internal carotid artery becomes significant when greater the 80 percent of the cross-sectional diameter of the artery is compromised. The typical clinical syndrome is characterized as the Transient Ischemic Event. A number of different syndromes can occur and the common syndromes include a short lived paralysis, sensory abnormalities, visual disturbances or a language disturbance. These events are not permanent with the patient returning to normal in a number of minutes to a few hours. When TIAs first start it is impossible to know whether the symptoms will clear or if the patient has suffered a cerebral thrombosis. The more TIAs a person has suffered the less likely it is to evolve into a completed stroke. Evaluation of the stenosis is readily made by flow DOPPER ultrasonography examination and either CT angiography or MRI angiography. **(Figure)** is an example of a high grade stenosis of the Internal Carotid Artery.

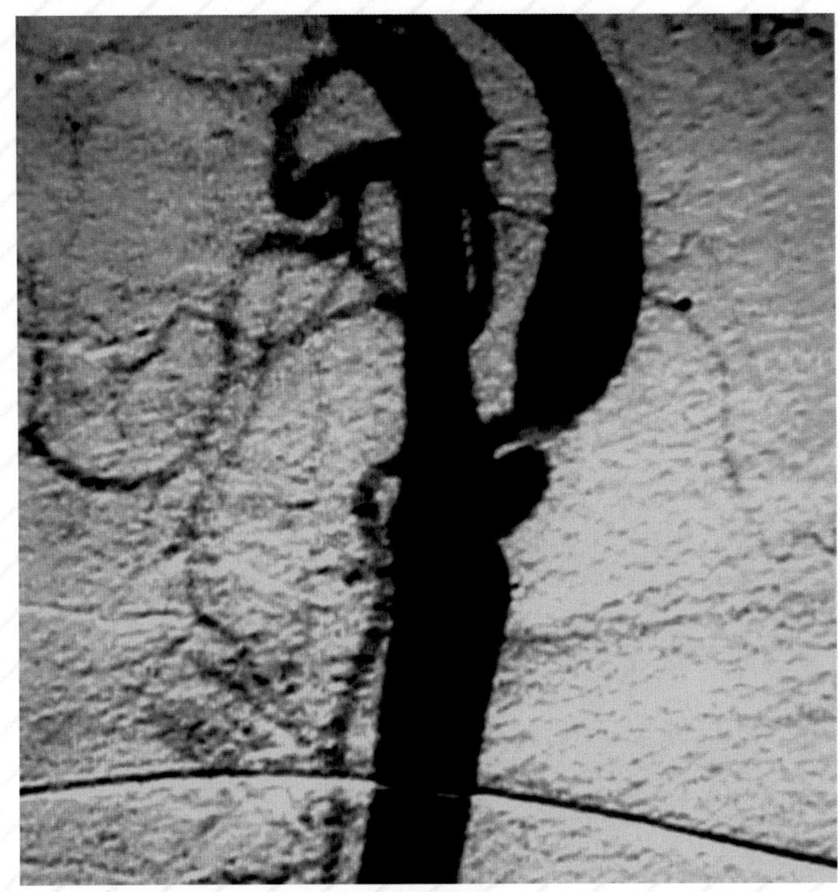

Figure 1. *Angiogram of carotid artery showing a very focal narrowing Note the post-stenotic dilatation of the internal carotid artery secondary to turbulent blood flow caused by the stenosis.*

The goal of this surgery is to prevent a stroke to the hemisphere supplied by this stenotic artery. **(Fig. 2).**

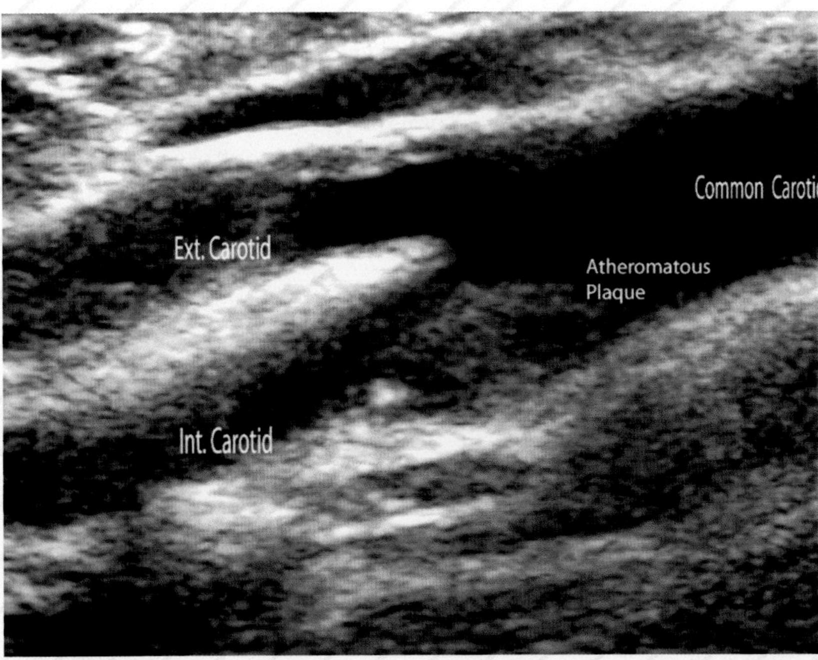

Figure 2. *Ultrasound of right caroted artery. Note the large atheromatous plaque nearly Occluding the Internal Carotid Artery at the bifurcation. This non-invasive test has almost Replaced angiography for the evaluation of carotid disease.*

The EEG has been utilized for a number of years to make a decision whether a vascular shunt will be required or if collateral circulation is adequate to support the at risk hemisphere during the surgery. Because of the anatomic variations in the Circle of Willis a number of patients do not have adequate collateral circulation to the operated side of the brain. The EEG should be monitored before clamping of the internal carotid artery to gather a baseline recording. The EEG is typically monitored during clamping of the internal carotid artery by the surgeon. If adequate collateral circulation is present no changes in the EEG pattern occur. If however, the collateral circulation is not adequate, a series of EEG changes occur very rapidly after clamping. First, there is an increase in beta activity (13-22 Hz) followed in seconds by slowing of the EEG into the Theta and Delta frequencies. All of this will occur within 10-20 seconds. If this sequence occurs then the clamp must be removed and a shunt placed for safe surgery. With the advent of fast computers the methods of analysis changed for monitoring the clamping of the internal carotid artery. The development of fast Fourier analysis allowed the raw EEG data to be separated into its power spectra and displayed as continuous bands of activity divided into the classic 4 bandwidths used in standard EEG analysis. The slowest frequency is the delta bandwidth with a frequency of 0.5-4 Hz; the theta activity has a bandwidth of 5-7 Hz, alpha activity 8-12 Hz, and beta activity 13-22 Hz. **(Fig. 3).**

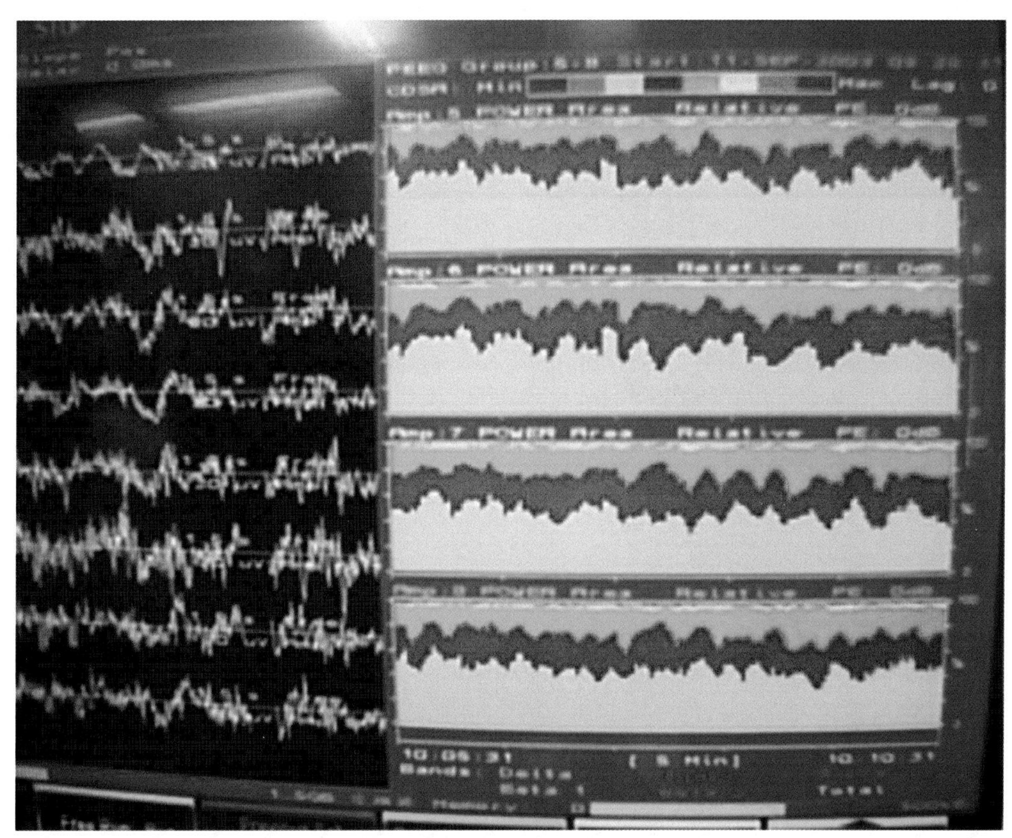

Figure 3. *Photograph of a computer screen which demonstrates raw EEG data on left side and computerized spectral analysis during Carotid artery surgery. Each power band is color coded. Light blue = delta activity, Red = theta activity, Green = alpha activity, and Yellow = beta activity. Patient is under anesthesia so the slower frequencies dominate.*

As one can see there are minor changes that occur over short periods of time, but with carotid test occlusion the changes will be over a longer period of time and persist. Spectral-edge analysis remains a powerful tool for monitoring this type of surgery. Spectral-edge represents an indirect measurement of blood flow.

Over the past 10 years a new technology has virtually replaced EEG and the processed EEG monitoring for carotid surgery. Flow DOPPLER ultrasonography is able to directly monitor the blood flow in the distal internal carotid artery system over the fronto-temporal cortex. Blood flow is measured over the temporal region. An adequate bone window is necessary to carry out this examination and is found in the majority of patients. **(Figure 4)** shows a patient with the ultrasonography equipment in place for surgery. This equipment allows for continuous blood flow monitoring.

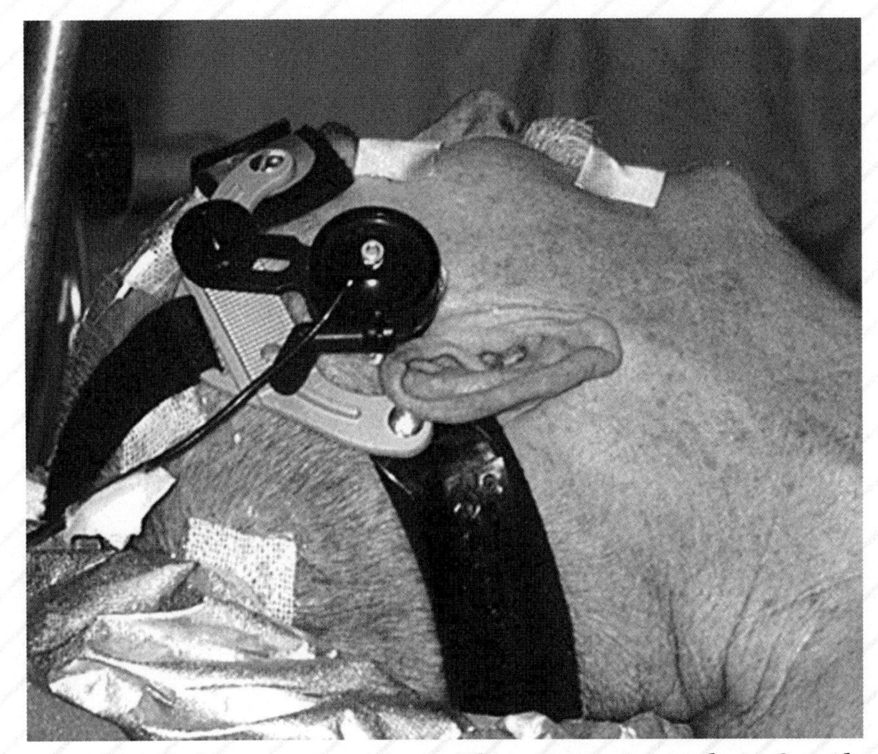

Figure 4. *Transcranial Doppler Intra-operative equipment. This equipment Is worn during Carotid surgery to monitor the blood flow in the middle cerebral artery.*

If the bone window is not adequate, then computerized spectral analysis can be used as a replacement. Using continuous TCD at the time of clamping the common carotid artery, any drop in blood flow greater than 30 percent will require a shunt being placed for continuation of the surgery. **(Figure 5)** shows the ultrasonogram before carotid clamping (A) and during clamping where the signal virtually disappears (B). This particular patient required a shunt being placed for safe endarterectomy.

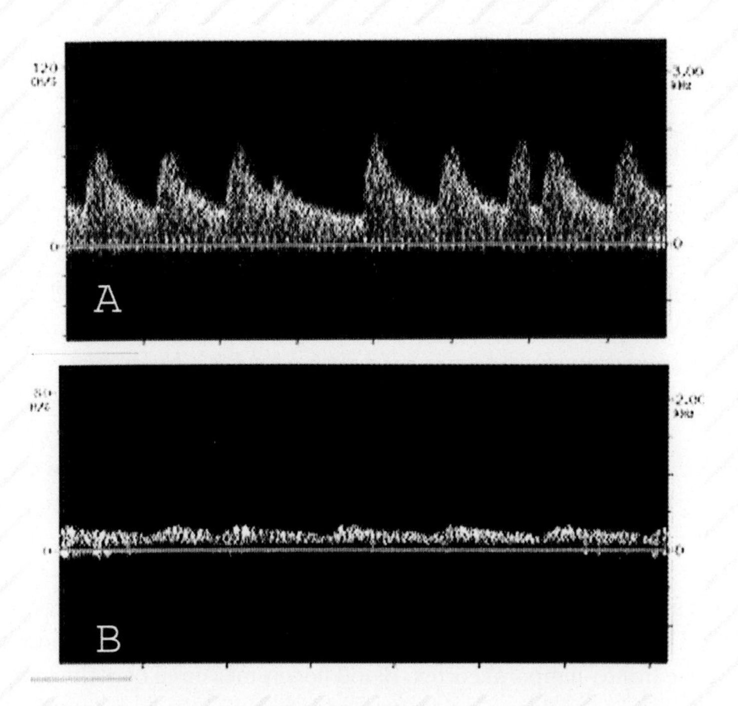

Figure 5. *Figure 5A shows the Doppler ultrasonogram before the test carotid occlusion. Figure 5B shows the same patient during the test occlusion of the Common Carotid Artery. The signal has virtually disappeared. This patient required a shunt for surgery.*

DOPPLER ultrasonography also has additional advantages in that it can readily detect particulate emboli **(Fig.6)** and air emboli **(Fig. 7).**

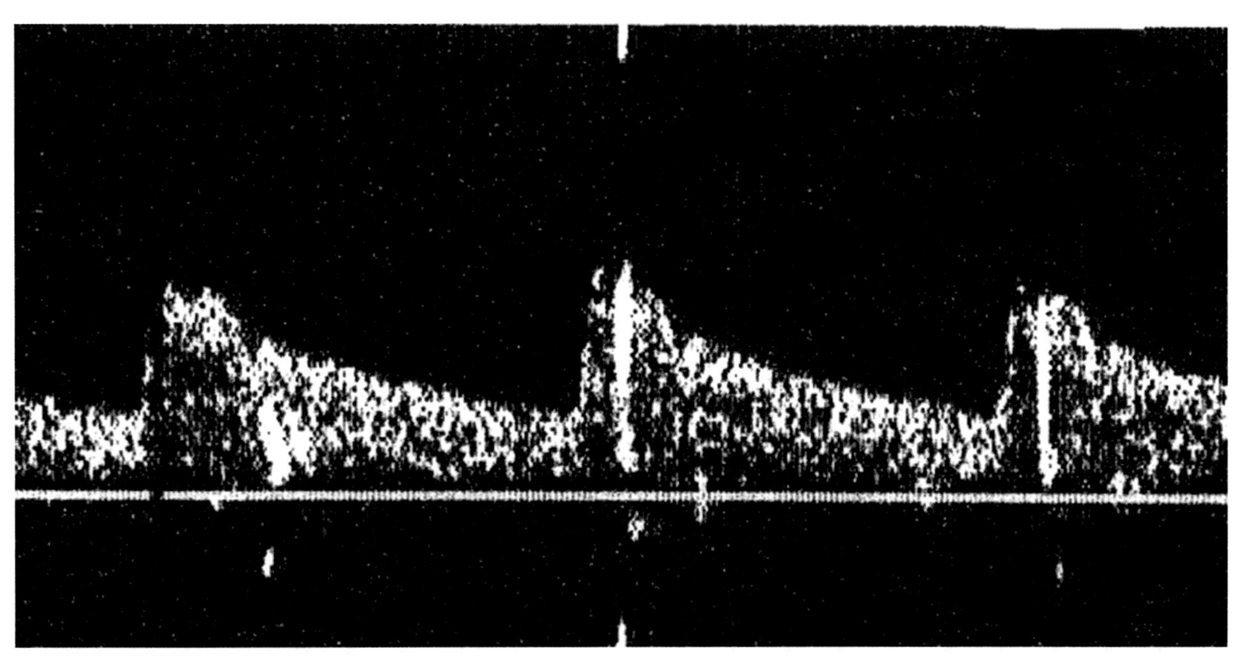

Figure 6. *Particulate emboli are demonstrated in this ultrasonogram. Note the very bright areas during each pulse. These bright areas represent the particulate emboli.*

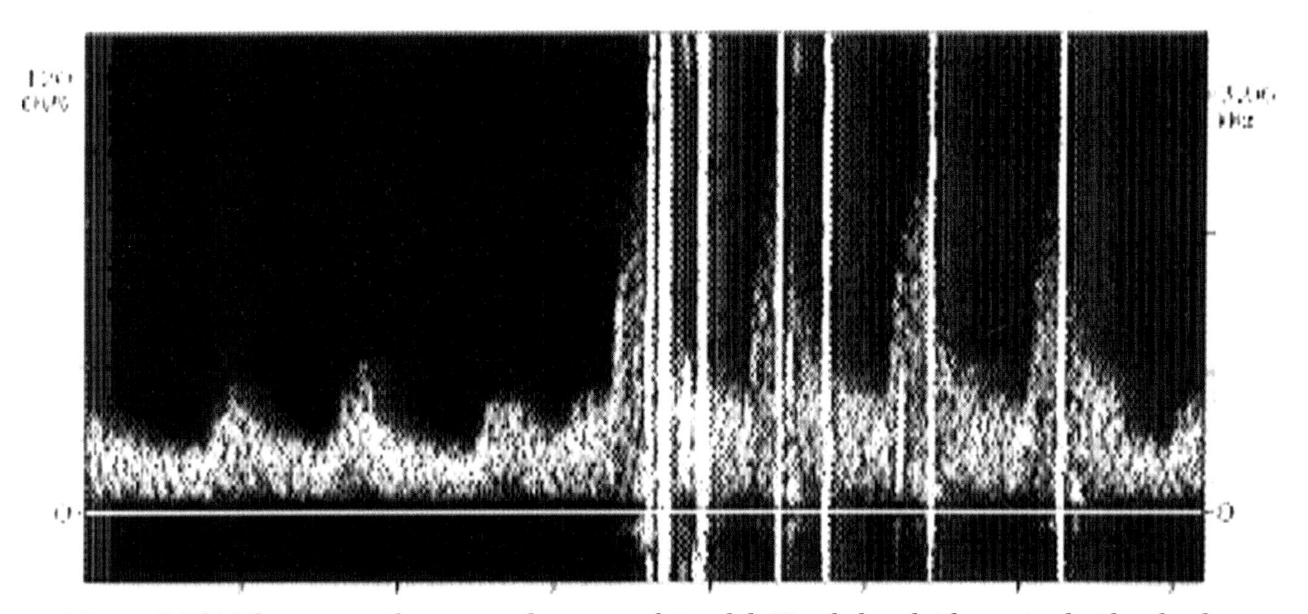

Figure 7. *This Ultrasonogram demonstrates the presence of air emboli. Note the large bright associated with each pulse. These large bright areas represent air emboli.*

Transcranial ultrasonography will most likely serve as the standard for monitoring of this type of vascular surgery.

Several other uses of TCD have been developed. The administration of Carbogen (5% CO_2) causes dilatation of the cerebral arteries and the changes that occur with this gas document the arteries have retained their automatic regulation. A failure to increase flow with 5 percent CO_2 shows that the cerebral arteries have lost their ability to auto-regulate. The loss of auto-regulation is commonly seen in severe hypertension, including hypertensive encephalopathy and eclampsia.

TCD is a rapid method for establishing brain death and ranks with nuclear medicine scan for speed and accuracy. Both EEG and Cortical evoked potentials (MNSSEPs) are helpful, but have significant problems if the patient has received sedative medications or if electrical artifacts are excessive. The advent of transplantation medicine makes it imperative that a comatose patient who will not recover be identified as early as possible so their vital organs can be utilized to save other patient's lives.

Selected Bibliography

Moore, W.S., et. al. Guideline for carotid endarterectomy: a multidisciplinary consensus statement from the ad hoc committee. American Heart Association. Stroke (1995) 28 (1): 188-201.

Stockard, J.J. and Bickford, R.G. The neurophysiology of anesthesia. In A Basis and Practice of Neuroanesthesia. Edited by E. Gordon. Pp. 3-46, (1975) Excerpta Medica. Amsterdam.

Chiappa, K. H., Burke, S.R., and Young, R.R. Results of electroencephalographic monitoring during 367 carotid endarterectomies: Use of a dedicated minicomputer. Stroke (1979) 10: 381-388.

Sundt, T.M. Jr., Sharbrough, F.W., Anderson, R.E. and Michenfelder, H.D. Cerebral blood flow measurements and electroencephalograms during carotid endarterectomy. J Neurosurg (1974) 41: 310-320.

Sundt, T.M. Jr., Sharbrough, F.W., Piepgras. D.G., et. al. Correlation of cerebral blood flow and electroencephalographic change during carotid endarterectomy. Mayo Clin Proc (1981) 56: 533-543.

Gordon, J.K., et. al. Correlation of intraoperative electroencephalographic and transcranial Doppler monitoring with post endarterectomy neurologic events. Presented at the 23rd International Joint Conference on stroke and Cerebral Circulation. Omni Rosen Hotel, Orlando, FL. Feb 5-7, 1998. (abstract)

Gaunt, M.E., et. al. Clinical relevance of intraoperative embolization detected by transcranial Doppler untrasonography during carotid endarterectomy : a prospective study of 100 patients. Brit J Surg (1993) 81 (10): 1435-1439.

Gaunt, M.E., et. al. Microembolism and hemodynamic changes in the brain during carotid endarterectomy. Stroke (1994) 25 (12): 2503-2505.

Smith, J.L., Evans, D.H., and Gaunt, M.E. Experience with transcranial Doppler monitoring reduces the incidence of particulate embolization during carotid endarterectomy. Bri J. Surg (85 (11): 56-59.

Aaslid, R., Huber,, P. and Nornes, H. Evaluation of cerebrovascular spasm with transcranial Doppler untrasound. J Neurosurg (1984) 60: 37-41.

Lindegaard, K.F., Normes, H., Bakke, S.J., et. al. Cerebral vasospasm after subarachnoid haemorrhage investigated by means of transcranial Doppler ultrasound. Act Neurotic supple (Wean) (1980 421 81-84.

Roomer, B., Brandt, and L. Bern man, L., ET. al. Simulaneous transcranial Doppler sonography and cerebral blood flow measurements of cerebrovascular CO2-reactivity in patients with aneusrysmal subarachnoid haemorrhage. Bri J. Neurosurg (1991) 5: 31-37.

Lampl, Y., Y., Gilad, R., Eschel, Y., et.al. Diagnosing Brain Death using the transcranial Doppler with a transorbital approach. Arch Neurol (2002) 59: 58-60.

Appendix

1. **Score Sheet for Language Mapping. Both the Surgeon and the Technologist should score the results using the codes below.**

Scoring codes:

Correct response (<)

Error (+)

Delays (D)

Hesitations (H)

Repetitions (R)

After language mapping is complete, the surgeon and technician should compare their score sheets for accuracy.

Score Sheet

CURRENT

PICTURES

CAR			
SOCK			
CARROT			
EAR			
CAT			
CHAIR			
CHICKEN			
CIGARETTE			
AIRPLANE			
ARM			
ARROW			
SAW			
BALL			
BARN			
BEAR			
BED			
BELT			
BICYCLE			
SQUIRREL			
BOOK			
BOWL			
BUS			
DRESSER			
EYE			
DRUM			
CANDLE			
COAT			
CORN			
COW			
TRUCK			
TRUMPET			
COUCH			
DESK			
DOLL			
DOG			
DOOR			
DRESS			
ELEPHANT			
FINGER			
FISH			

FOOT			
FORK			
GLASSES			
GUITAR			
HAMMER			
HAT			
HORSE			
KNIFE			
SHOE			
LAMP			
LEG			
LION			
MOON			
MOTORCYCLE			
MOUSE			
STAR			
TABLE			
NAIL			
NOSE			
PAINTBRUSH			
PIG			
PANTS			
PIANO			
POT			
TELEPHONE			
ROLLERSKATE			
RULER			
SCREWDRIVER			
SHIRT			
SPOON			
STOOL			
STOVE			
SUITCASE			
TELEVISION			
TIGER			
BRUSH			
TOMATO			
TRAIN			
VIOLIN			
WINDOW			

NAME...

DATE...

A- Electrical stimuli are applied at random at each of the labelled sites and indicated on the score sheet with a 'correct' (<) or 'error' (+) label. Delays (D), hesitations (H), repetitions (R) or other errors are also noted . A variation of similar names for the same picture are acceptable as correct. The surgeon gives the number of each site stimulated and this is indicated on the score sheet.

7

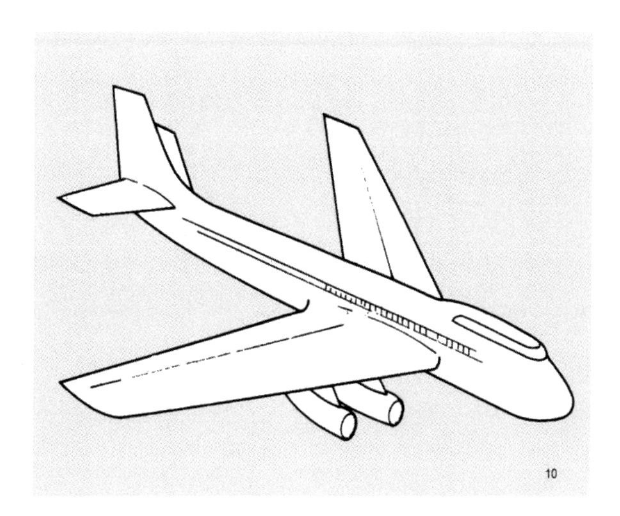
10

8

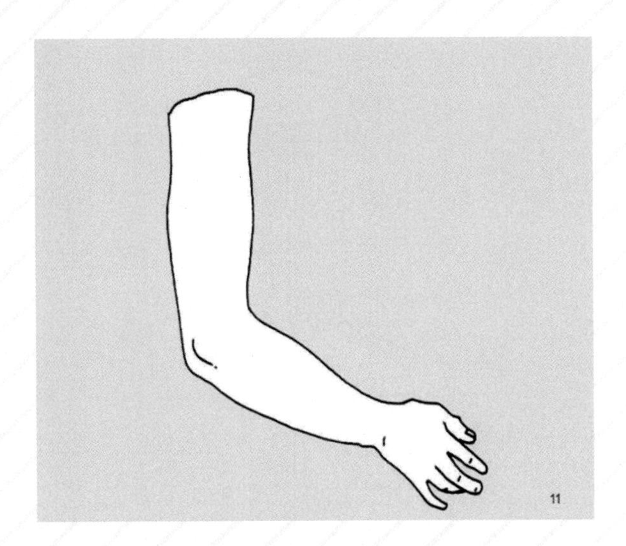
11

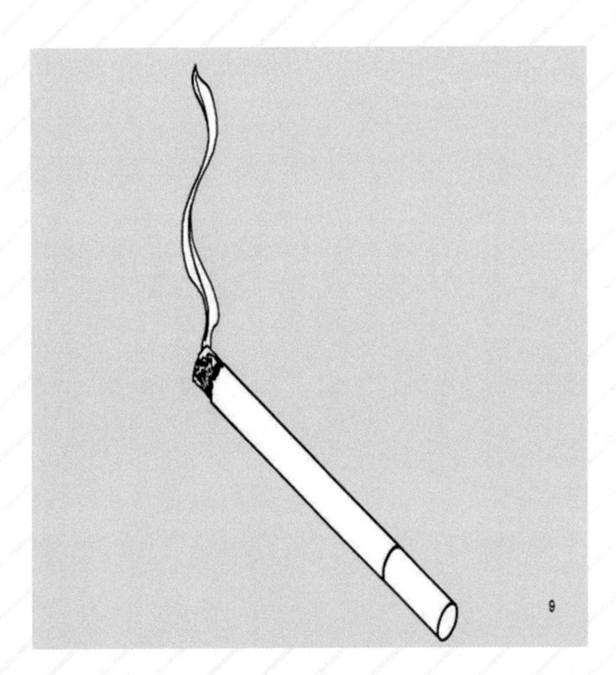
9

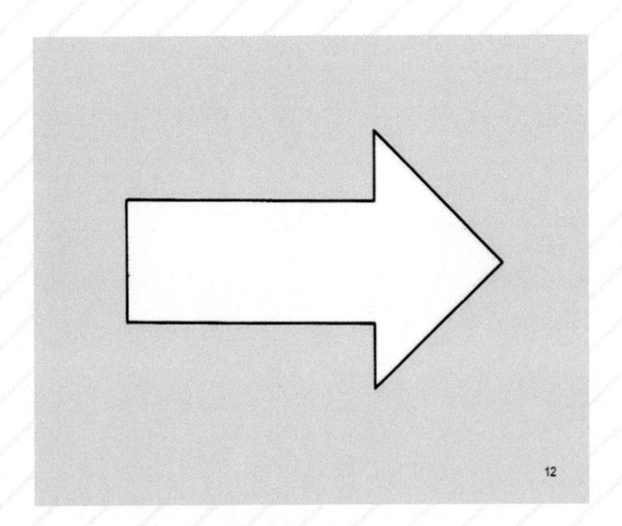
12

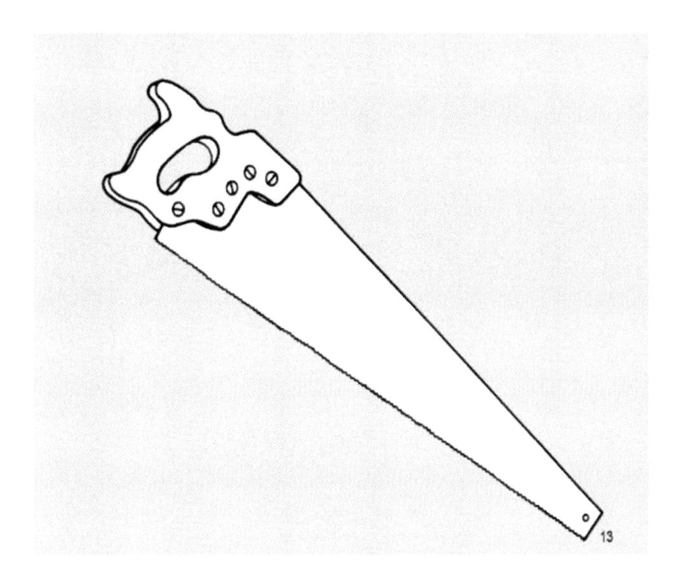
13

16

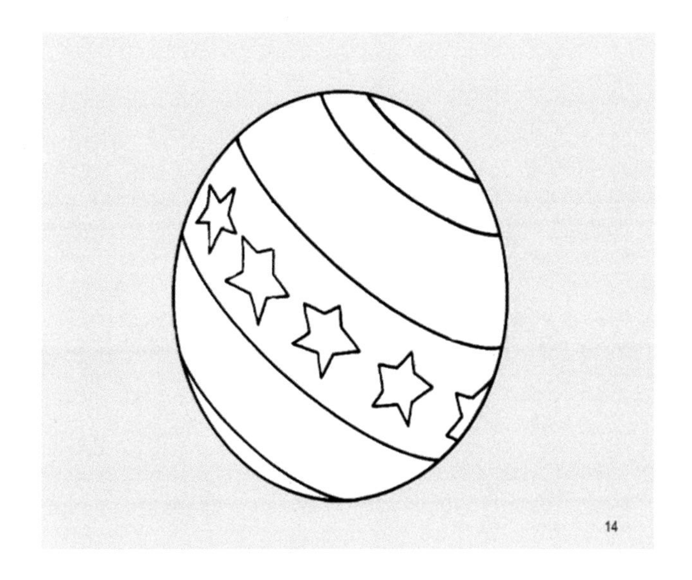
14

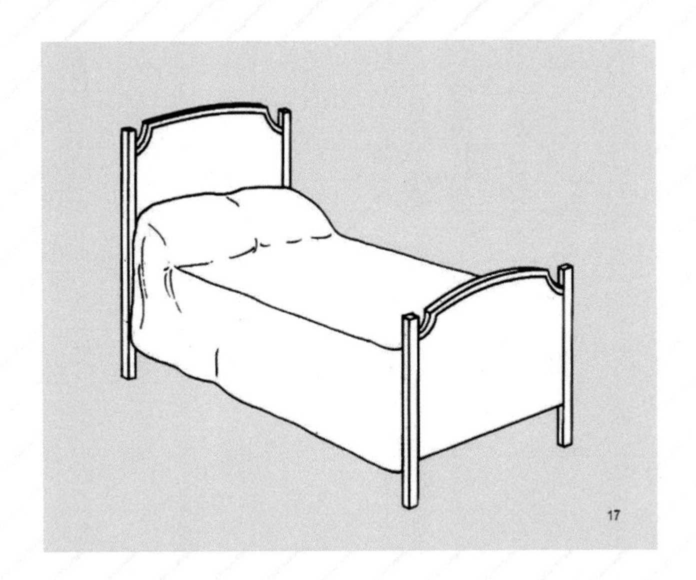
17

15

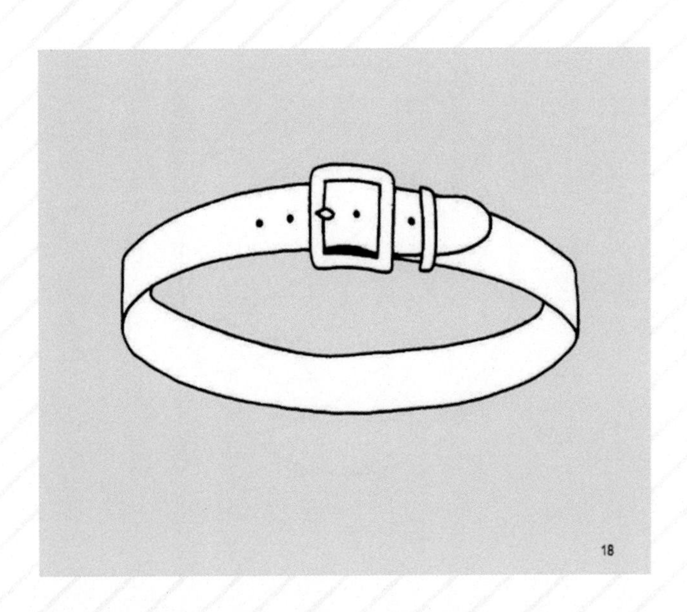
18

19

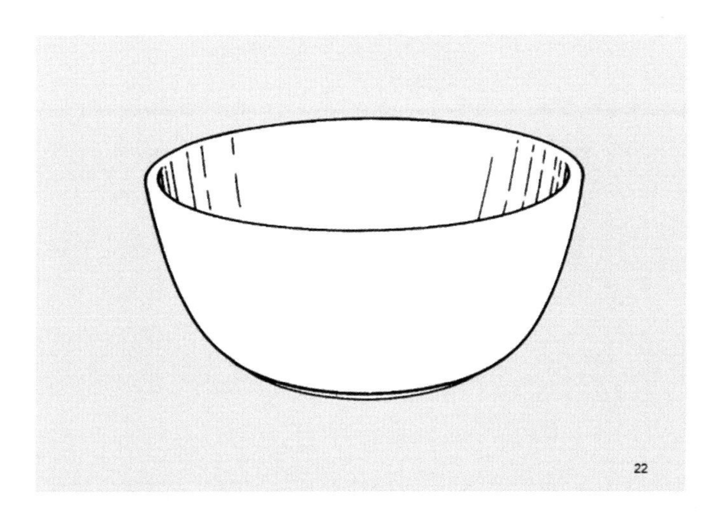
22

20

23

21

24

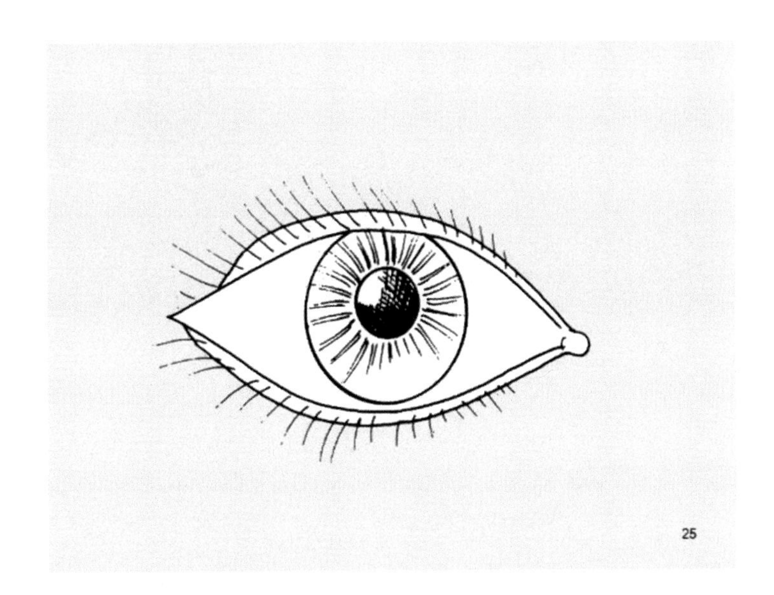
25

28

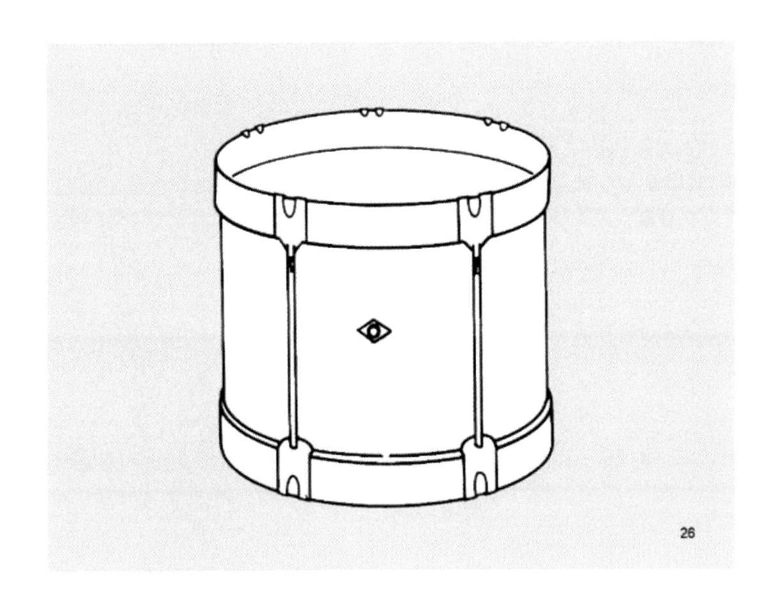
26

29

27

30

31

34

32

35

33

36

37

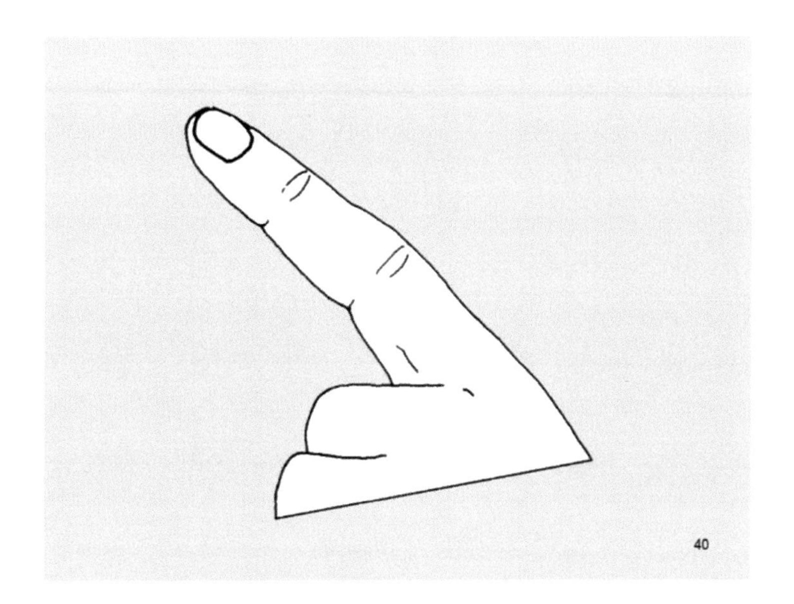
40

38

41

39

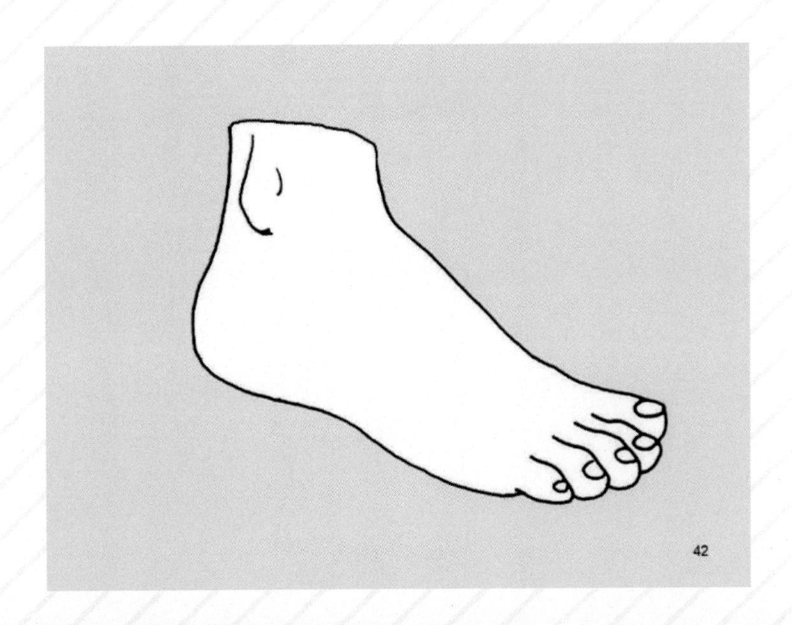
42

43

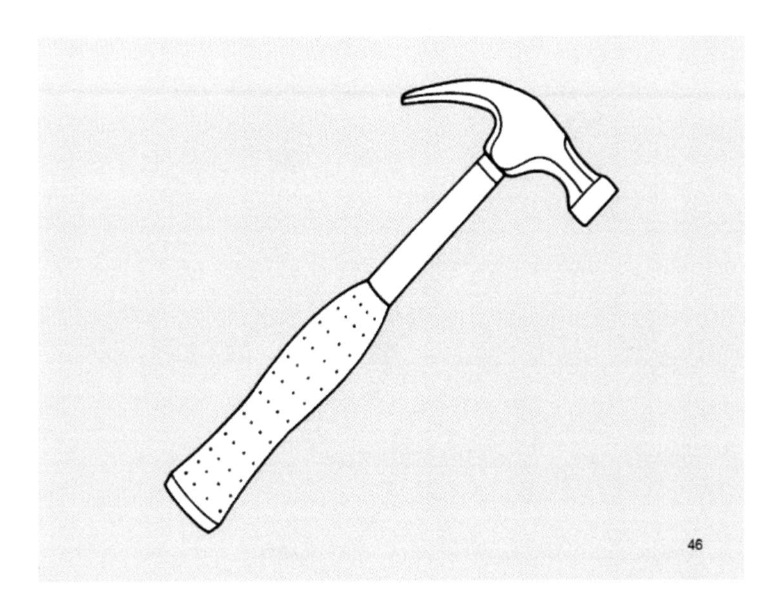
46

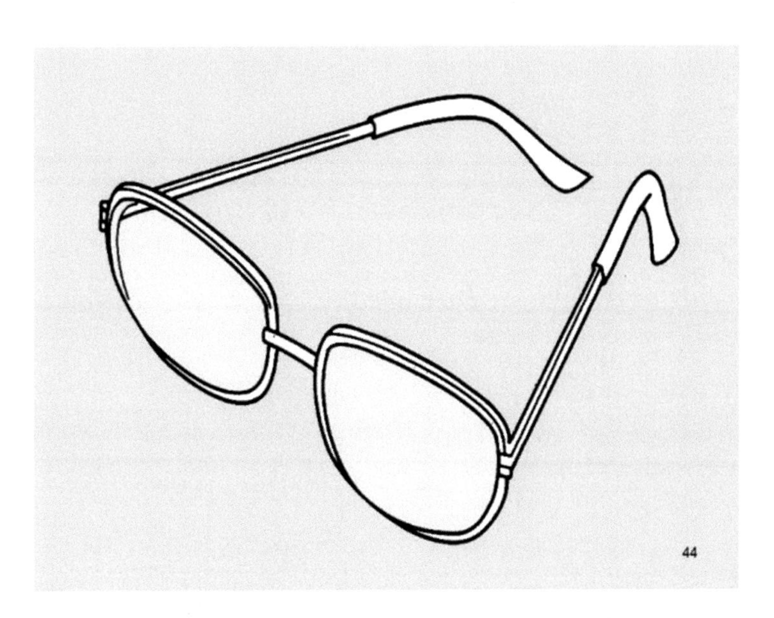
44

47

45

48

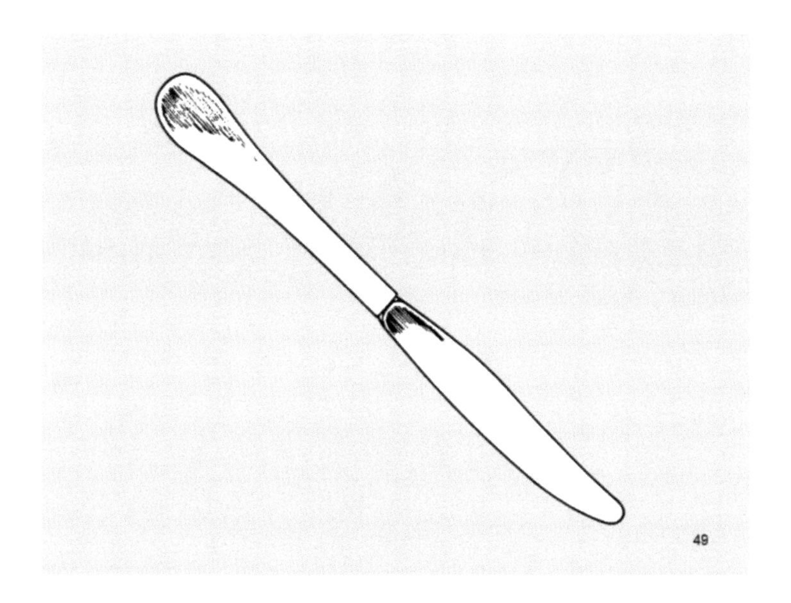
49

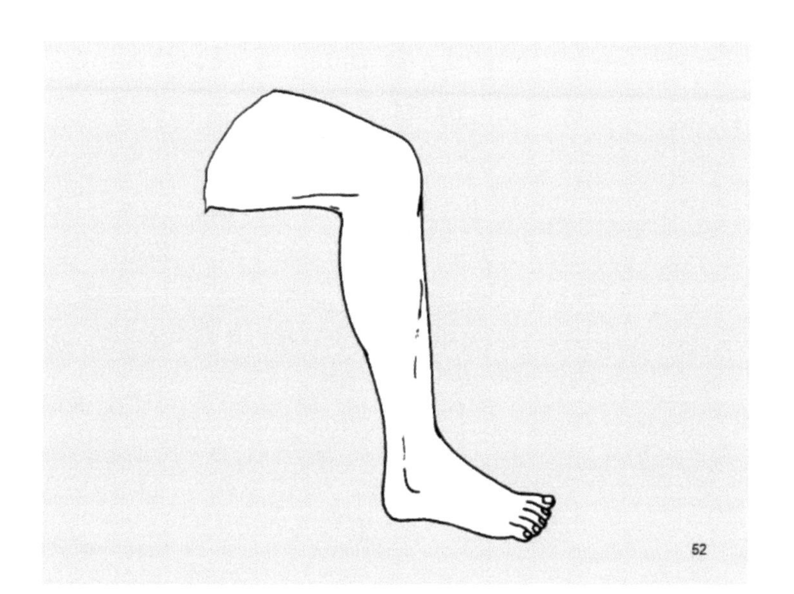
52

50

53

51

54

55

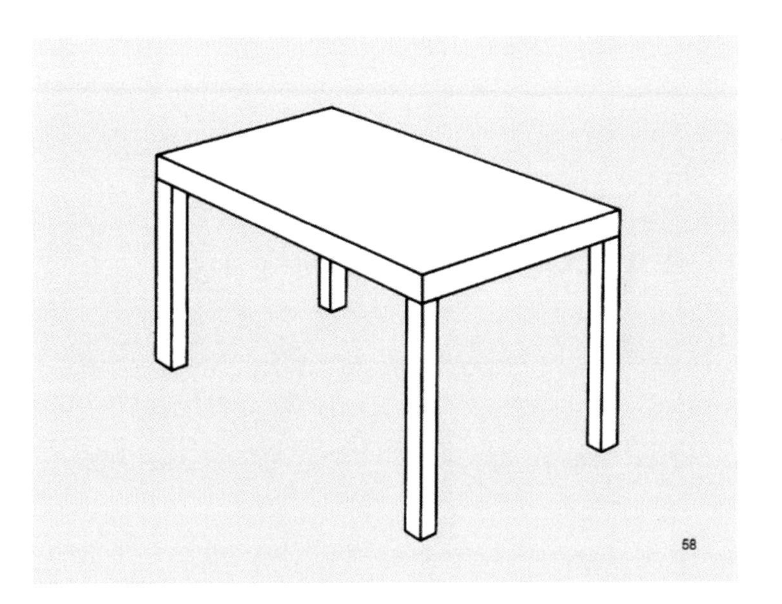
58

56

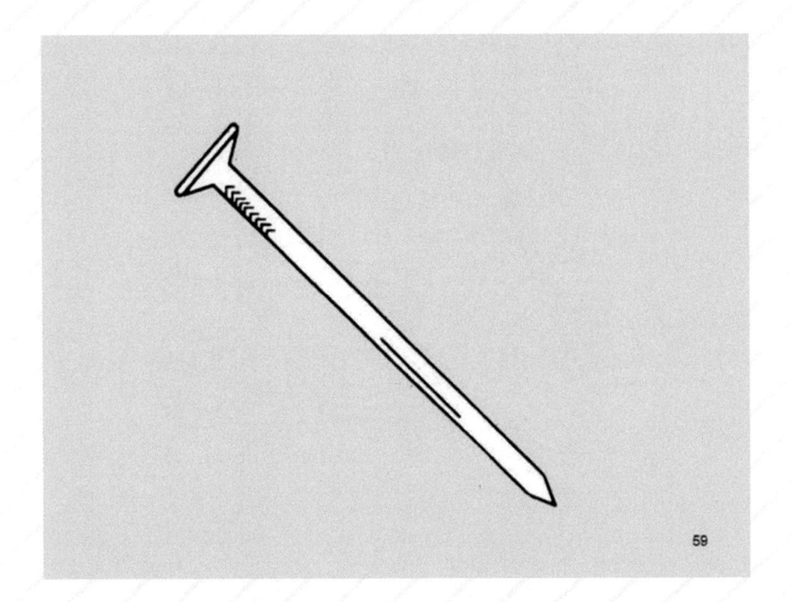
59

57

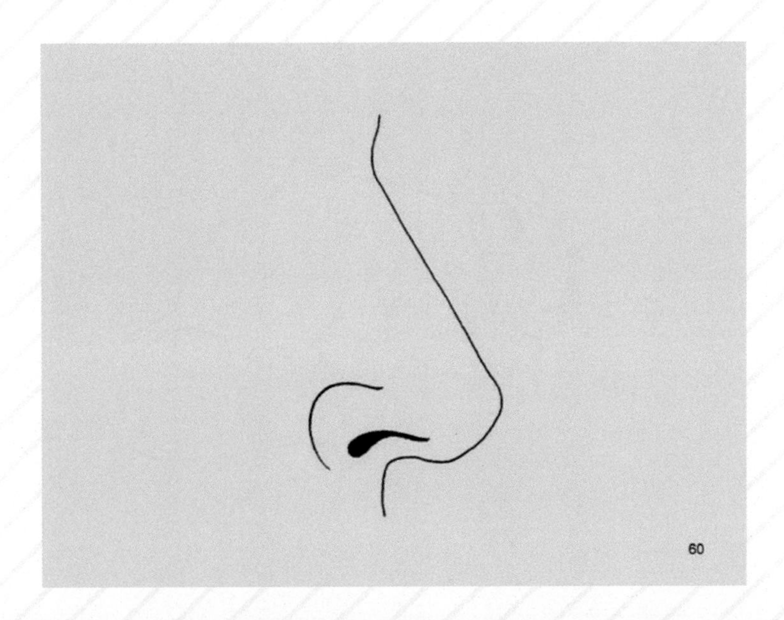
60

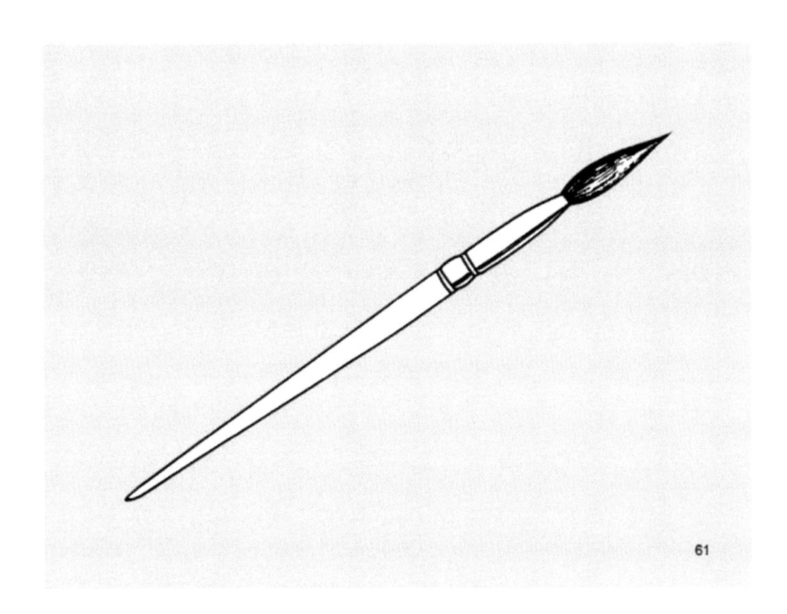
61

64

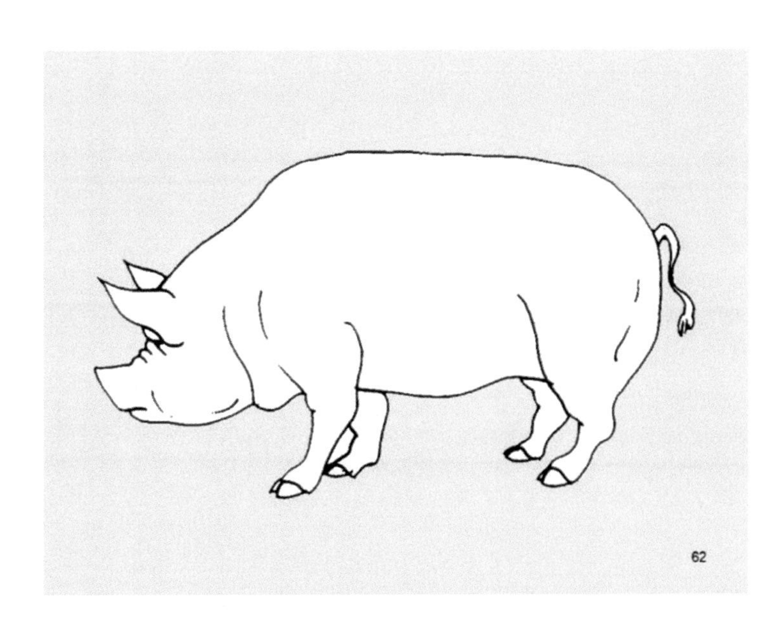
62

65

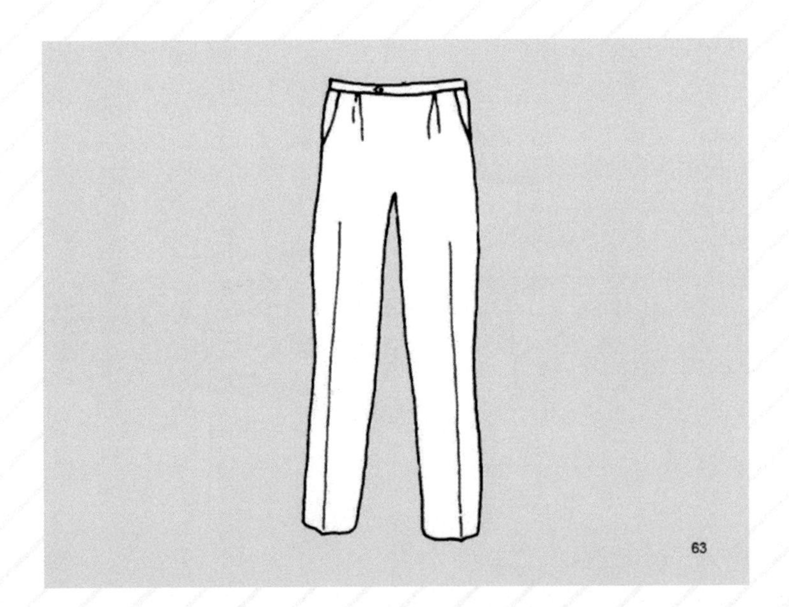
63

66

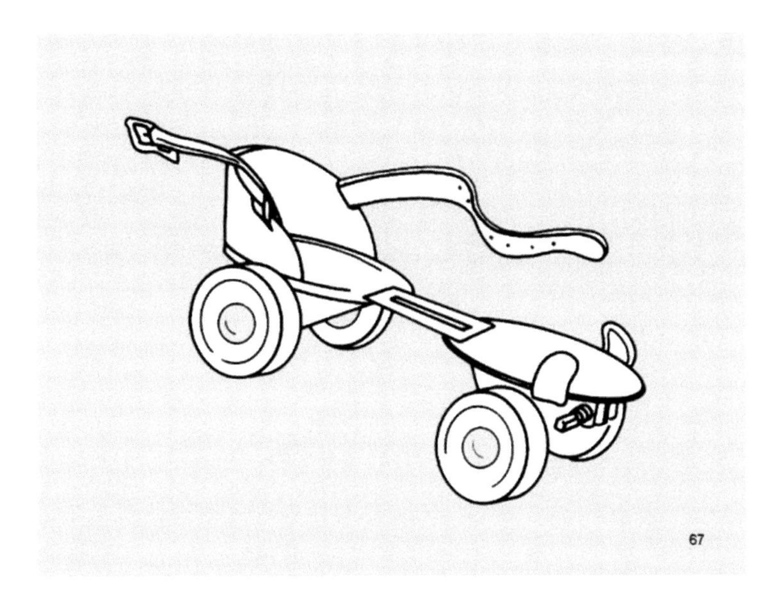
67

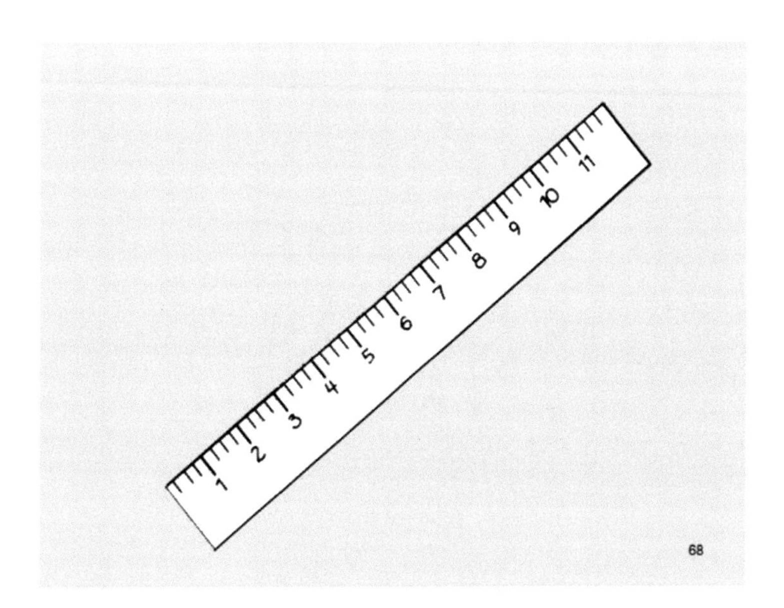
1 2 3 4 5 6 7 8 9 10 11
68

71

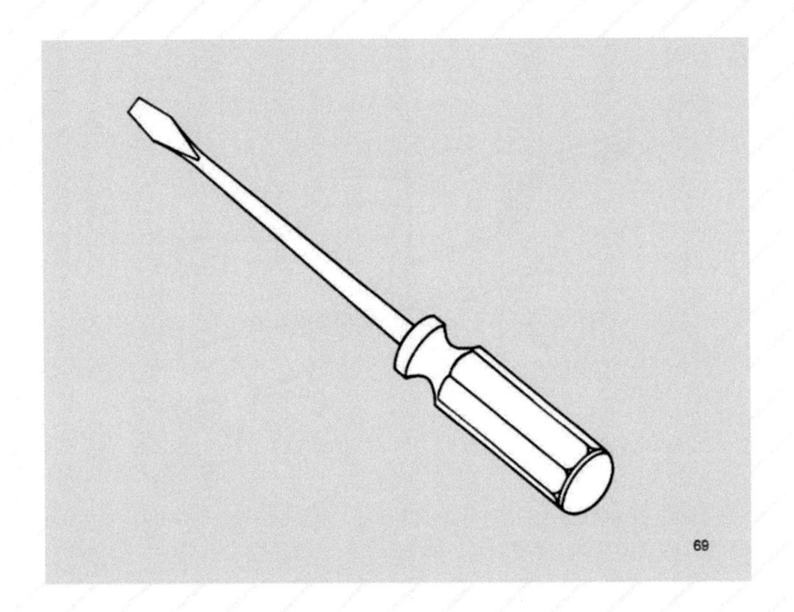
69

72

73

76

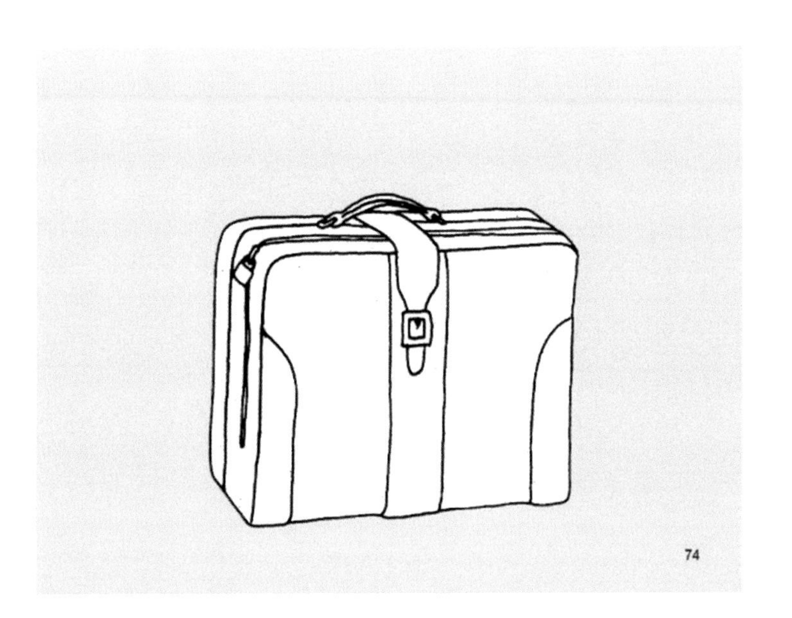
74

77

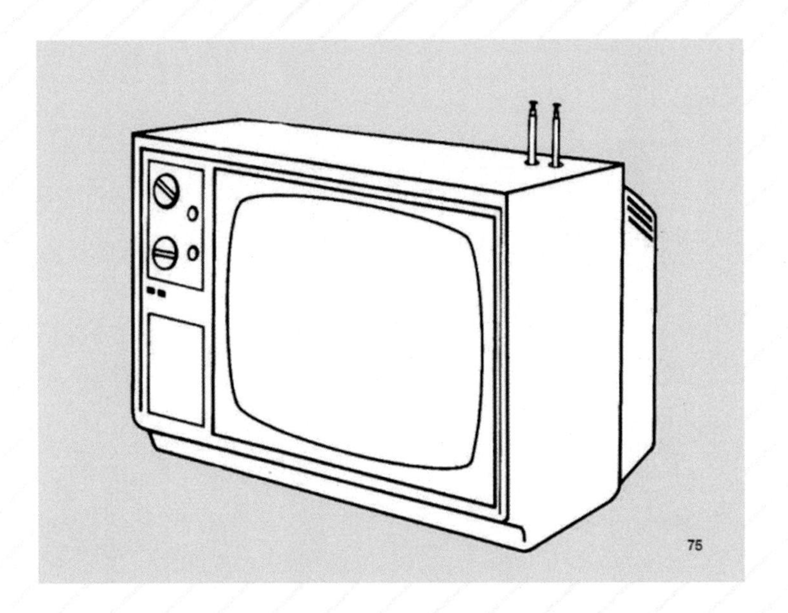
75

78

79

80

81

The following pages include line-drawings of connectors and maps used in Language Mapping at the time of Electrocorticography.

The following line drawings of Electrocorticography accessories were done by the late, Mr. Hector Lettich, Research Eleacroencephalography technologist.

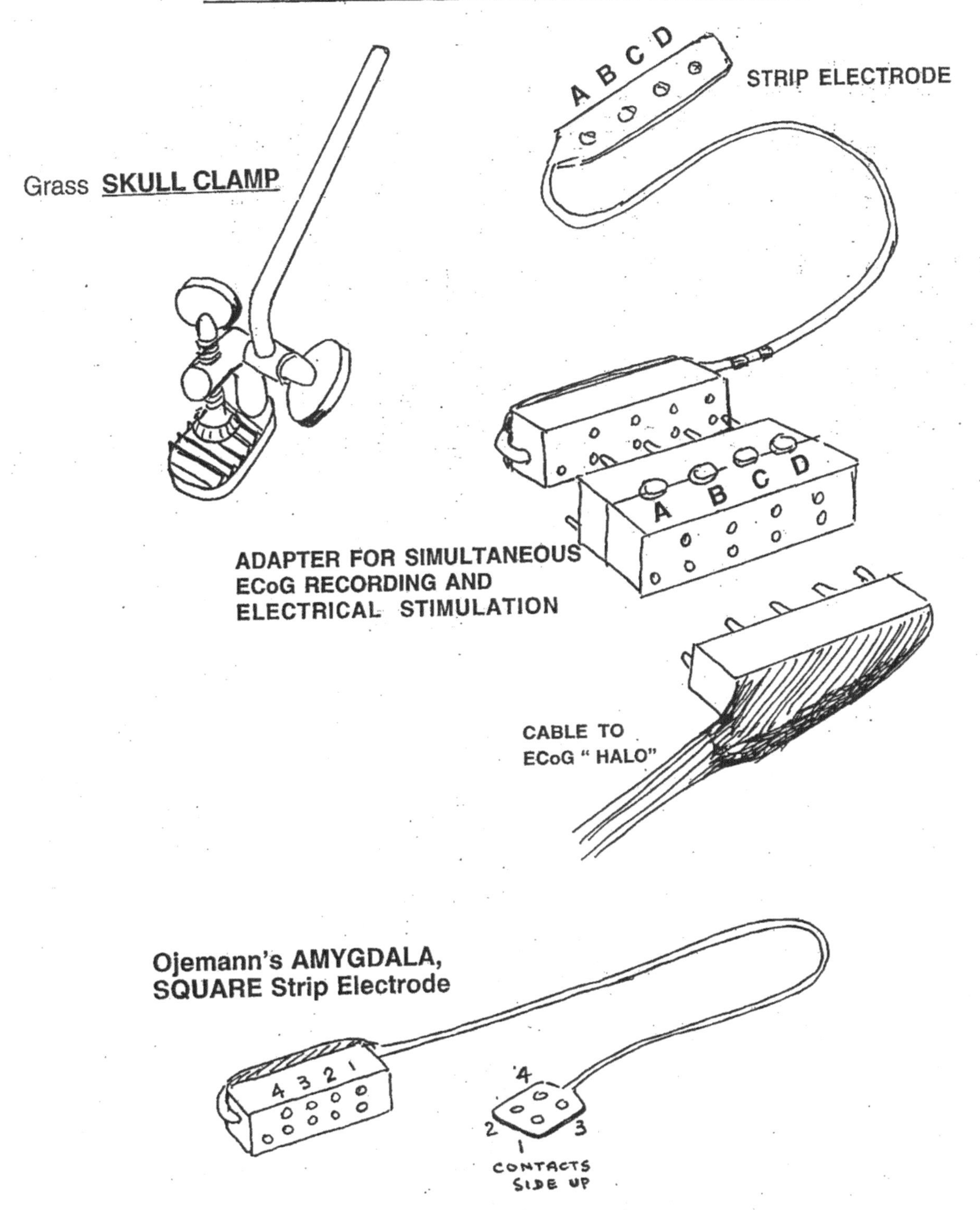
EPILEPSY SURGERY ACCESSORIES
A B C D
STRIP ELECTRODE
Grass SKULL CLAMP
A B C D
ADAPTER FOR SIMULTANEOUS
ECoG RECORDING AND
ELECTRICAL STIMULATION
CABLE TO
ECoG " HALO"
Ojemann's AMYGDALA,
SQUARE Strip Electrode
4 3 2 1
4
2
3
1
CONTACTS
SIDE UP

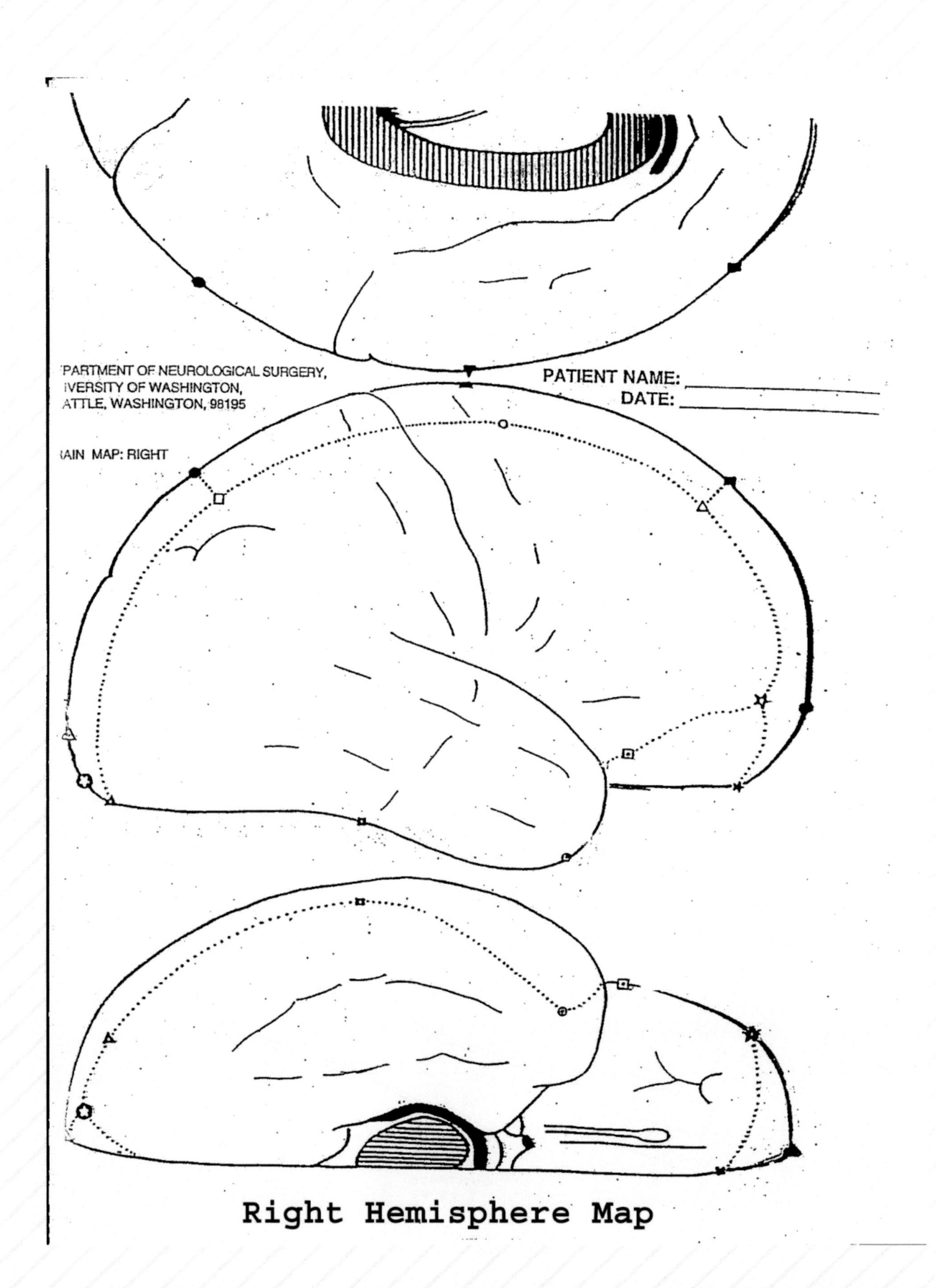

Right Hemisphere Map

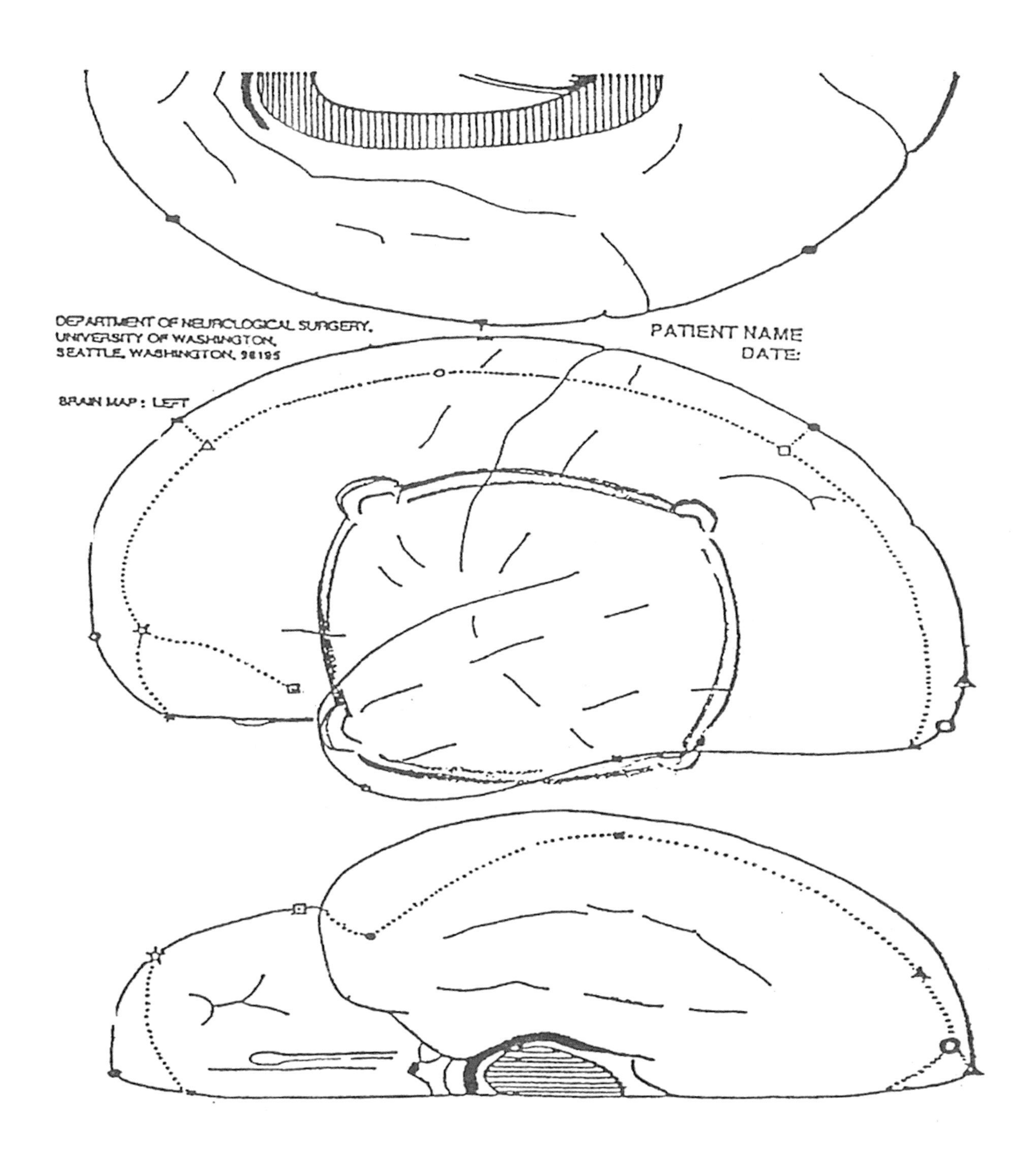

Brain Map Left Hemisphere

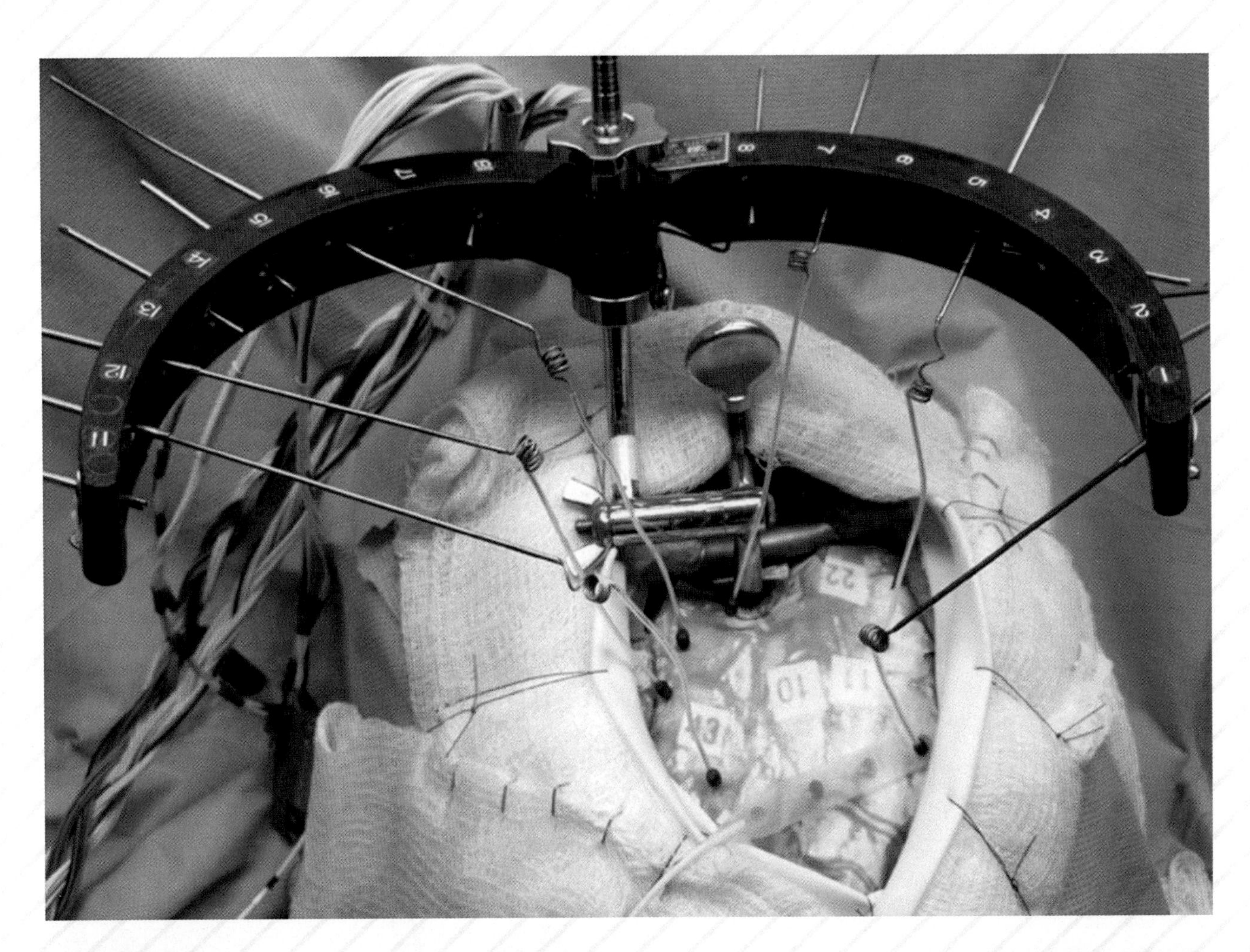
18
17
16
15
14
13
12
11
8
7
6
5
4
3
2
1
22
10
11

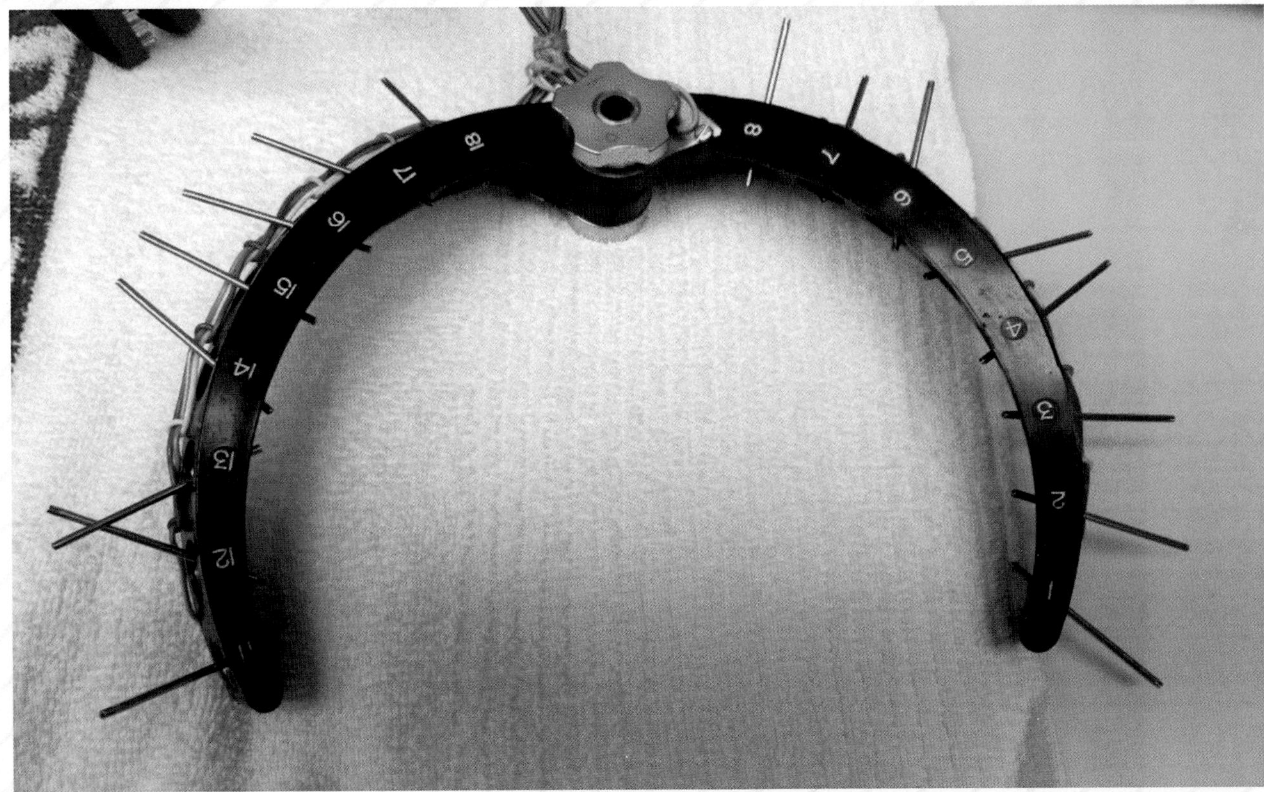
18
17
16
15
14
13
12
11
8
7
6
5
4
3
2
1

Grass Halo

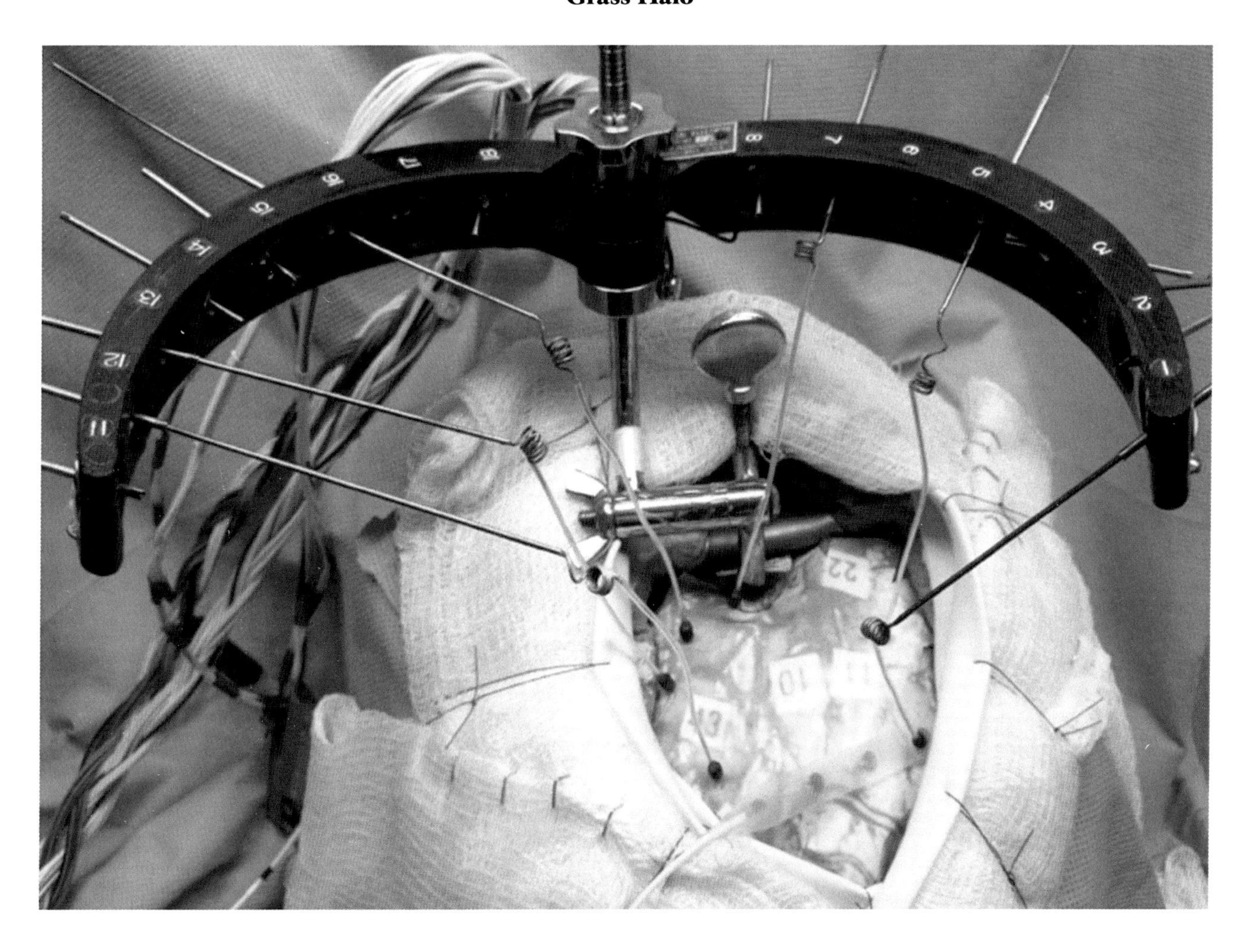

Model showing electrodes in Halo

Electrode Strips and Grids with their Code Charts

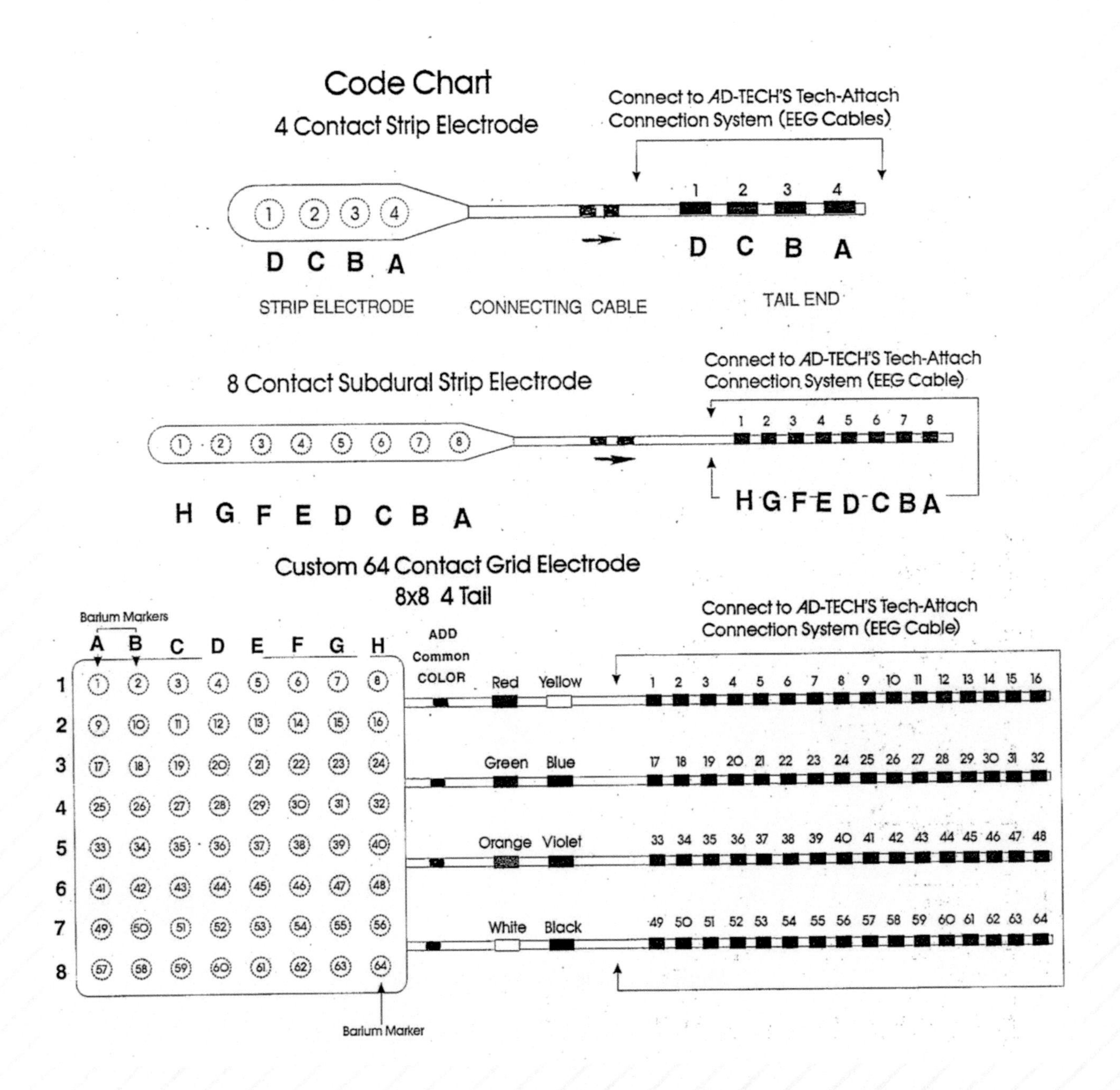

A CD of a Microsoft Powerpoint presentation for Language Mapping is provided. This is to be used with a laptop computer. Built into this program is a projector sound between slides so the surgeon will know when a new slide is presented.

Printed in the United States
By Bookmasters